Frontiers in Computational Chemistry

Frontiers in Computational Chemistry

Computer Applications for Drug Design and Biomolecular Systems
(Volume 1)

Edited by

Zaheer Ul-Haq
Panjwani Center for Molecular Medicine & Drug Research
International Center for Chemical & Biological Sciences
University of Karachi
Pakistan

&

Jeffry D. Madura
Department of Chemistry & Biochemistry
Center for Computational Sciences
Duquesne University, Pittsburgh
USA

AMSTERDAM • BOSTON • HEIDELBERG • LONDON • NEW YORK • OXFORD
PARIS • SAN DIEGO • SAN FRANCISCO • SINGAPORE • SYDNEY • TOKYO
ELSEVIER

Elsevier
Radarweg 29, PO Box 211, 1000 AE Amsterdam, Netherlands
The Boulevard, Langford Lane, Kidlington, Oxford OX5 1GB, UK
225 Wyman Street, Waltham, MA 02451, USA

Notices
Knowledge and best practice in this field are constantly changing. As new research and experience broaden our understanding, changes in research methods, professional practices, or medical treatment may become necessary.

Practitioners and researchers must always rely on their own experience and knowledge in evaluating and using any information, methods, compounds, or experiments described herein. In using such information or methods they should be mindful of their own safety and the safety of others, including parties for whom they have a professional responsibility.

To the fullest extent of the law, neither the Publisher nor the authors, contributors, or editors, assume any liability for any injury and/or damage to persons or property as a matter of products liability, negligence or otherwise, or from any use or operation of any methods, products, instructions, or ideas contained in the material herein.

ISBN: 978-1-60805-865-5

British Library Cataloguing in Publication Data
A catalogue record for this book is available from the British Library

Library of Congress Cataloging-in-Publication Data
A catalog record for this book is available from the Library of Congress

CONTENTS

PREFACE

Computational chemistry is a very diverse field spanning from the development and application of linear free energy relationships (*e.g.* QSAR, QSPR), to electronic structure calculations, molecular dynamics simulations, and to solving coupled differential equations (*e.g.* drug metabolism). The focus of ***Frontiers in Computational Chemistry*** is to present material on molecular modeling techniques used in drug discovery and the drug development process. Topics falling under this umbrella include computer aided molecular design, drug discovery and development, lead generation, lead optimization, database management, computer and molecular graphics, and the development of new computational methods or efficient algorithms for the simulation of chemical phenomena including analyses of biological activity. In this volume, we have collected eight different perspectives in the application of computational methods towards drug design.

In chapter 1 "Computational Strategies to Incorporate GPCR Complexity in Drug Design" the authors review various computational approaches to G protein-coupled receptors (GPCRs). They review the use of GPCR databases to extract starting information about the structure and function of these systems. The authors also review different strategies currently being probe the molecular mechanisms of drug action as well as the development of new drugs.

The topic of chapter 2 "Knowledge-Based Drug Repurposing: A Rational Approach Towards the Identification of Novel Medical Applications of Known Drugs" is of current interest in the pharmaceutical industry. As we learn more about the biochemical pathways and the interactions of compounds with proteins of these pathways, one can gain an appreciation of how current and previous drugs can be used for other medical uses. This chapter discusses the use of cheminformatics and bioinformatics in identifying new insights about known drugs.

Chapter 3, "Tuning the Solvation Term in the MM-PBSA/GBSA Binding Affinity Predictions" focuses on the development and application of a computational tool. A widely used method, Molecular Mechanics Poisson-Boltzmann (Generalized Born) Surface Area (MM-PBSA, MM-GBSA), is discussed in terms of applying the method to calculate accurate binding affinities. The authors point out that in order to obtain good, reliable results the MM-PBSA or MM-GBSA methods need to be tuned for a particular system. In particular, they focus on interior dielectric constant as well as the PB and GB solvers.

A very active area of experimental and computational research is protein-protein interactions that is the topic of Chapter 4, "Recent Advances in the Discovery and Development of Protein-Protein Interaction Modulators by Virtual Screening". In particular, the application of virtual screening methods to find compounds that modulate protein-protein interactions. This is a very challenging task since protein interfaces are flat, large, and lack distinct features. The authors provide a review of the use of virtual screening in protein-protein interactions as its role in drug discovery.

Across the scientific field, we come across the term "big data." In particular, that data generated from genomic projects is overwhelming. In Chapter 5 "Computational Design of Biological Systems: From Systems to Synthetic Biology" the authors describe the development and use of computational methods on large biological data sets to potentially engineer circuits. This systems biology approach to understanding biological function is being used to develop synthetic biological systems. Such developments have potential uses in biotechnology and in the development of strategies to treat various diseases such as cancer.

Biological systems are complex systems to study. In Chapter 6, "Considering the Medium when Studying Biologically Active Molecules: Motivation, Options and Challenges" we are reminded that when studying biological systems not to forget the environment surrounding the system. Most of the time, the environment is left out due to its complexity; however, one must keep in mind that the environment may play a significant role in biological activity. The authors review some insight into how to appropriately include the environment into the study of a particular biological system.

As computational power, hardware and software, continue to increase so do the systems, both temporally and spatially. One approach to address the increase in systems is presented in Chapter 7 "New frontiers of coarse-grained approach to protein folding." Coarse-graining involves the reduction in the number of particles of the system by representing a small group of particles, *e.g.* an amino side-chain by a single particle. This reduction in the number of particles to represent a biological system has the potential to allow for greater exploration of the free energy landscape as well as simulation increased timescales. The authors review the use of coarse-graining in the study of protein folding.

The last chapter "Computational chemistry strategies-tackling function and inhibition of pharmaceutically relevant targets" reviews the various computational methods used to identify pharmaceutically relevant targets. The authors illustrate the application of various tools from first principles to empirical methods in the discovery and development of new compounds that potentially lead or become the next drug. They appropriately point out that it is through the combination of experiment and computations that lead to significant advancement in molecular medicine.

Zaheer Ul-Haq
Panjwani Center for Molecular Medicine & Drug Research
International Center for Chemical & Biological Sciences
University of Karachi
Pakistan

&

Jeffry D. Madura
Department of Chemistry & Biochemistry
Center for Computational Sciences
Duquesne University, Pittsburgh
USA

List of Contributors

Adam K. Sieradzan Faculty of Chemistry, University of Gdańsk, Wita Stwosza 63, 80-952 Gdańsk, Poland and Department of Physics and Astronomy and Science for Life Laboratory, Uppsala University, P.O. Box 803, S-75108 Uppsala, Sweden

Adam Liwo Faculty of Chemistry, University of Gdańsk, Wita Stwosza 63, 80-952 Gdańsk, Poland

Agnieszka A. Kaczor University of Eastern Finland, School of Pharmacy, Department of Pharmaceutical Chemistry, Kuopio, Finland and Department of Synthesis and Chemical Technology of Pharmaceutical Substances, Faculty of Pharmacy with Division for Medical Analytics, Medical University of Lublin, Lublin, Poland

Alan Talevi Medicinal Chemistry, Department of Biological Sciences, Faculty of Exact Sciences, National University of La Plata, Buenos Aires, Argentina

Alessandro Contini Dipartimento di Scienze Farmaceutiche - Sezione di Chimica Generale e Organica "Alessandro Marchesini", Università degli Studi di Milano, Via Venezian, 21 20133 Milano, Italy

Antti Niemi Department of Physics and Astronomy and Science for Life Laboratory, Uppsala University, P.O. Box 803, S-75108 Uppsala, Sweden and Laboratoire de Mathematiques et Physique Theorique CNRS UMR 6083, Fédération Denis Poisson, Université de Tours, Parc de Grandmont, F37200 Tours, France and Department of Physics, Beijing Institute of Technology, Haidian District, Beijing 100081, People's Republic of China

Carolina L. Bellera Medicinal Chemistry, Department of Biological Sciences, Faculty of Exact Sciences, National University of La Plata, Buenos Aires, Argentina

Chung-Hang Leung State Key Laboratory of Quality Research in Chinese Medicine, Institute of Chinese Medical Sciences, University of Macau, Macao, China

Daniel Shiu-Hin Chan Department of Chemistry, Hong Kong Baptist University, Kowloon Tong, Hong Kong, China

Dik-Lung Ma Department of Chemistry, Hong Kong Baptist University, Kowloon Tong, Hong Kong, China

Eduardo A. Castro Institute of Physicochemical Theoretical and Applied Research (INIFTA), National Council of Scientific and Technical Research (CONICET) CCT La Plata, Buenos Aires, Argentina

Irene Maffucci Dipartimento di Scienze Farmaceutiche - Sezione di Chimica Generale e Organica "Alessandro Marchesini", Università degli Studi di Milano,

	Via Venezian, 21 20133 Milano, Italy
Jana Selent	Research Programme on Biomedical Informatics (GRIB), Department of Experimental and Health Sciences, Universitat Pompeu Fabra, IMIM (Hospital del Mar Medical Research Institute), Barcelona, Spain
Li-Juan Liu	State Key Laboratory of Quality Research in Chinese Medicine, Institute of Chinese Medical Sciences, University of Macau, Macao, China
Liliana Mammino	Department of Chemistry, University of Venda, South Africa
Luis E. Bruno-Blanch	Medicinal Chemistry, Department of Biological Sciences, Faculty of Exact Sciences, National University of La Plata, Buenos Aires, Argentina
Marco De Vivo	Drug Discovery and Development, Italian Institute of Technology, Genoa, Italy
María L. Sbaraglini	Medicinal Chemistry, Department of Biological Sciences, Faculty of Exact Sciences, National University of La Plata, Buenos Aires, Argentina
Maria Marti-Solano	Research Programme on Biomedical Informatics (GRIB), Department of Experimental and Health Sciences, Universitat Pompeu Fabra, IMIM (Hospital del Mar Medical Research Institute), Barcelona, Spain
Matteo Dal Peraro	Institute of Bioengineering, School of Life Sciences, ÉcolePolytechniqueFédérale de Lausanne (EPFL), Lausanne, Switzerland and Swiss Institute of Bioinformatics (SIB), Lausanne, Switzerland
Mauricio E. Di Ianni	Medicinal Chemistry, Department of Biological Sciences, Faculty of Exact Sciences, National University of La Plata, Buenos Aires, Argentina
Michele Cascella	Department of Chemistry and Centre for Theoretical and Computational Chemistry (CTCC), University of Oslo, Oslo, Norway;
Milsee Mol	National Centre for Cell Science, NCCS Complex, Ganeshkhind, Pune University Campus, Pune 411007, India
Modi Wang	Department of Chemistry, Hong Kong Baptist University, Kowloon Tong, Hong Kong, China
Mwadham M. Kabanda	Department of Chemistry, North-West University (Mafikeng Campus), South Africa
Rafik Karaman	Bioorganic Chemistry Department, Faculty of Pharmacy Al-Quds University, P.O. Box 20002, Jerusalem, Palestine
Ramon Guixà-González	Research Programme on Biomedical Informatics (GRIB), Department of Experimental and Health Sciences, Universitat Pompeu Fabra, IMIM (Hospital del Mar Medical Research Institute), Barcelona, Spain

Shailza Singh	National Centre for Cell Science, NCCS Complex, Ganeshkhind, Pune University Campus, Pune 411007, India
Sheng Lin	Department of Chemistry, Hong Kong Baptist University, Kowloon Tong, Hong Kong, China
Xubiao Peng	Department of Physics and Astronomy and Science for Life Laboratory, Uppsala University, P.O. Box 803, S-75108 Uppsala, Sweden

CHAPTER 1

Computational Strategies to Incorporate GPCR Complexity in Drug Design

Maria Marti-Solano[1], Agnieszka A. Kaczor[2,3,*], Ramon Guixà-González[1] and Jana Selent[1,*]

[1]Research Programme on Biomedical Informatics (GRIB), Department of Experimental and Health Sciences, Universitat Pompeu Fabra, IMIM (Hospital del Mar Medical Research Institute), Barcelona, Spain; [2]University of Eastern Finland, School of Pharmacy, Department of Pharmaceutical Chemistry, Kuopio, Finland and [3]Department of Synthesis and Chemical Technology of Pharmaceutical Substances, Faculty of Pharmacy with Division for Medical Analytics, Medical University of Lublin, Lublin, Poland

Abstract: G protein-coupled receptors (GPCRs) represent the most important family of drug targets to date. However, state-of-the-art experimental procedures, able to characterize in deep both GPCR modulation in health and disease and the molecular mechanisms of drug action at these receptors, have provided a more nuanced picture than previously expected. Several aspects of GPCR function, which are currently being characterized, clarify some regulatory processes regarding these receptors and, at the same time, introduce novel levels of complexity which should be taken into consideration for rational drug design. In this scenario, computational approaches can help in several ways rationalize the increasing amount of data on GPCRs and their ligands. On the one hand, a set of databases devoted to these receptors provide excellent starting points for data mining. On the other, exploitation of the ever-increasing ligand and structure-based information by novel computational techniques can help addressing emerging questions in the GPCR field. Some of these questions comprise the refined modulation of GPCR signaling states by biased agonists, the exploitation of GPCR oligomers as drug targets, the analysis of polypharmacology in GPCR ligands, the development of strategies for receptor deorphanization or the prediction of off-target interactions of known drugs targeting these receptors. In this chapter, we will cover some of these strategies for knowledge-based rational design for GPCRs and will discuss the main hurdles which they may need to overcome to yield novel, safer and more efficacious drugs possessing polished mechanisms of action.

***Corresponding author Jana Selent:** Research Programme on Biomedical Informatics (GRIB), Department of Experimental and Health Sciences, Universitat Pompeu Fabra, IMIM (Hospital del Mar Medical Research Institute), Dr. Aiguader 88, E-08003 Barcelona (Spain) / University of Eastern Finland, School of Pharmacy, Department of Pharmaceutical Chemistry, Yliopistoranta 1, P.O. Box 1627, FI-70211 Kuopio, Finland; Tel/Fax: +39 933 160 648/+34 933 160 550; E-mail: jana.selent@upf.edu

Agnieszka A. Kaczor: Medical University of Lublin, Faculty of Pharmacy with Division for Medical Analytics, Department of Synthesis and Chemical Technology of Pharmaceutical Substances, 4A Chodzki St., 20093 Lublin, Poland; Tel: +48 81448 7270; Fax: +48 81448 7272; E-mail: agnieszkakaczor@umlub.pl

Zaheer Ul-Haq and Jeffry D. Madura (Eds). **10.1016/B978-1-60805-865-5.50001-0**

Keywords: Allosteric modulation, biased agonism, chemogenomics, drug design, G protein-coupled receptors, homology modeling, ligand promiscuity, molecular dynamics, oligomerization, virtual screening,

1. INTRODUCTION

G protein-coupled receptors (GPCRs) are transmembrane proteins responsible for the transmission of signals to the intracellular milieu upon detection of a wide variety of extracellular stimuli. About a thousand human genes code for this type of receptors [1], which are implicated in most physiological processes involving communication between cells or detection of exogenous signals such as light, odorants or flavors. Due to their importance in cell physiology, these receptors have historically received special attention in drug discovery, even before they were thoroughly characterized. Indeed, GPCRs are considered the most important drug targets to date [2].

All GPCRs are characterized by a set of common structural features: they have an extracellular N-terminal domain and an intracellular C-terminal domain, connected by seven helices which cross the plasma membrane. Classification according to phylogenetic criteria yields the following five GPCR families: Glutamate, Rhodopsin, Adhesion, Frizzled/Taste2 and Secretin [3]. The Rhodopsin family (also termed class A) has been the most exploited one in drug discovery and it is estimated that drugs targeting this family of receptors represent approximately a 25% of marketed small molecules [4]. Structural information on GPCR topology has dramatically increased during the past years, in part thanks to the development of original crystallographic strategies [5]. In this sense, crystallization of the ternary complex formed by the active β_2-adrenergic receptor in complex with an agonist and coupled to a G protein, represented a major advance for the understanding of the structural basis of GPCR functioning [6] (Fig. **1**). Besides that, several other representatives of class A GPCRs have been crystallized in complex with ligands covering a wide range of activities, thus providing detailed insight into the nature of specific ligand-receptor interactions [7]. Finally, recent crystallization of the Smoothened [8] and Corticotropin Releasing Factor [9] receptors has shed light into structural receptor diversity beyond the class A GPCR family. This structural information is helping in the understanding of GPCR functioning and, at the same time, contributing to unravel subtle receptor differences which point to a higher degree of complexity in receptor modulation than previously expected.

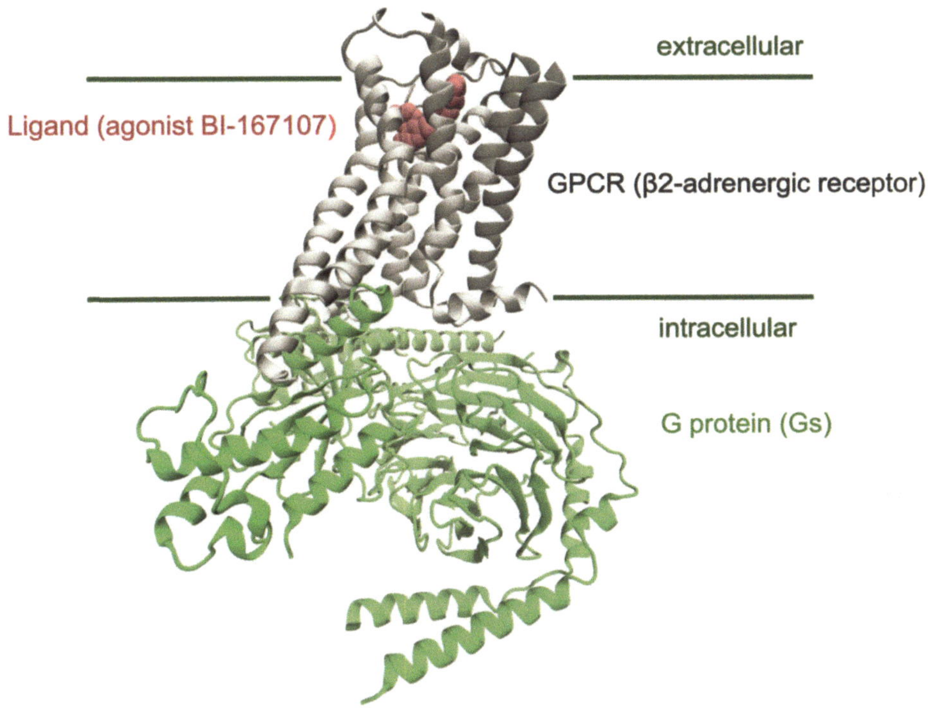

Figure 1: The ternary complex model. The original ternary complex model contemplates a ligand (mauve), a GPCR (white) and a G protein capable of signal transduction (green) as the basic signaling complex that is responsible for receptor function. The publication of the structure of the β_2-adrenergic receptor in complex with an agonist and coupled to a G protein (PDB code 3SN6), which we see in this Figure, helped clarifying the structural basis of interactions between these different signaling components.

1.1. On Complex GPCR Modulation

The ternary complex model describes GPCR function as the interplay of three basic components: a receptor, an agonist and a G protein (Fig. **1**). In this model, receptor activation is favored by interaction with an agonist, which translates into the activation of a particular G protein in the intracellular compartment that, in turn, is capable of initiating particular signaling cascades. However, the increasing experimental evidence on GPCR functioning has revealed that GPCR signaling can be modulated in ways much more complicated than the ones contemplated in the ternary complex model (see Fig. **2**). On the one hand, the discovery that a single receptor could couple to more than one G protein type and, besides that, that GPCRs could trigger G protein-independent pathways, stimulated a more nuanced characterization of GPCR ligands. This characterization led researchers to realize that there are ligands, which would later be named biased agonists, capable of preferentially activating one receptor-associated pathway over another. This has been related to the existence of multiple receptor states, with different propensities to couple to G proteins or other signaling partners, and which can be differentially stabilized by biased compounds. This complex receptor modulation, which has been termed functional selectivity [10], has opened a new avenue for the interrogation of specific GPCR-activated pathways and their impact on health and disease, as well as for the subsequent detection of pathway-selective drugs with a refined mechanism of action [11]. In this way, characterization of the importance of particular pathways associated with a given receptor can provide insight into the optimal functional selectivity profile for the treatment of a particular disease. This is the case for niacin [12], a compound which binds to the GPR109A receptor and, by lowering cAMP levels through stimulation of $G_{\alpha i}/G_{\alpha o}$ proteins, reduces the production of triglycerides and LDL. However, niacin can also lead to dermatological side effects mediated by β-arrestin signaling, such as skin itching and flushing, which limit its clinical use. For this reason, the development of a biased agonist capable of binding to the GPR109A receptor, and specifically stimulating G protein-mediated pathways, could lead to a safer alternative treatment which could still preserve the therapeutic effects of niacin.

In addition, accumulation of experimental evidence from cross-linking experiments and radioligand binding determinations reporting negative and positive cooperativity, suggested the possibility that GPCRs may be capable of oligomerization [13]. In the past years, both homo- and hetero-dimerization have been described for an increasing amount of these receptors and, in some cases, these associations have been related to particular functional outcomes [14]. For this reason, GPCR oligomers have also been described as potential drug targets which, due to their restricted tissue distribution, could provide a new source of drug specificity. Despite the increasing amount of described functional interactions between dimers, the development of drugs with the ability to target receptor oligomers is still very challenging. Therefore, a deeper characterization of the basis of receptor dimerization and of its impact on signaling, together with the development of original treatment strategies, will be necessary for the pharmacological exploitation of this phenomenon [15].

The current nature of ligand screening campaigns, which incorporate functional readouts as well as binding affinity data, has facilitated the detection of an additional class of GPCR ligands [16]. Such ligands possess the ability to modulate GPCR function by binding to receptor regions away from the orthosteric binding site. Allosteric modulators usually bind to receptor areas with a low degree of conservation between GPCR subtypes. This binding specificity could also be the basis for the design of more selective drugs. Additionally, the fact that allosteric modulators can function together with ligands interacting at the orthosteric binding site, makes drugs exploiting this phenomenon especially useful when treatment can be achieved by enhancing an endogenous signal. As an example, Cinacalcet is a positive allosteric modulator of the CaS calcium-sensing receptor. This drug, which is currently commercialized for the treatment of hyperparathyroidism, potentiates activation of the CaS receptor, a class C GPCR. In particular, Cinacalcet interacts with the receptor at the level of the transmembrane helix bundle and, by promoting receptor activation upon calcium binding to the orthosteric binding site, inhibits parathyroid hormone secretion [17].

Finally, in depth characterization of GPCR ligands, has revealed that known drugs targeting GPCRs often present a high degree of promiscuity [18]. The ability of GPCR drugs to bind to more than one receptor subtype at low concentrations was first envisaged as a drawback for GPCR drug discovery. However, nowadays, the efficacy of certain drugs targeting GPCRs is considered to be mediated by their capacity to regulate several targets at the same time – for instance, in the case of drugs related to the treatment of CNS diseases [19]. In the case of antipsychotic drugs, for instance, searching for drugs with selectivity for a specific GPCR subtype did not yield more efficacious drugs than the existing first and second generation treatments, which present promiscuous binding profiles for a variety of receptor families. At present, efforts are devoted to the identification of targets responsible for therapeutic efficacy and to the search of drugs capable of preserving affinity for these targets and, at the same time, avoiding targets mediating side effects.

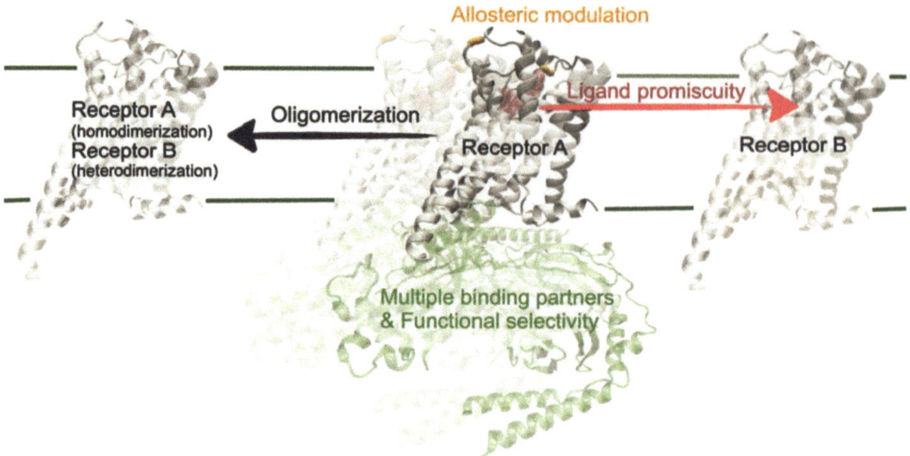

Figure 2: Different sources of complexity in GPCR modulation. GPCRs and their ligands present multiple layers of complexity such as ligand promiscuity towards different receptors (mauve), the ability of some compounds - allosteric modulators - to bind receptors away from their orthosteric binding pocket (orange), the capacity of receptors to function as homo and heteromers (black), and the existence of multiple signal transducers capable of binding different receptor activation states (green) which can be stabilized by biased agonists promoting functional selectivity.

1.2. Discovering Available Data on GPCRs

As previously mentioned, interest about the implication of GPCR function in health and disease, along with their established importance as drug targets, has led to the generation of large amounts of experimental data on these receptors. This data characterizes, among others, the binding affinities of GPCR ligands, their capacity to activate particular signaling pathways, the effect of mutations on GPCR function and their impact on ligand binding, the potential of particular receptors to form oligomers and the functional impact of oligomerization, and the structural atomic details of ligand-receptor interaction and of different levels of receptor activation. This information is, at present, available, among others, at the following online databases:

1. GPCRDB (http://www.gpcr.org/7tm/): this database contains experimental and computational data covering GPCR sequences, available receptor mutagenesis data, structural information at an atomistic level, receptor homology models and a series of sequence alignment tools allowing building alignments, predicting the impact of particular mutations or finding specific GPCR sequence motifs.

2. IUPHAR-DB (http://iuphar-db.org/): it provides peer-reviewed data on pharmacological, functional and pathophysiological information on GPCRs. Information in this database covers different features including structural information on ligands and their affinity and efficacy data, detailed information on the capability of GPCRs to couple to different intracellular mediators, data on tissue distribution and receptor physiological functions, and genetic information on receptor variants.

3. GPCR SARfari (https://www.ebi.ac.uk/chembl/sarfari/gpcrsarfari/): this database, which is integrated in ChEMBL, provides information on GPCR sequence and structure and contains screening information for a large set of structurally-characterized compounds (at present >140000). Notably, queries in this database allow discriminating natural ligands, clinical candidates and FDA approved GPCR drugs together with their trade names and chemical structures.

4. GPCR-OKB (http://data.gpcr-okb.org/gpcr-okb/): information in this database covers computational and experimental evidence on GPCR oligomerization. In particular, it allows assessing the potential of a given receptor to oligomerize by analyzing publications using different characterization methods. Besides, it provides information on the cases in which physiologically relevant effects of oligomerization have been reported and on the proposed structural details of receptor-receptor interactions.

5. GPCRSD (http://zhanglab.ccmb.med.umich.edu/GPCRSD/): this resource provides up to date information on experimentally-solved GPCR crystal structures along with their PDB codes and related citations and together with information on the ligands they are bound to.

The amount and variety of data presented in these repositories represent an excellent starting point for the development of computational models to extract relevant information on GPCR modulation and functioning. In fact, *in silico* approaches have been historically used to analyze this type of receptors. Originally, the absence of crystal structure information led to the development of ligand-based techniques dealing with the prediction of GPCR binding affinities and efficacies. Then, crystallization of rhodopsin first, and of several other class A GPCRs later, made possible the application of structure-based techniques for the selection of new GPCR ligands either by using the crystal structures themselves or by the construction of homology models. Nowadays, the wealth of GPCR experimental data allows for the construction of complex models addressing issues such as polypharmacology. Besides, from a structural perspective, the amount of crystallographic, mutagenesis and biophysical data available, together with the constant increase of computational power, allows building models and performing simulations for the analysis of GPCR conformational space, which can, for instance, incorporate information on the effects of biased agonists or oligomeric interaction partners. In this book chapter, we will cover some of these computational techniques and will analyze the potential they hold for rationalizing the increasing amount of evidence on novel GPCR regulation mechanisms, as well as for discovering new drugs exploiting this GPCR complex modulation.

2. LIGAND-BASED APPROACHES TO EXPLORE GPCR COMPLEXITY

Historically, the difficulty to model GPCR structures due the absence of available templates made ligand-based approaches the method of choice to rationalize receptor binding determinants. These approaches mainly relied on the analysis of structure-activity relationships (SAR) of previously characterized ligands [20]. Output from this SAR analysis allowed generating pharmacophore models (see Fig. **3**), which would serve as the basis for ligand-based virtual screening or, alternatively, it could serve to derive quantitative SAR (QSAR) relationships from the 2D or 3D description of compounds. This description could later be used to generate linear models with the potential to predict the behavior of previously uncharacterized compounds. An interesting example of the use of QSAR for the rationalization of ligand behavior can be found in the study of long-acting dual dopaminergic D_2/β_2-adrenoceptor agonists [21]. Using a 3D-QSAR approach based on the calculation of molecular interaction fields with the GRID software, Austin *et al.* were able to determine that both compound lipophilicity and basicity at the level of their secondary amine were key for their effect duration. Lessons learned in this study have been associated to the development of long-acting β_2-

adrenergic receptor agonists by Pfizer and also of indacaterol, a recently approved long-acting β_2-adrenergic receptor agonist from Novartis [22].

Nowadays, and especially for class A GPCRs, these techniques, despite being still widely used, are generally complemented with receptor 3D structural information derived either by direct analysis of the crystal structure or by the generation of homology models. However, approaches based on the analysis of ligand properties are still applied to address issues on GPCR complex modulation. Some authors, for instance, have used this analysis to explore the structural basis of ligand agonism and antagonism at different GPCRs. As an example, Custodi *et al.* developed decision trees by using relatively simple 2D and 3D molecular descriptors, with the capacity to discriminate between ligands with agonistic and antagonistic effects in different class A GPCR sub-classes [23].

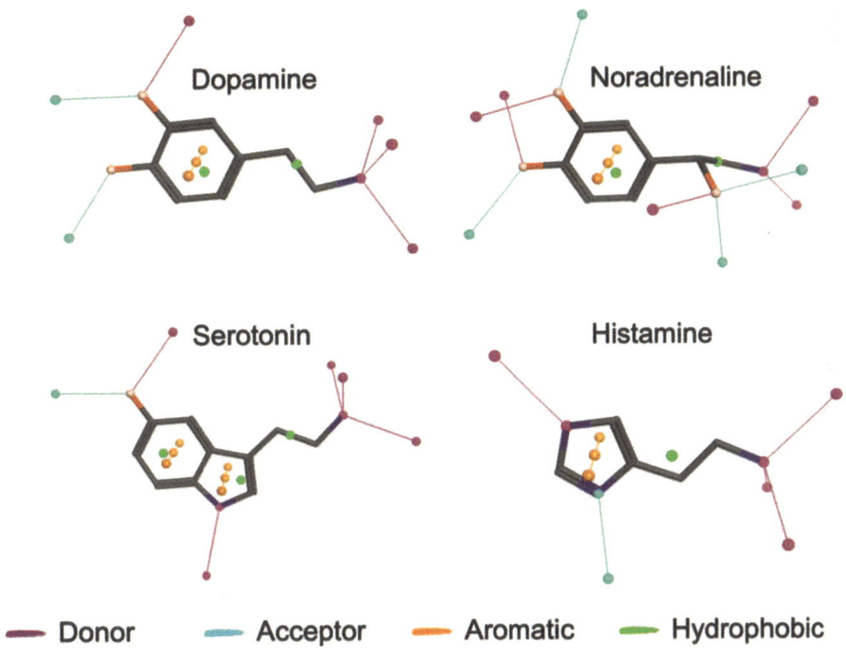

Figure 3: Example of a pharmacophoric description for a series of aminergic GPCR receptor ligands.

Another aspect of GPCR complexity, which is still being addressed by ligand-based techniques, is the common existence of multi-target affinity profiles in compounds binding to these receptors. In line with the observation that GPCR ligands tend to show promiscuous binding affinity behaviors, these molecules have also received an especial attention from the chemogenomics field [24]. In the case of drug promiscuity, this approach intends to predict relationships between the chemical structures of ligands and the receptors they are able to target [25]. In this context, several strategies have been explored including approaches solely based on ligand structure and others which incorporate some level of binding pocket representation [26].

Recently, in an interesting approach, Lin and coworkers developed a pharmacological organization of GPCRs based on the similarity between their ligands [27]. In their work, they compare this classification to others based on receptor sequence similarity at the level of the GPCR binding pocket [28]. Analysis of the resulting dendrogram allowed the authors to select high similarity ligands targeting receptors with low sequence conservation, which presented polypharmacology upon experimental characterization. For instance, they identified previously undescribed ligand associations between opioid and serotoninergic receptors or between receptors with molecularly diverse endogenous ligands such as neuropeptide and cannabinoid receptors. However, probably one of the most interesting outcomes of this study relates to the capacity of this characterization to expand to non-GPCR targets. Due to the sequence independence of the method, it turns possible to select ligands known to bind to other target classes, which present a high degree of similarity to GPCR binders. Using this approach, some compounds were identified that bind both GPCRs and other protein classes such as kinases, phospholipases or hydrolases.

Prediction of complex GPCR ligand pharmacology was also used by Besnard *et al.* to obtain compounds with particular polypharmacological profiles [29]. In this case, the authors used an automated, adaptive design approach to evolve the chemical structure of the acetylcholinesterase inhibitor Donepezil. Their first goal was to improve activity of this drug, indicated for Alzheimer disease, for the dopaminergic D_2 receptor and, at the same time, to increase its likelihood of crossing the blood-brain barrier. After applying their models, eight drug-like novel analogues were synthesized and tested, and all of them showed substantial D_2 receptor affinities. After testing the most potent compound, it was both observed that it possessed blood brain barrier penetration and that it presented a polypharmacological profile. Analysis of the receptors targeted by this newly identified ligand revealed that it showed affinity for α_1-adrenergic receptors. Due to the fact that these receptors were considered potential anti-targets, the adaptive models were again used to evolve the eight new compounds towards a polypharmacological profile with improved selectivity over these anti-targets. According to the authors, their approaches could be extended to other drug–target classes, provided that enough ligand structure–activity data is available to create useful models.

Incorporation of different levels of receptor structural detail to ligand definitions has also been the basis for the analysis and detection of GPCR ligands with non-traditional mechanisms of action. As an example, Gloriam *et al.* mapped the binding sequence motifs of three known privileged structures targeting Family A GPCRs and analyzed the sequence of the class C GPRC6A receptor to select ligands acting as allosteric antagonists at this receptor [30]. In a different study, Nijmeijer *et al.* used FLAP (Fingerprints for Ligands and Proteins) 3D-QSAR to describe ligand and receptor features responsible for β-arrestin biased signaling at the human histaminergic H_4 receptor [31]. These examples highlight the importance that new GPCR structural information has on our understanding of ligand-mediated GPCR functioning. However, analysis of the chemogenomics approaches presented so far, also shows that combining information on the structures of compounds and their binding preferences can improve our understanding on the basis of GPCR modulation.

3. STRUCTURE-BASED METHODS FOR THE STUDY OF GPCRs

The last decade has been marked by a rapid growth in experimentally solved structures of GPCRs with more than 100 structures available to-date covering 19 different GPCR types (http://zhanglab.ccmb.med.umich.edu/GPCRSD/). These structures have provided an unprecedented level of insight into the basis of ligand-receptor interaction and also into the structural basis of receptor activation and coupling (please refer to reference [7] for a comprehensive review). However, given the amount of GPCRs encoded in the human genome (approximately a thousand [1]), computational modeling is still a highly relevant tool for exploring functional complexity and selective targeting for the majority of GPCRs, which have not yet been crystallized. The potential of computational modeling, but also its limitations, has been systematically assessed in three community-wide GPCR Dock competitions (2008, 2010 and 2013, http://gpcr.scripps.edu). These evaluations show, in detail, to what extent the GPCR modeler community is able to predict ligand-receptor interactions combining available structural information and state-of-the-art modeling protocols (homology modeling and docking). This assessments are a highly informative evaluation of the contribution that modeling can have on rational drug design, as successful construction of an accurate ligand-GPCR complex is of high value for optimizing lead structures in terms of binding affinity, efficacy and safety.

3.1. Homology Modeling

Homology modeling refers to the construction of an all-atom model of the target receptor using its sequence and experimentally-derived high-resolution data of a phylogenetically-close receptor (template). Particular modeling care has to be taken for regions which potentially interact with ligands such as large parts of the transmembrane (TM) domain and the extracellular loop 2 (ECL2). The modeling assessments DOCK 2008 [32], 2010 [33] and 2013 [34] have demonstrated that GPCR modeling strongly depends on the available template (Table **1**): thus the higher the sequence identity between target structure and template, the better the structural prediction. In particular, this holds true for the TM regions comprising the seven helical domains of high structural conservation within the GPCR family. Thus, sequence identities greater than 40% resulted in homology models with excellent TM RMSDs < 2 Å as observed for the class A GPCRs, like the dopamine D_3 receptor (D_3R) and the serotoninergic receptors 5-HT_{1B} and 5-HT_{2B} (5-$HT_{1B}R$ and 5-$HT_{2B}R$) (Table **1** (a)). Greater structural deviations of the predicted model to the experimental structure (TM RMSDs > 2 Å) were obtained for sequence identities lower than 30% between target receptor and template. Among them, one of the biggest challenges was found in the prediction of the human smoothened homolog receptor (SMO) - a class frizzled (class F) GPCR with a TM sequence identity as low as 14% to available templates in the modeling assessment DOCK 2013 [34]. Low sequence identity bears the risk of alignment inaccuracies between target sequence and available structural templates: even inaccuracies as small as one-residue shift in a single TM helix result in a dislocation of residues and impairment of important interhelical and ligand-receptor contacts.

In contrast to the TM domains, extracellular loop regions such as the ECL2 are far more challenging receptor sections to model. The same loop sequence of the same receptor can exist in different conformational states depending on the ligand type bound to it and therefore the loop has to be modeled in the presence of the ligand. The difficulties of modeling the ECL2 in comparison to the structurally better conserved TM region are reflected in the three modeling assessments 2008 [32], 2010 [33] and 2013 [34]: none of the predicted complexes reached an ECL2 RMSD < 2Å (Table **1** (b)). Clearly, more sophisticated modeling techniques [35] have to be applied to produce better predictions of such flexible receptor regions, which, due to their diversity between receptor sub-classes, represent interesting targets for selective allosteric modulators.

3.2. Docking: Predicting Ligand-Receptor Interaction

The recent advances in GPCR crystallization and homology modeling also condition the accurate prediction of ligand-receptor interactions by docking. The main objectives of docking in drug discovery campaigns are (i) to identify the ligand binding pocket, (ii) to dock promising structures into this binding pocket and (iii) to predict ligand contacts with surrounding key residues of the target receptor.

The definition of the binding pocket and thus ligand placement is easier for aminergic receptors (such as, D_3R, 5-$HT_{1B}R$ or 5-$HT_{2B}R$). The binding pocket is characterized by a highly conserved Asp3.32 in TM3 [36] that typically interacts with a positively charged nitrogen of the ligand by electrostatic interactions. In addition, a hydrophobic pocket between TM3 and TM6 accommodates hydrophobic ligand fragments. Such structural knowledge facilitates the definition of the binding pocket for aminergic receptors (D_3, 5-HT_{1B} or 5-HT_{2B}) when compared to non-aminergic receptors (A_{2A}, CXCR4 and SMO) (see pocket RMSD, Table **1**(b)). Moreover, this structural information is enormously supportive to accurately place the ligand during the docking procedure. This is impressively demonstrated in the modeling assessments for the best submitted complexes of the aminergic receptors 5-HT_{1B}, 5-HT_{2B} and D_3 [37] which showed a ligand RMSD $\leq$ 1.51 Å (Table **1** (b)). Such a predictive potential is ideal for structure-based drug discovery programs. However, it seems to be limited to aminergic GPCRs. Thus, non-aminergic receptors are often characterized by larger and less defined binding pockets. This widens the amount of possible solutions when placing the ligand in the binding pocket making the prediction of correct ligand poses extremely difficult. In fact, this is reflected by the large ligand RMSDs (from 4.42 to 8.88 Å) obtained for the best complexes of the chemokine receptor CXCR4 and the smoothened receptor (SMO) (Table **1** (b)). Besides the definition of the binding pocket, another key issue that contributes to easiness of docking is the size of the ligand. Large ligands such as peptides (*e.g.* CVX15) have numerous rotatable bonds. This is the source of a vast amount of possible conformations and makes it exceptionally challenging to identify the biologically active one. Hence, no submitted model of the CXCR4-CVX15 complex in the DOCK 2010 achieved a ligand RMSD below 8 Å to the target structure (Table **1** (b)) [33].

Table 1: Target receptors and complexes in the modeling assessment DOCK 2008, 2010 and 2013

(a) Homology Model[3]					(b) Receptor - Ligand Complex[4]					
Target Receptor[1]	**Class**	**Template[2]**	**TM Sequence Identity**	**TM RMSD**	**Ligand in Target Complex**	**TM RMSD**	**ECL2 RMSD**	**Pocket RMSD**	**Ligand RMSD**	**Rank[5]**
SMO (2013)	F	M_3R	14%	2,78	LY-2940681	5,3	14,34	13,85	4,42	3
SMO (2013)	F	M_3R	14%	2,78	SANT-1	3,9	11,27	5,61	4,31	5
CXCR4 (2010)	A	β_1AR	25%	2,05	IT1t	2,21	7,42	3,04	4,88	5
CXCR4 (2010)	A	β_1AR	25%	2,05	CVX15	2,88	8,19	4,11	8,88	5
A_{2A} (2008)	A	β_2AR	36%	2,00	ZM241385	2,5	3,8	3,4	2,7	2
5-HT_{2B} (2013)	A	β_1AR	41%	1,52	ERG	2,21	5,67	2,69	1,05	3
D_3 (2010)	A	β_1AR	43%	1,26	Eticlopride	1,38	2,87	1,5	0,96	3
5-HT_{1B} (2013)	A	β_1AR	48%	1,52	ERG	1,82	4,34	1,41	1,51	2

[1] in parentheses year of the modeling competition
[2] best available template
[3] best submitted homology model
[4] best submitted receptor-ligand complex
[5] modeler group`s rank for the best model (max. 5 models per target complex were submitted)

Another level of complexity in modeling GPCRs is added by the fact that these receptors exist in different activation states. The first two modeling assessments (2008 and 2010) had only considered inactive structures in the blind prediction [32, 33]. Structurally, the binding pocket of the inactive state is wider when compared to the one of the active receptor [38-40]. Cross-docking experiments of inverse agonists into active crystal structures or agonists into inactive crystal structures highlight the importance of selecting a template with a correct activation state (inactive/active) for accurate predictions of orthosteric ligand binding [38].

Even more challenging for GPCR modeling is the finding that there are nuanced conformational states between classically defined active and inactive GPCR structures – namely, conformational states linked to different propensities for G protein or β-arrestin signaling. Wang *et al.* captured such conformational differences by crystallizing the complexes of two serotonin receptors with ergotamine [41], a compound used for the treatment of acute migraine attacks. Interestingly, the 5-HT_{1B} receptor is capable to signal through both G protein and β-arrestin pathways when interacting with ergotamine. Conversely, the 5-HT_{2B} receptor is biased towards β-arrestin signaling when interacting with this drug [42]. These complexes represented a new challenge in the last edition DOCK2013 [34]. Encouragingly, most submitted models reproduced accurately the activation state of the 5-HT_{1B}R, which adopts a classical active state in TM5-6 and TM7 region. In contrast, the ergotamine/5-HT_{2B} complex adopts a so far unseen conformational state with an active TM7 rotation and a TM6 rotation that is more consistent with an inactive state. Unfortunately, none of the submitted models to the DOCK2013 competition has captured this important structural feature. As biased agonism is gaining increasing relevance for drug development, modeling efforts have to be devoted to improve the performance of predictions of biased ligand-GPCR complexes.

To summarize, the three assessments of the current state of GPCR modeling (2008, 2010 and 2013) clearly demonstrate that the increasing amount of experimental data facilitates the prediction of ligand binding to a receptor target reaching atomic resolution, provided that a closely-related template is available.

3.3. Taking Advantage of New Structural Information for the Discovery of New GPCR Ligands by Virtual Screening

Virtual screening is nowadays a standard tool in drug discovery used to identify new compounds targeting a protein of interest [43]. Computational screening techniques have gained acceptance due to the fact that, compared to high-throughput screening approaches, they are able to reduce both time and cost by limiting the number of compounds which have to be experimentally tested [44].

There are two main strategies for *in silico* screening: ligand-based and structure-based virtual screening. This second approach can be applied when the 3D structure of a drug target is available from experimental studies (for instance, from X-ray crystallography) or accessible by molecular modeling (homology modeling). When a 3D structure of the target is available, high-throughput docking is the method of choice [45]. During this process, each of the screened compounds is docked in several possible conformations using a combination of shape matching and predictions both of favorable hydrogen bonding and of charge-charge interactions [46]. Ligand-based virtual screening, conversely, is used when no reliable structural information of the target can be obtained. Nowadays, this strategy is less and less common for class A GPCRs due to the increasing availability of receptor crystal structures. However, this approach is still used for other GPCR families where structural information is still scarce.

In the case of structure-based virtual screening, this technique performs naturally best when it makes use of the X-ray structure of a molecular target. Thus, published crystal structures of GPCRs are commonly used in virtual screening. Furthermore, homology models of closely-related receptors have also been proven to be suitable for structure-based virtual screening [47]. Interestingly, docking and screening results against new GPCR structures resulted in considerably better hit rates than those of soluble proteins [48]. This high success rate may be a consequence of the fact that compound databases, such as ZINC [49], contain a disproportionate number of molecules that resemble GPCR ligands. Furthermore, the well-buried GPCR orthosteric sites can almost entirely sequester or complement a small organic molecule, allowing it to be recognized with high ligand efficiency while, in the case of soluble proteins, binding sites are more heterogeneous and their surface is often larger, flatter and less buried with respect to the solvent [48].

Another option when applying virtual screening is to screen for compound fragments. Fragment-based virtual screening considers binding of small chemical fragments, which may bind only weakly to the sub-pockets of a given target (Fig. **4**). The next step is growing the identified fragments or combining them to produce a lead with higher affinity. Fragment-based screening has several advantages over classical virtual screening approaches: (1) it enables identification of more hydrophilic hits with a higher number of hydrogen bonds, which favor enthalpically-driven binding, (2) it results in smaller ligands with higher efficacy, (3) it is more effective as screening of N fragments is estimated to be equal to screening between N^2 and N^3 compounds, (4) it results in less sterically crowded hits. Virtual fragment screening is a powerful and commonly used screening approach to look for GPCR ligands. As an example, the performance of an agonist-bound A_{2A} adenosine receptor structure was evaluated for the retrieval of known agonists and then employed to screen for new fragments optimally fitting the corresponding sub-pocket [50]. Remarkably, of the 16 predicted high scoring candidates, 15 compounds had sub-micromolar affinity for the receptor, several of which had affinities reaching approximately 10 nM.

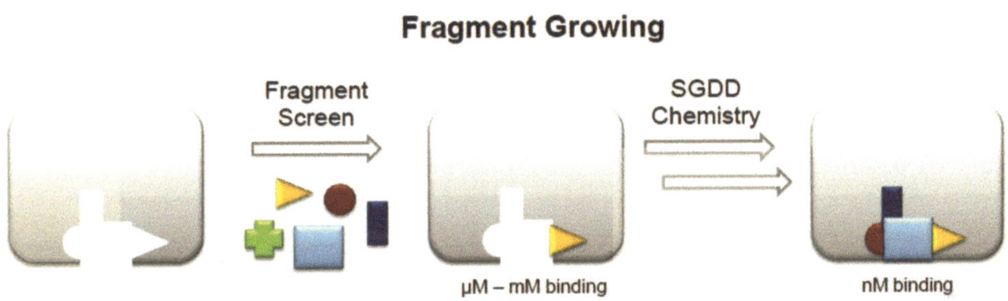

Figure 4: Fragment-based virtual screening. The method identifies fragments with micromolar activity which fit to parts of the binding pocket. The fragments are combined with linkers using structure-guided drug discovery (SGDD) yielding compounds with can reach nanomolar activity.

3.4. Advances in GPCR Virtual Screening

Aminergic GPCRs are particularly important drug targets. Thus, the resolution of a crystal structure of the β_2 adrenergic receptor [51] was especially welcome from the structure-based drug design GPCR community. This detailed receptor description enabled the first crystal structure-based virtual screening. Kolb *et al.* [52] docked about 1 million lead-like molecules against this receptor. Subsequently, they selected and tested 25 high-ranked molecules (considering chemical clustering, visual inspection, and favorable interaction with Asp3.32).

From these tested compounds, 6 were active with binding affinities below 4 μM, with the best molecule binding with a K_i of 9 nM. They also found that 5 of these molecules were inverse agonists, which is consistent with a carazolol-bound starting conformation. Furthermore, another interesting outcome of this study came as the predicted binding mode of the highest affinity hit was confirmed by X-ray studies. This degree of structural prediction shows that GPCR structure-based virtual screening may not only result in new ligands but also provide suitable starting points for the more challenging structure-based hit optimization [45]. Recently, a large library virtual screen against an activated β_2 adrenergic receptor structure resulted in the detection of GPCR binding compounds with a preferential retrieval of agonists over inverse agonists [53]. This study complements the previous virtual screening exercises against inverse agonist-bound GPCR structures which tended to yield inverse agonists. Thanks to their results, the authors conclude that the docking hits resulting from virtual screening campaigns are deeply related to the functional state of the conformation of the GPCR target. In the same line, and before the crystal structure of an activated β_2 adrenergic receptor was available, Schneider *et al.* [54] demonstrated that an homology model of this receptor based on the opsin crystal structure was better at retrieving active compounds in virtual screening experiments than the crystal structure of the inactivated form of the β_2 adrenergic receptor.

In another interesting approach, Kolaczkowski *et al.* [55] recently assessed how modifying receptor templates by induced fit docking could impact posterior virtual screening. In their study, they modeled dopaminergic D_1 and D_2 receptors. After constructing homology models of these receptors, they modified them by ligand-steered binding site optimization. According to the authors, these modified receptors performed better in the subsequent virtual screening experiments than typical homology models. In fact, they observed that the most important aspect determining success in virtual screening had to do with the ligand used in the induced fit docking. This ligand choice was, in fact, more determinant than the choice of crystal structure used to build the D_1 and D_2 receptor homology models (in this case, the β_2 adrenergic and the dopaminergic D_3 receptors).

Another example of the application of a GPCR X-ray structure in virtual screening used the purinergic adenosine A_{2A} receptor. Antagonists of these receptors may be used to treat a wide range of conditions including Parkinson disease, inflammation, cancer, ischemia reperfusion injury, sickle cell disease, diabetic nephropathy, infectious diseases or CNS disorders [56]. After the publication of the crystal structure of the adenosine A_{2A} receptor in complex with an antagonist, Katritch *et al.* [57] performed molecular docking and virtual screening of more than 4 million commercially available drug-like and lead-like compounds. After virtual screening, the highest 56 ranking compounds were tested *in vitro*. Of the tested compounds, 23 presented affinities under 10 μM, 11 of those with sub-μM affinities and two compounds with affinities under 60 nM. Moreover, these hits were characterized by their chemical diversity, as they belonged to at least 9 different chemical scaffolds and were characterized by very high ligand efficiency. For this reason, the authors conclude that their screening strategy could represent a starting point for the search of drug discovery leads. Furthermore, and as expected given the fact that the initial crystal structure was a complex featuring an antagonist, 11 out of 14 compounds tested in a functional assay were able to effectively block more than 75% cAMP generation at a concentration of 10 nM, which strongly supports their antagonistic activity.

Using an original strategy to find new ligands for the CXC chemokine receptor 7 (CXCR7), a potential drug target for cancer chemotherapy, Yoshikawa and co-workers [58] took advantage of their experience in the prediction of the ligand binding pocket of CXCR4 in GPCR Dock 2010. Using their method, they modeled the CXCR7 receptor structure using the CXCR4 receptor as a general template, but also incorporating information from other crystallized class A GPCR structures in order to cover a higher conformational space, and performed virtual screening of around 800000 commercially available drug-like compounds. From this screening experiment, 626 candidate compounds were selected, 21 of which presented IC_{50} values of 1.29-11.4 μM upon experimental characterization. In another approach, centered likewise in the evaluation of which X-ray structural information may be more suitable for modeling and virtual screening, Pala *et al.* [59] evaluated a set of MT_2 melatonin receptor models. Besides considering different structural templates for homology modeling, the authors also analyzed the impact of allowing for binding pocket readjustments by applying induced fit techniques. To do so, they used known MT_2 melatonin receptor ligands in which mutagenesis information on binding structural determinants was known. According to their results, the importance both of the template choice and model structural refinement for screening results was confirmed. In another interesting example, and in order to perform virtual screening at the GPR17 receptor, a GPCR responding to both uracil nucleotides and cysteinyl-leukotrienes which has been proposed as a target for neurodegenerative diseases, Eberini *et al.* [60] created a structural model of this receptor built as a chimera of four homology templates. In this way, the authors modeled GPR17 loops from different crystallized receptors according to their higher degree of homology to the receptor of interest, and then performed a high-throughput virtual screening exploration. To do so, they screened more than 130,000 lead-like compounds from which they were capable of identifying 4 full agonists, with a better potency than their reference ligand.

Virtual screening has not only been explored to identify orthosteric ligands of GPCRs but also to find allosteric modulators. As we have previously mentioned, due to the fact that allosteric modulators act by modifying physiological activation of receptors, these compounds can provide improved selectivity and safety, a ceiling effect preventing overdosage, high receptor selectivity, or even activation pathway selectivity along with maintenance of spatial and temporal determinants of GPCR signaling [61]. Allosteric drugs can thus help in the problem of drug dependence, overdose risk and other adverse effects of orthosteric drugs. However, structure-based virtual screening for allosteric GPCR ligands has been hampered by the lack of structural data for allosteric binding sites [62]. Despite this difficulty, some groups have successfully applied virtual screening to search for new GPCR allosteric modulators. Lane *et al.* [63] for instance, used two models of the dopaminergic D_3 receptor in its apo form and in complex with dopamine to screen a library of 4.1 million compounds. The top 150 compounds for each of the two receptor models were selected for further re-docking and assessment. After selecting chemically diverse scaffolds and discarding ligands with a high similarity to already described D_3 ligands, the authors selected 25 compounds per receptor model and purchased them from chemical vendors. Interestingly, the compounds derived from the model in complex with dopamine proved to have very attractive profiles in D_3, but also in D_2 dopaminergic receptors, by behaving as non-competitive negative allosteric modulators at these receptors.

In summary, the increasing number of GPCR crystal structures and more and more accurate homology models enable successful structure-based virtual screening as exemplified above. Yet, structure-based identification of novel ligands for GPCRs with low homology to the currently available GPCR crystal structures (*e.g.*, class B and class C GPCR allosteric ligands) is still a challenging task and, at this point, ligand-based virtual screening may be a useful alternative.

3.5. Successful Optimization of Lead Structures

Despite the increasing amount of structural data for GPCR targets with atomic resolution, lead optimization has been rarely reported in the literature. Nevertheless, a few reported studies indicate that GPCR models can be efficiently applied in lead optimization obtaining compounds with improved affinity or physicochemical parameters (reviewed in [64]).

For example, one very recent study by Andrews *et al.* demonstrates the successful optimization of affinities and selectivity of antagonists at the adenosine A_{2A} receptor *versus* the A_1 receptor while preserving a balanced, drug-like profile [65]. In a first step, GRID maps using different molecular probes (*e.g.* sp^3 carbon, sp^2 carbon, NH or C=O) were constructed in order to obtain a comparison of both receptors with respect to their shape, size and electrostatics. In a second step, inspection of the GRID maps as well as of the size of the binding pocket suggested the addition of a small lipophilic substituent into the 1,2,4-triazine antagonist series for obtaining higher affinity and selectivity at the A_{2A} *versus* the A_1 receptor. Remarkably, the success of this strategy was later demonstrated in experimental binding studies and stresses the potential of lead optimization using structure-based approaches.

3.6. A Dynamic View on Different Receptor Conformational States

All the aforementioned structure-based techniques tend to look at receptor-ligand interactions from a static point of view. However, we are nowadays aware that GPCRs are inherently flexible and that understanding drug action on these receptors requires considering their ability to explore different conformational states. In this sense, a new view has originated, which understands these receptors as flexible structures capable of transitioning between multiple conformational states with different capabilities to couple to signaling partners [66] (Fig. **5**). This change of paradigm raised awareness on the importance of characterizing these different conformational states. Besides being necessary to understand the structural basis of GPCR function, description of these receptor states could be the basis for developing ligands capable of stabilizing particular receptor conformations associated to a precise signaling outcome (biased agonists). In this scenario, molecular dynamics have proven to be an informative method to interrogate receptor conformation and to analyze the impact of ligand binding on receptor structural stabilization.

Ligand-mediated receptor stabilization has been the focus of several MD simulation studies trying to correlate receptor signaling levels to ligand-induced receptor states. Lee *et al.*, for instance analyzed the dynamic behavior of both adenosine and UK432097 at the A_{2A} adenosine receptor starting from their crystal structures [67]. In their simulations, they observed that UK432097 was capable of forming a wider hydrogen bond network with the receptor than the natural agonist adenosine. UK432097 also showed a higher degree of stability in the receptor binding pocket. The authors relate this higher amount of interaction of UK432097, which possesses a higher potency than adenosine, to the stabilization of a decreased number of receptor conformations characterized by a high G protein activation capacity. Studies like this one are however scarce due to the fact that receptor crystal structures in complex with differently behaving ligands are unfortunately not available. Therefore, several methods to explore conformational changes related to receptor association with particular ligands have been developed. One of such methods is the *ligand-induced transmembrane rotational conformational changes* (LITiCon) method created by Bhattacharya and coworkers [68]. In order to sample receptor conformational space, the LITiCon method generates a set of conformations by systematically rotating transmembrane helices in the vicinity of the ligand. Out of all these conformations the ones with the lowest ligand-binding energy and higher degree of ligand-receptor interactions are selected. This method was applied for the first time to analyze β_2-adrenergic receptor conformations in complex with compounds with activities covering from the full agonist to the inverse agonist spectrum. These ligands proved to be able to differentially engage several receptor switches implicated in receptor activation and inactivation like the ionic lock formed between positions R3.50 and E6.30 or the rotamer toggle switch of W6.48 on TM6. Using a different approach, Provasi *et al.* took advantage of the availability of the active state of the β_2-adrenergic receptor in complex with a nanobody to apply an adaptive biasing method. Using this technique, they were able to determine ligand-specific conformations along the transition between activated and inactivated receptor states [69]. Analysis of the free energy landscapes of receptors in complex with ligands inducing different behaviors led the authors to conclude that, while inverse agonists tend to stabilize a single receptor state, complexes with agonists could sample more easily different receptor conformations.

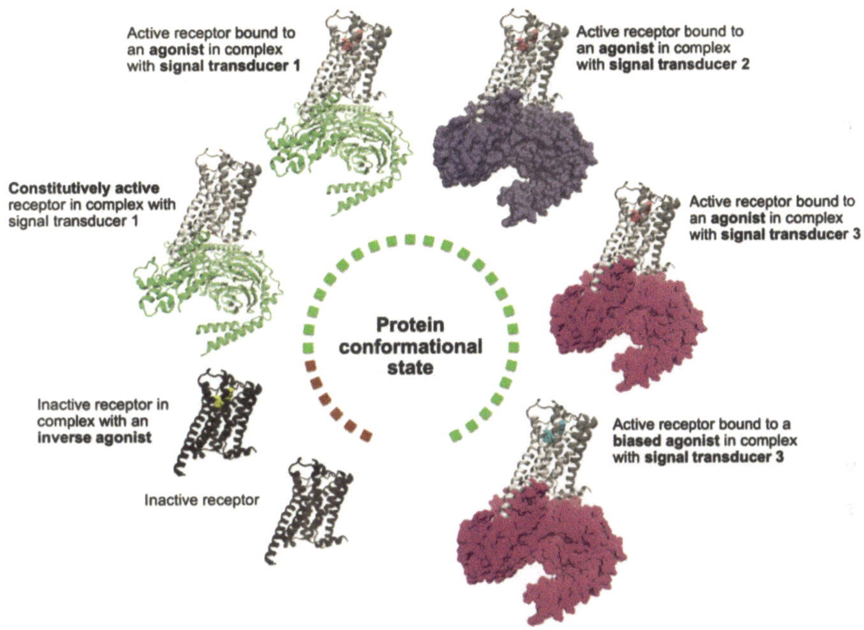

Figure 5: Different receptor conformational states couple preferentially to particular signal transducers. In this schematic picture, we can see how a single receptor can potentially exist in different conformational states associated with particular signaling outcomes. In this way, receptors can adopt inactive states both in the absence of ligands and when in complex with an inverse agonist. However, some unliganded receptors can also signal when they are in their apo form, thus presenting constitutive activity. Besides that, a single receptor can also exist in different conformational states with a particular preference to couple to intracellular signal transducers. These different receptor conformations can in turn be stabilized by compounds which either bind indiscriminately to all of these receptor states - in the case of (unbiased) agonists - or stabilize a particular conformation with a preference to couple to one signal transducer - in the case of biased agonists.

3.7. Monitoring the Receptor Activation Process

At present, there exists the conviction that understanding the process by which receptors activate and inactivate can yield information of special value for the rational design of GPCR drugs. However, simulating these transitions is still a challenge due to the computational cost of simulating large conformational changes and also to the relative scarcity of crystal structures of receptors in different activation states. In an effort to overcome the difficulty to analyze receptor activation by classical MD, which has been until now computationally unamenable due to the fact that, according to experimental evidence, this process may require milliseconds to take place [70], different groups have explored original solutions to understand this process.

One of these approaches consisted in the use of a Monte Carlo algorithm by Bhattacharya *et al.* to derive ligand-dependent β_2-adrenergic receptor activation pathways from previously-generated LITiCon poses [71]. Some ligand-receptor complexes derived from these activation pathways were used in a subsequent study to perform all-atom MD [72]. In the case of rhodopsin, crystallization of its active state allowed Provasi *et al.* to perform biased MD simulations which pointed to the existence of at least four metastable states between its photoactivated state and its opsin-like conformation [73]. The subsequent crystallization of the β_2-adrenergic receptor in complex with a G protein-mimetic nanobody was also used as the basis to analyze the process of receptor activation. In this case, Nygaard *et al.* performed an unbiased MD study of unprecedented computational magnitude [74]. In order to analyze receptor activation, they evaluated in detail the opposite process: conversion from an active conformation to an inactive one. Their simulations pointed to the singular conclusion that receptor activation may begin in the receptor region implicated in G protein binding. In this way, agonists would be responsible for stabilizing receptor states possessing an activated G protein binding site but would not be able to lock the receptor in its active state by themselves. In a different approach, Miao *et al.* applied accelerated molecular dynamics to study the process of receptor activation at the muscarinic M_2 receptor [75]. Using this simulation technique, they compared the receptor in complex with the antagonist 3-qui-nuclidinyl-benzilate, present in the crystal structure, to the receptor in its apo form. Observation of these two systems confirms the inability of the receptor in complex with an antagonist to activate and also the ability of the unbound receptor to transition to the active state upon application of dual-boost accelerated MD. According to these simulations, receptor activation is associated to the formation of a hydrogen bond between residues Tyr206$^{5.58}$ and Tyr440$^{7.53}$ in the G protein binding site and to an outward tilt of the cytoplasmic end of TM6 (associated to the disruption of the TM3/6 ionic lock). Besides, by analyzing community networks across the receptor structure, the authors observed that, at an intracellular level, the strength of the overall network between helices in the active apo receptor is significantly weaker when compared to the inactive and intermediate states. In particular, TM6 becomes loosely connected to TM3, TM5, and TM7, which probably allows for its tilting and adoption of an active conformation.

Finally, Kohlhoff and coworkers [76] recently made use of Google's Exacycle cloud-computing platform to perform tens of thousands of independent simulations of the β_2-adrenergic receptor. To do so, they started their simulations using inactive and active receptor crystal structures in their apo form and in complex with a partial inverse agonist and a full agonist. In order to characterize the transition between active and inactive receptor states, the authors built 3,000-state Markov State Models (MSM), using clustering along four structural metrics representing structural activation and inactivation features, and mapped out the transitions between all states. Using these models, they were able to generate 150 μs activation trajectories, which highlight the ability of the agonist (BI-167107) to strengthen correlations between extracellular and intracellular residue groups to stabilize active states. In contrast, these correlations become disconnected in the presence of an inverse agonist (carazolol) and appear indiscriminate in the case of the apo receptor. They also observe that docking to MSM states can facilitate the detection of receptor ligands acting both as agonists and antagonists, as well as expand ligand chemical space, an observation which could be especially an advantage in virtual screening approaches.

Albeit limited, at present, there exists some information on the structural basis of G protein coupling to different receptors. In this sense, crystallization first of opsin in complex with the C terminus of the transducin Gα subunit, and later of the β-adrenergic receptor in complex with Gs, have opened new opportunities for the study of GPCR / G protein structural crosstalk. Goetz *et al.*, for example, studied how G protein binding to the β-adrenergic receptor was affected by binding of either the inverse agonist carazolol or the agonist isoprenaline at the orthosteric binding site [77]. In this case, they took the C terminus of the transducin Gα subunit as a template and used it as a surrogate of the Gs protein. According to their results, the presence of the Gα fragment was able to induce an enlargement of the agonist binding pocket. Besides, presence of carazolol in the receptor binding pocket seemed to destabilize G protein binding, a phenomenon which was not observable in the complex with isoprenaline. Subsequently, the publication of the structure of the β_2-adrenergic receptor in complex with Gs led Feng *et al.* [78] to analyze the stability of this complex. According to their results, removal of the nanobody which had been used for the crystallization of this receptor, led to a structural reorganization which started at the agonist binding pocket and was transmitted to the G protein coupling region of the receptor to finally reach the G protein alpha subunit. More recently, Kling and coworkers used the crystallized structure of the ternary complex, together with a homology model of the dopaminergic D_2 receptor in complex with the Gαi subunit, to perform MD simulations [79]. Their analysis, which also included free energy calculations after computational alanine-scanning mutagenesis of the receptor / G protein interface, identified distinct hot-spots important for receptor / G protein selectivity. Interestingly, they observed that hydrophobic interactions could be crucial for coupling of the β_2-adrenergic receptor in complex with Gs, while dopaminergic D_2 receptor coupling to Gi could be mainly determined by ionic interactions between basic amino acids of receptor and negatively charged amino acids of this G protein subtype.

3.8. Analyzing Ligand-GPCR Binding Paths

Another interesting phenomenon of GPCR modulation whose study is becoming increasingly available computationally has to do with ligand binding. Understanding this process is especially attractive from a drug design perspective as it can help rationalize determinants of ligand kinetics and binding, as well as help pinpoint possible receptor hotspots capable of binding allosteric modulators. Dror and coworkers have also analyzed this process using their unbiased molecular dynamics simulation techniques [80]. In particular, they analyzed binding of three antagonists - propranolol, alprenolol and dihydroalprenolol - and the agonist isoproterenol to the

β_2-adrenergic receptor. Interestingly, their simulations detected a receptor vestibular region - located between the extracellular loops 2 and 3 and the helices 5, 6 and 7 – which was visited by all the compounds tested. The authors postulate that this intermediate region could correspond to the binding site of some allosteric modulators such as gallamine, which would exert their effects by blocking the entrance and exit of ligands targeting the orthosteric binding pocket. A similar approach was later used to study the process of tiotropium binding to the muscarinic M_2 and M_3 receptors [81]. In this study, tiotropium was also capable of binding an intermediate *vestibule* region which could help rationalizing the experimental observation that some orthosteric ligands can also act as allosteric modulators of muscarinic receptors. Another interesting conclusion arising from these simulations was that the different rate of dissociation of this drug at the M_2 and M_3 receptors could help explaining clinically important 'kinetic selectivity' of thiotropium for M_3 receptors despite similar equilibrium binding affinities at the two types of receptors. In a very recent publication, Dror *et al.* also used MD simulations to clarify the structural determinants of allosterism at muscarinic M_2 receptors [82]. They performed unbiased simulations that allowed characterizing the binding pathways of both positive and negative allosteric modulators. Their results, which were further validated by mutagenesis studies, suggest a common mechanism of binding for the structurally divergent allosteric modulators. These compounds would be capable of establishing cation-π interactions with two pairs of tyrosine residues which would form two 'binding centers' in the extracellular vestibule of the receptor. On the other hand, the authors also assessed the interplay between orthosteric and allosteric binding in these receptors. To do so, they simulated systems including negative and positive allosteric modulators together with the orthosteric antagonist N-methylscopolamine. Using this setup they identified two major drivers of allosteric modulation: i) the electrostatic repulsion between allosteric and orthosteric ligands, which depended both on their charges and on charge spatial proximity and ii) the stabilization of open or closed allosteric and orthosteric binding pockets by positive and negative allosteric modulators respectively.

Finally, some MD simulations have specifically focused on the capacity of ions to function as allosteric modulators of receptor structure. As an example, Selent and coworkers [83] investigated how allosteric binding of sodium ions to the dopaminergic D_2 receptor could impact receptor structure. According to their results, binding of sodium ions into a deep allosteric site near Asp2.50 could be responsible for locking the rotamer toggle switch W6.48 on TM6 in a distinct conformational state. Notably, the existence of this sodium binding site was later observed in the high resolution crystal structure of the human A_{2A} adenosine receptor [84]. In the same line, this time starting from the structure of the A_{2A} adenosine receptor, Gutiérrez-de-Terán *et al.* recently analyzed the allosteric effects of sodium and the allosteric small molecule amiloride in receptor activation [85]. In this study, the authors took advantage of the availability of crystal structures of the receptor in complex with both agonists and antagonists. According to their simulations, which they complemented with binding and thermostability assays, they suggest that, when either a sodium ion or amiloride binds to the allosteric pocket of the A_{2A} adenosine receptor, they are capable of stabilizing an inactive conformation which hampers agonist binding.

As we have seen, molecular dynamics simulations can yield revealing information on receptor stability and on the mechanisms by which different modulators can modify the equilibrium between different receptor populations. The rapidly increasing computational resources together with new crystal structure information on GPCRs will surely allow getting a deeper understanding on the basis of receptor transitions and help guide the design of ligands stabilizing particular conformational states.

3.9. Studying Higher Order Receptor Complexes to Search for New GPCR Modulators

In the past years, the characterization of GPCRs forming dimers or higher-order oligomers has challenged the classical view in which these receptors were believed to function as monomeric units. As a result, during the last decade, the organization of GPCRs in cell membranes has been a matter of intense study. On the one hand, while certain class A GPCRs can effectively function as monomers [86], these proteins still have the ability to exist as higher-order complexes. For instance, a monomeric unit of rhodopsin, the first purified GPCR, is sufficient to fully activate transducin [87], its cognate G protein. And yet, atomic force microscopy experiments have demonstrated that rhodopsin molecules organize as dense arrays of dimers in native disc membranes [88]. On the other hand, the quaternary structure of GPCRs does not display the same level of stability across different families. Thus, whereas most class C GPCRs form dimers, stably linked by a covalent disulphide bridge [89], class A GPCRs can engage in both stable and transient interactions [90-92]. Although the biological significance of these findings still needs a deeper characterization, modulating the stability of GPCR dimers or oligomers may become soon subject to the development of new drugs.

In order to better characterize receptor-receptor interactions and to detect compounds capable of modulating them, several computational approaches are helping to guide and complement available pharmacological evidence. In the past years, the recent resolution of different crystal structures of GPCR homodimers [93, 94] and homooligomers [95, 96] has helped in the difficult task of modeling GPCR complexes. These crystal structures are revealing new data on potential dimerization interfaces, which further enrich the computational modeling techniques used to study GPCR dimers and oligomers. These techniques can be generally divided in sequence- and structure-based methods (see [97] for a complete and well-organized review on this topic). In the common scenario where structural data is lacking, sequence-based techniques exploit the vast amount of information contained on protein sequences to predict residues and/or domains involved in putative dimerization interfaces. Due to the advent of new GPCR crystal structures during the last years, structure-based methods are being intensely used instead to unravel new features of GPCR dimerization. In this respect, some GPCR crystal structures deposited in the Protein Data Bank (PDB) have provided the first hints on GPCR dimerization modes. This has allowed advancing receptor-receptor docking, as most protein-protein docking methods involve an initial searching step followed by sampling and refinement phases to, respectively, disregard decoy poses and filter out non-desired dimers. In the past years, protein-protein docking studies have considered the interfaces present in the new crystal structures of GPCR dimers to select relevant interfaces for dimers of related receptors [97]. Furthermore, the improvement of traditional docking algorithms in an effort to reflect, for example, the membrane environment surrounding GPCRs, has converted docking into one of the structure-based methods of choice to study GPCR dimers. In this line, a collection of docking servers offer nowadays an overall automation to predict new receptor interfaces by protein-protein docking methods. Several studies have already taken advantage of these tools to characterize different GPCR dimers (reviewed in [97]).

Protein-protein docking methods can only yield a static representation of the receptor dimerization phenomenon. In contrast, other techniques, such as MD, attempt to study the dynamic nature of this process. In a recent example, Rodríguez *et al.* analyzed the dynamics of the CXCR4 taking advantage of the crystallization of the inactive state of its homodimer [98]. In their work, the authors analyze the impact of binding of a co-crystallized small molecule, the antagonist IT1t, and of the cyclic peptide CVX15. Comparison of ligand effects on binding site stability reveals that, in contrast to the peptidic ligand, IT1t produces a negligible effect at the binding pocket level. Hence, the conformation resulting from the simulation of the apo form of the dimer could be appropriate for evaluating the binding of small organic molecules exploring the same binding site region.

However, in the cases in which the dimeric interface has not been crystallographically determined, prediction of relevant dimerization interfaces is still a challenge which can be explored by MD simulations. In this case, the complexity needed to represent systems including several receptors in biologically-relevant conditions, and the necessary length of simulations studying GPCR association, have forced computational scientists to search for alternatives to all-atom molecular representations in MD simulations. Thus, coarse-grained MD simulations are nowadays the preferred tool to study GPCR oligomerization as these simulations are in the range of 2-3 times faster than all-atom MD [99] (Fig. **6**). In this way, the use of such methods could serve as a first approximation to detect relevant receptor-receptor interfaces for the development of ligands exploiting dimerization. As an example, Filizola's group has elegantly exploited a combination of biased, non-biased, all-atom and coarse-grained simulations techniques to study the interface of different GPCRs. Whereas Provasi *et al.* [100] predicted dimer association constants by studying the lifetime of δ-opioid receptor homodimers by umbrella sampling coarse-grained simulations, Johnston *et al.* [101] later worked on two different arrangements of the same dimer by coarse-grained well-tempered metadynamics. Recently, Johnston *et al.* [102] compared the relative stability of two dimerization interfaces of β_1 and β_2-adrenergic receptor homodimers, two closely related receptors. By a combined approach using both coarse-grained and all-atom simulations, the authors conclude that H1/H8 interface is the most stable in both receptors, a finding described for rhodopsin homodimers almost simultaneously by Periole *et al.* [103]. These studies illustrate the important effort that has been made in recent years to understand the thermodynamics and kinetics of GPCR oligomerization by combining state-of-the-art computation techniques.

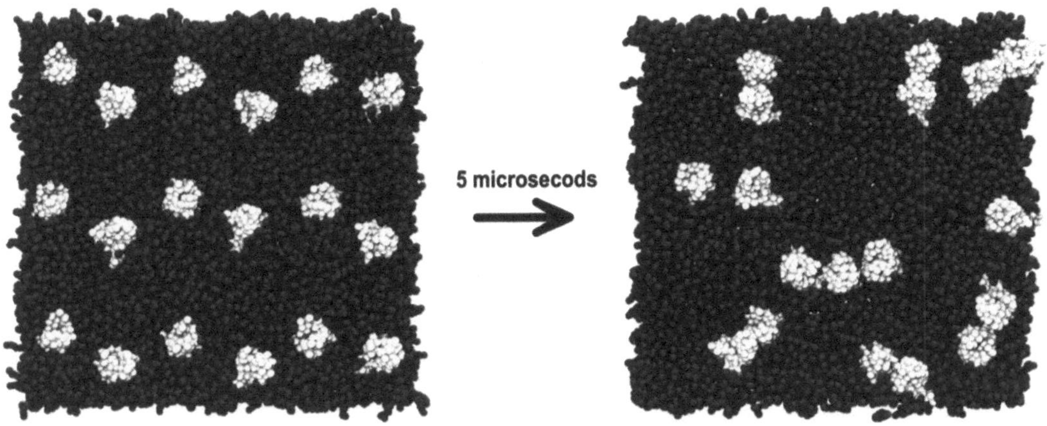

Figure 6: Coarse-grained simulations of GPCRs. The figure represents a top view of two snapshots of a CG simulation of 18 GPCR monomers (white) embedded in a membrane (grey). Water and ions are omitted for clarity. The simulation time, 5 μs, is enough for GPCR monomers to freely diffuse and rotate so that they find each other and ultimately form dimers. These type of approaches allow simulating the formation of GPCR dimers and oligomers, which would not be amenable using classical all-atom molecular dynamics [104].

3.10. Strategies for the Design of Ligands Targeting Receptor Complexes

The above mentioned techniques can provide physiologically-relevant receptor dimer models that represent important starting structures for developing ligands targeting GPCR complexes *in silico*. These compounds have received a special attention due to the fact that binding to heteromers could lead to safer drug candidates with potential higher tissue selectivity. Based on the aforementioned crosstalk phenomenon, where the function of one GPCR protomer can be affected by the other protomer [105], one treatment strategy could be based in targeting just one GPCR subtype using monovalent ligands. This strategy could either be based on the interaction of the ligand with one of the orthosteric or allosteric sites of the complex or, alternatively, on the stabilization or disruption of the receptor-receptor interface. In parallel, the rational design of ligands consisting of two pharmacophoric entities (bivalent ligands), which are able to bind both protomers of the complex at the same time, is also an active research field. This 'dual-mode' strategy enables a selective targeting of GPCR heteromers by assembling two

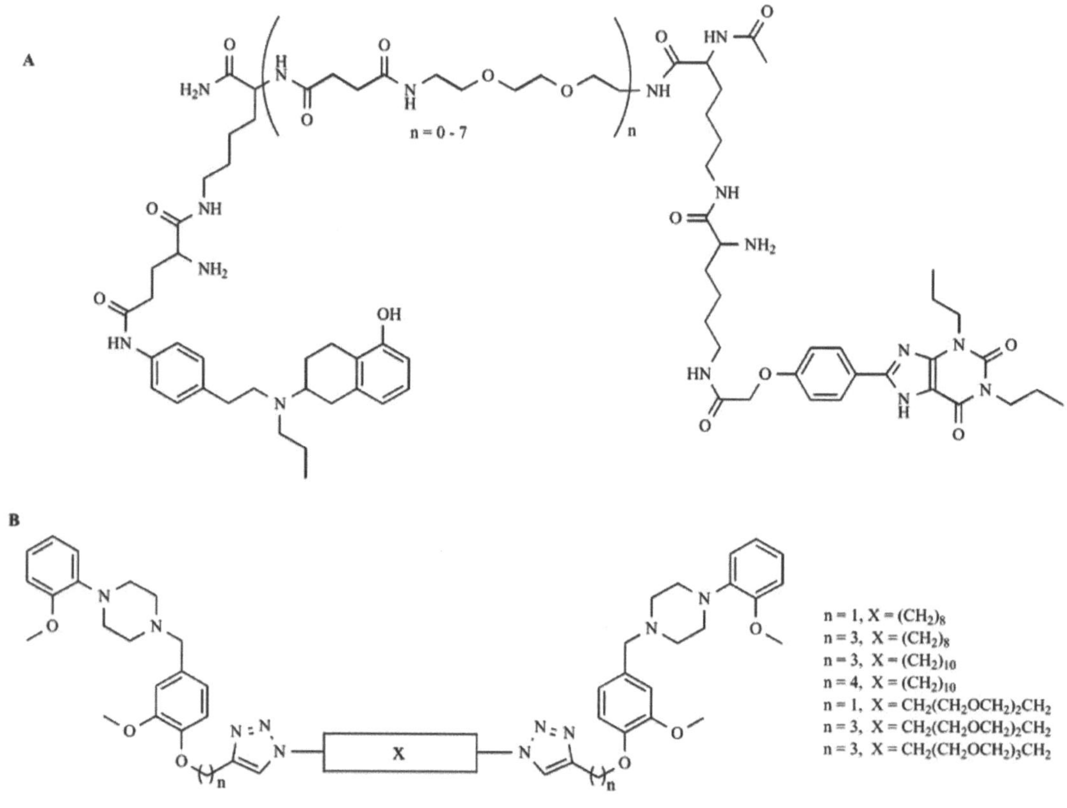

Figure 7: Examples of GPCR bivalent ligands. The figure, reprinted with permission from [109], shows an example of hetero- (A) and homobivalent ligands (B) targeting the A_{2A}-D_2 heteromer and D_2-D_2 homomers, respectively.

different pharmacophores within the same molecule or even of GPCR homomers by simply using the same pharmacophore twice. For example, homobivalent ligands consisting on identical pharmacophores (*e.g.* 1,4-disubstituted aromatic piperidines or piperazines) can be used to target D_2 receptor homomers [106]. In contrast, heterobivalent ligands comprised of one D_2 agonist (*e.g.* a PPHT derivative) and one A_{2A} antagonist (*e.g.* a xanthine derivative) have been designed to target the A_{2A}-D_2 heteromer as a potential therapy in Parkinson's disease [107]. The design of these ligands involves the tethering of both pharmacophoric entities by a spacer that is able to provide both a particular length and enough conformational flexibility to allow the accommodation of the ligand in both binding pockets (Fig. 7) [108]. The major drawback of this type of ligands comes from their high molecular weight and hydrophobicity, provided by the long alkyl spacers needed to bridge the receptor-receptor interface. As a result, poor absorption properties frequently hamper the druggability of bivalent ligands and, at present, they are generally used as chemical tools to study dimer behavior.

All in all, computational techniques for modeling GPCR dimer- and oligomerization have undergone an important development over the last decade. From more classical sequence-based approaches to the evolving field of MD simulations, computer modeling holds promise for guiding the rational design of new molecular probes, and also of new drug candidates targeting GPCR dimers and oligomers.

4. COMPUTATIONAL GPCR DRUG DISCOVERY: CHALLENGES AND CONCLUSIONS

As we have seen along this chapter, GPCRs are not only the most important known drug targets, but they also constitute an open area of research to obtain new solutions for unmet medical needs. This continuing importance of GPCRs as drug targets can be seen in the analysis of recent approvals by the Food and Drug Administration. In this sense, between 2010 and 2012, almost 20% of new approved drugs targeted these receptors [110].

The increasing amount of public data on GPCRs, their ligands and their binding partners has opened new opportunities for *in silico* approaches capable of using this information to gain a deeper understanding on these receptors. In this sense, for instance, information on ligand binding affinities for different GPCRs can be used in chemogenomics approaches capable of selecting new compounds for receptor deorphanization. In parallel, the increasing amount of experimentally-determined structural information, together with recent advances in computational power and simulation software, allow obtaining more accurate models for virtual screening, as well as gaining a dynamic view on unknown receptor conformational states and their modulation by different interaction partners (ligands, signal transducers or GPCRs). In addition, new information on activity outcomes induced by particular ligands has created the possibility to study the basis of biased agonism, and also to design compounds capable of interrogating the importance of particular pathways in health and disease from a systems pharmacology perspective.

From our viewpoint, new knowledge on GPCR functioning, despite adding new layers of complexity with regard to drug action at this type of targets, will finally help obtaining safer and more effective therapies exploiting phenomena such as drug promiscuity, biased agonism, allosteric modulation or receptor oligomerization.

AKNOWLEDGEMENTS

M M-S is supported by a doctoral fellowship from the University and Research Secretariat of the Catalan Government and the European Social Fund (2013FI_B00143). AAK contributed to this chapter during a postdoctoral stay at University of Eastern Finland, Kuopio, Finland, under a Marie Curie fellowship. She also wants to acknowledge the project "The equipment of innovative laboratories doing research on new medicines used in the therapy of civilization and neoplastic diseases" within the Operational Program Development of Eastern Poland 2007-2013, Priority Axis I Modern Economy, operations I.3 Innovation promotion. This work was also funded by La MARATÓ de TV3 Foundation; Grant number: 091010; and the Ministerio de Educación y Ciencia; Grant number: SAF2009-13609-C04-04. JS acknowledges support from the Instituto de Salud Carlos III FEDER (CP12/03139) and from the GLISTEN European Research Network.

CONFLICT OF INTEREST

The authors confirm that this chapter contents have no conflict of interest.

REFERENCES

[1] Takeda, S.; Kadowaki, S.; Haga, T.; Takaesu, H.; Mitaku, S. Identification of G Protein-Coupled Receptor Genes from the Human Genome Sequence. *FEBS Lett.* **2002**, *520*, 97–101.

[2] Overington, J. P.; Al-Lazikani, B.; Hopkins, A. L. How Many Drug Targets Are There? *Nat. Rev. Drug Discov.* **2006**, *5*, 993–996.

[3] Fredriksson, R.; Lagerström, M. C.; Lundin, L.-G.; Schiöth, H. B. The G-Protein-Coupled Receptors in the Human Genome Form Five Main Families. Phylogenetic Analysis, Paralogon Groups, and Fingerprints. *Mol. Pharmacol.* **2003**, *63*, 1256–1272.

[4] Hopkins, A. L.; Groom, C. R. The Druggable Genome. *Nat. Rev. Drug Discov.* **2002**, *1*, 727–730.

[5] Audet, M.; Bouvier, M. Restructuring G-Protein-Coupled Receptor Activation. *Cell* **2012**, *151*, 14–23.

[6] Rasmussen, S. G. F.; DeVree, B. T.; Zou, Y.; Kruse, A. C.; Chung, K. Y.; Kobilka, T. S.; Thian, F. S.; Chae, P. S.; Pardon, E.; Calinski, D.; Mathiesen, J. M.; Shah, S. T.; Lyons, J. A.; Caffrey, M.; Gellman, S. H.; Steyaert, J.; Skiniotis, G.; Weis, W. I.; Sunahara, R. K.; Kobilka, B. K. Crystal Structure of the B2 Adrenergic Receptor-Gs Protein Complex. *Nature* **2011**, *477*, 549–555.

[7] Venkatakrishnan, A. J.; Deupi, X.; Lebon, G.; Tate, C. G.; Schertler, G. F.; Babu, M. M. Molecular Signatures of G-Protein-Coupled Receptors. *Nature* **2013**, *494*, 185–194.

[8] Wang, C.; Wu, H.; Katritch, V.; Han, G. W.; Huang, X.-P.; Liu, W.; Siu, F. Y.; Roth, B. L.; Cherezov, V.; Stevens, R. C. Structure of the Human Smoothened Receptor Bound to an Antitumour Agent. *Nature* **2013**, 1–8.

[9] Hollenstein, K.; Kean, J.; Bortolato, A.; Cheng, R. K. Y.; Doré, A. S.; Jazayeri, A.; Cooke, R. M.; Weir, M.; Marshall, F. H. Structure of Class B GPCR Corticotropin-Releasing Factor Receptor 1. *Nature* **2013**, *499*, 438–443.

[10] Urban, J.; Clarke, W.; von Zastrow, M.; Nichols, D. E.; Kobilka, B.; Weinstein, H.; Javitch, J. A. Functional Selectivity and Classical Concepts of Quantitative Pharmacology. *J. Pharmacol. Exp. Ther.* **2007**, *320*, 1–13.

[11] Martí-Solano, M.; Guixà-González, R.; Sanz, F.; Pastor, M.; Selent, J. Novel Insights into Biased Agonism at G Protein-Coupled Receptors and Their Potential for Drug Design. *Curr. Pharm. Des.* **2013**, *19*, 5156–5166.

[12] Walters, R.; Shukla, A.; Kovacs, J. J.; Violin, J. D.; Dewire, S. M.; Lam, C. M.; Chen, J. R.; Muehlbauer, M. J.; Whalen, E. J.; Lefkowitz, R. J. β-Arrestin1 Mediates Nicotinic Acid–induced Flushing, but Not Its Antilipolytic Effect, in Mice. *J. Clin. Invest.* **2009**, *119*, 1321–1321.

[13] Salahpour, A.; Angers, S.; Bouvier, M. Functional Significance of Oligomerization of G-Protein-Coupled Receptors. *Trends Endocrinol. Metab.* **2000**, *11*.

[14] Milligan, G. G Protein-Coupled Receptor Hetero-Dimerization: Contribution to Pharmacology and Function. *Br. J. Pharmacol.* **2009**, *158*, 5–14.

[15] Filizola, M. Increasingly Accurate Dynamic Molecular Models of G-Protein Coupled Receptor Oligomers: Panacea or Pandora's Box for Novel Drug Discovery? *Life Sci.* **2010**, *86*, 590–597.

[16] Christopoulos, A. Allosteric Binding Sites on Cell-Surface Receptors: Novel Targets for Drug Discovery. *Nat. Rev. Drug Discov.* **2002**, *1*, 198–210.

[17] Brown, E. M. Clinical Utility of Calcimimetics Targeting the Extracellular Calcium-Sensing Receptor (CaSR). *Biochem. Pharmacol.* **2010**, *80*, 297–307.

[18] Allen, J. a; Roth, B. L. Strategies to Discover Unexpected Targets for Drugs Active at G Protein-Coupled Receptors. *Annu. Rev. Pharmacol. Toxicol.* **2011**, *51*, 117–144.

[19] Roth, B. L.; Sheffler, D. J. Magic Shotguns *Versus* Magic Bullets: Selectively Non-Selective Drugs for Mood Disorders and Schizophrenia. *Nat. Rev. Drug Discov.* **2004**, *3*, 3–9.

[20] Klabunde, T.; Hessler, G. Drug Design Strategies for Targeting G-Protein-Coupled Receptors. *Chembiochem* **2002**, *3*, 928–944.

[21] Austin, R. P.; Barton, P.; Bonnert, R. V; Brown, R. C.; Cage, P. A.; Cheshire, D. R.; Davis, A. M.; Dougall, I. G.; Ince, F.; Pairaudeau, G.; Young, A. QSAR and the Rational Design of Long-Acting Dual D 2 -Receptor / 2 -Adrenoceptor Agonists J. Med. Chem. **2003**, 3210–3220.

[22] Cumming, J. G.; Davis, A. M.; Muresan, S.; Haeberlein, M.; Chen, H. Chemical Predictive Modelling to Improve Compound Quality. *Nat. Rev. Drug Discov.* **2013**, *12*, 948–962.

[23] Custodi, C.; Nuti, R.; Oprea, T. I.; Macchiarulo, A. Fitting the Complexity of GPCRs Modulation into Simple Hypotheses of Ligand Design. *J. Mol. Graph. Model.* **2012**, *38*, 70–81.

[24] Brianso, F.; Carrascosa, M. Cross-Pharmacology Analysis of G Protein-Coupled Receptors. *Curr. Top. Med. Chem.* **2011**, *11*, 1956–1963.

[25] Klabunde, T. Chemogenomic Approaches to Drug Discovery: Similar Receptors Bind Similar Ligands. *Br. J. Pharmacol.* **2007**, *152*, 5–7.

[26] Gloriam, D. E. Chemogenomics of Allosteric Binding Sites in GPCRs. *Drug Discov. Today. Technol.* **2013**, *10*, e307–13.

[27] Lin, H.; Sassano, M. F.; Roth, B. L.; Shoichet, B. K. A Pharmacological Organization of G Protein-Coupled Receptors. *Nat. Methods* **2013**, *10*, 140–146.

[28] Gloriam, D. E.; Foord, S. M.; Blaney, F. E.; Garland, S. L. Definition of the G Protein-Coupled Receptor Transmembrane Bundle Binding Pocket and Calculation of Receptor Similarities for Drug Design. *J. Med. Chem.* **2009**, *52*, 4429–4442.

[29] Besnard, J.; Ruda, G. F.; Setola, V.; Abecassis, K.; Rodriguiz, R. M.; Huang, X.-P.; Norval, S.; Sassano, M. F.; Shin, A. I.; Webster, L. A; Simeons, F. R.; Stojanovski, L.; Prat, A.; Seidah, N. G.; Constam, D. B.; Bickerton, G. R.; Read, K. D.; Wetsel, W. C.; Gilbert, I. H.; Roth, B. L.; Hopkins, A. L. Automated Design of Ligands to Polypharmacological Profiles. *Nature* **2012**, *492*, 215–220.

[30] Gloriam, D. E.; Wellendorph, P.; Johansen, L. D.; Thomsen, A. R. B.; Phonekeo, K.; Pedersen, D. S.; Bräuner-Osborne, H. Chemogenomic Discovery of Allosteric Antagonists at the GPRC6A Receptor. *Chem. Biol.* **2011**, *18*, 1489–1498.

[31] Nijmeijer, S.; Vischer, H. F.; Sirci, F.; Schultes, S.; Engelhardt, H.; de Graaf, C.; Rosethorne, E. M.; Charlton, S. J.; Leurs, R. Detailed Analysis of Biased Histamine H_4 Receptor Signalling by JNJ 7777120 Analogues. *Br. J. Pharmacol.* **2013**, *170*, 78–88.

[32] Michino, M.; Abola, E.; Brooks, C. L.; Dixon, J. S.; Moult, J.; Stevens, R. C. Community-Wide Assessment of GPCR Structure Modelling and Ligand Docking: GPCR Dock 2008. *Nat. Rev. Drug Discov.* **2009**, *8*, 455–463.

[33] Kufareva, I.; Rueda, M.; Katritch, V.; Stevens, R. C.; Abagyan, R. Status of GPCR Modeling and Docking as Reflected by Community-Wide GPCR Dock 2010 Assessment. *Struct. (London, Engl. 1993)* **2011**, *19*, 1108–1126.

[34] Kufareva, I.; Katritch, V.; Participants of GPCR Dock 2013; Stevens, R. C.; Abagyan, R. Advances in GPCR Modeling Evaluated by the GPCR Dock 2013 Assessment: Meeting New Challenges. *Structure* **2014**, 22 (cursive), 1120-1139.

[35] Goldfeld, D. A.; Zhu, K.; Beuming, T.; Friesner, R. A. Successful Prediction of the Intra- and Extracellular Loops of Four G-Protein-Coupled Receptors. *Proc. Natl. Acad. Sci. U. S. A.* **2011**, *108*, 8275–8280.

[36] Ballesteros, J.; Weinstein, H. Integrated Methods for the Construction of Three-Dimensional Models and Computational Probing of Structure-Function Relations in G Protein-Coupled Receptors. *Methods Neurosci.* **1995**, *25*, 366–428.

[37] Obiol-Pardo, C.; López, L.; Pastor, M.; Selent, J. Progress in the Structural Prediction of G Protein-Coupled Receptors: D3 Receptor in Complex with Eticlopride. *Proteins* **2011**, *79*, 1695–1703.

[38] Beuming, T.; Sherman, W. Current Assessment of Docking into GPCR Crystal Structures and Homology Models: Successes, Challenges, and Guidelines. *J. Chem. Inf. Model.* **2012**, *52*, 3263–3277.

[39] De Graaf, C.; Rognan, D. Selective Structure-Based Virtual Screening for Full and Partial Agonists of the Beta2 Adrenergic Receptor. *J. Med. Chem.* **2008**, *51*, 4978–4985.

[40] Vilar, S.; Karpiak, J.; Berk, B.; Costanzi, S. In Silico Analysis of the Binding of Agonists and Blockers to the β2-Adrenergic Receptor. *J. Mol. Graph. Model.* **2011**, *29*, 809–817.

[41] Wang, C.; Jiang, Y.; Ma, J.; Wu, H.; Wacker, D.; Katritch, V.; Han, G. W.; Liu, W.; Huang, X.-P.; Vardy, E.; McCorvy, J. D.; Gao, X.; Zhou, X. E.; Melcher, K.; Zhang, C.; Bai, F.; Yang, H.; Yang, L.; Jiang, H.; Roth, B. L.; Cherezov, V.; Stevens, R. C.; Xu, H. E. Structural Basis for Molecular Recognition at Serotonin Receptors. *Science* **2013**, 340 (cursive), 610-614.

[42] Wacker, D.; Wang, C.; Katritch, V.; Han, G. Structural Features for Functional Selectivity at Serotonin Receptors. *Science* **2013**, *469*, 175–180.

[43] Lill, M. Virtual Screening in Drug Design. *Methods Mol. Biol.* **2013**, *993*, 1–12.

[44] Senderowitz, H.; Marantz, Y. G Protein-Coupled Receptors: Target-Based in Silico Screening. *Curr. Pharm. Des.* **2009**, *15*, 4049–4068.

[45] Kooistra, A. J.; Roumen, L.; Leurs, R.; de Esch, I. J. P.; de Graaf, C. From Heptahelical Bundle to Hits from the Haystack: Structure-Based Virtual Screening for GPCR Ligands. *Methods Enzymol.* **2013**, *522*, 279–336.

[46] Congreve, M.; Marshall, F. The Impact of GPCR Structures on Pharmacology and Structure-Based Drug Design. *Br. J. Pharmacol.* **2010**, *159*, 986–996.

[47] Ananthan, S.; Zhang, W.; Hobrath, J. V. Recent Advances in Structure-Based Virtual Screening of G-Protein Coupled Receptors. *AAPS J.* **2009**, *11*, 178–185.

[48] Shoichet, B. K.; Kobilka, B. K. Structure-Based Drug Screening for G-Protein-Coupled Receptors. *Trends Pharmacol. Sci.* **2012**, *33*, 268–272.

[49] Irwin, J. J.; Sterling, T.; Mysinger, M. M.; Bolstad, E. S.; Coleman, R. G. ZINC: a Free Tool to Discover Chemistry for Biology. *J. Chem. Inf. Model.* **2012**, *52*, 1757–1768.

[50] Tosh, D. K.; Phan, K.; Gao, Z.-G.; Gakh, A. A.; Xu, F.; Deflorian, F.; Abagyan, R.; Stevens, R. C.; Jacobson, K. A.; Katritch, V. Optimization of Adenosine 5'-Carboxamide Derivatives as Adenosine Receptor Agonists Using Structure-Based Ligand Design and Fragment Screening. *J. Med. Chem.* **2012**, *55*, 4297–4308.

[51] Cherezov, V.; Rosenbaum, D. M.; Hanson, M. a; Rasmussen, S. G. F.; Thian, F. S.; Kobilka, T. S.; Choi, H.-J.; Kuhn, P.; Weis, W. I.; Kobilka, B. K.; Stevens, R. C. High-

Resolution Crystal Structure of an Engineered Human Beta2-Adrenergic G Protein-Coupled Receptor. *Science* **2007**, *318*, 1258–1265.

[52] Kolb, P.; Rosenbaum, D. M.; Irwin, J. J.; Fung, J. J.; Kobilka, B. K.; Shoichet, B. K. Structure-Based Discovery of Beta2-Adrenergic Receptor Ligands. *Proc. Natl. Acad. Sci. U. S. A.* **2009**, *106*, 6843–6848.

[53] Weiss, D. R.; Ahn, S.; Sassano, M. F.; Kleist, A.; Zhu, X.; Strachan, R.; Roth, B. L.; Lefkowitz, R. J.; Shoichet, B. K. Conformation Guides Molecular Efficacy in Docking Screens of Activated β-2 Adrenergic G Protein Coupled Receptor. *ACS Chem. Biol.* **2013**, *8*, 1018–1026.

[54] Schneider, M.; Wolf, S.; Schlitter, J.; Gerwert, K. The Structure of Active Opsin as a Basis for Identification of GPCR Agonists by Dynamic Homology Modelling and Virtual Screening Assays. *FEBS Lett.* **2011**, *585*, 3587–3592.

[55] Kołaczkowski, M.; Bucki, A.; Feder, M.; Pawłowski, M. Ligand-Optimized Homology Models of D1 and D2 Dopamine Receptors: Application for Virtual Screening. *J. Chem. Inf. Model.* **2013**.

[56] De Lera Ruiz, M.; Lim, Y.-H.; Zheng, J. The Adenosine A2A Receptor as a Drug Discovery Target. *J. Med. Chem.* **2013**.

[57] Katritch, V.; Jaakola, V.-P.; Lane, J. R.; Lin, J.; Ijzerman, A. P.; Yeager, M.; Kufareva, I.; Stevens, R. C.; Abagyan, R. Structure-Based Discovery of Novel Chemotypes for Adenosine A(2A) Receptor Antagonists. *J. Med. Chem.* **2010**, *53*, 1799–1809.

[58] Yoshikawa, Y.; Oishi, S.; Kubo, T.; Tanahara, N.; Fujii, N.; Furuya, T. Optimized Method of G-Protein-Coupled Receptor Homology Modeling: Its Application to the Discovery of Novel CXCR7 Ligands. *J. Med. Chem.* **2013**, *56*, 4236–4251.

[59] Pala, D.; Beuming, T.; Sherman, W.; Lodola, A.; Rivara, S.; Mor, M. Structure-Based Virtual Screening of MT2 Melatonin Receptor: Influence of Template Choice and Structural Refinement. *J. Chem. Inf. Model.* **2013**, *53*, 821–835.

[60] Eberini, I.; Daniele, S.; Parravicini, C.; Sensi, C.; Trincavelli, M. L.; Martini, C.; Abbracchio, M. P. In Silico Identification of New Ligands for GPR17: a Promising Therapeutic Target for Neurodegenerative Diseases. *J. Comput. Aided. Mol. Des.* **2011**, *25*, 743–752.

[61] Burford, N. T.; Watson, J.; Bertekap, R.; Alt, A. Strategies for the Identification of Allosteric Modulators of G-Protein-Coupled Receptors. *Biochem. Pharmacol.* **2011**, *81*, 691–702.

[62] Ivetac, A.; McCammon, J. A. Mapping the Druggable Allosteric Space of G-Protein Coupled Receptors: a Fragment-Based Molecular Dynamics Approach. *Chem. Biol. Drug Des.* **2010**, *76*, 201–217.

[63] Lane, J. R.; Chubukov, P.; Liu, W.; Canals, M.; Cherezov, V.; Abagyan, R.; Stevens, R. C.; Katritch, V. Structure-Based Ligand Discovery Targeting Orthosteric and Allosteric Pockets of Dopamine Receptors. *Mol. Pharmacol.* **2013**, *84*, 794–807.

[64] Tautermann, C. S. The Use of G-Protein Coupled Receptor Models in Lead Optimization. *Future Med. Chem.* **2011**, *3*, 709–721.

[65] Andrews, S. P.; Brown, G. A; Christopher, J. A. Structure-Based and Fragment-Based GPCR Drug Discovery. *ChemMedChem* **2013**, 1–21.

[66] Park, P. S. H. Ensemble of G Protein-Coupled Receptor Active States. *Curr. Med. Chem.* **2012**, *19*, 1146–1154.

[67] Lee, J. Y.; Lyman, E. Agonist Dynamics and Conformational Selection During Microsecond Simulations of the A(2A) Adenosine Receptor. *Biophys. J.* **2012**, *102*, 2114–2120.

[68] Bhattacharya, S.; Hall, S. E. E.; Li, H.; Vaidehi, N. Ligand-Stabilized Conformational States of Human Beta2 Adrenergic Receptor: Insight into G-Protein-Coupled Receptor Activation. *Biophys. J.* **2008**, *94*, 2027–2042.

[69] Provasi, D.; Artacho, M. C.; Negri, A.; Mobarec, J. C.; Filizola, M. Ligand-Induced Modulation of the Free-Energy Landscape of G Protein-Coupled Receptors Explored by Adaptive Biasing Techniques. *PLoS Comput. Biol.* **2011**, *7*, e1002193–e1002193.

[70] Vilardaga, J.-P.; Bünemann, M.; Krasel, C.; Castro, M.; Lohse, M. J. Measurement of the Millisecond Activation Switch of G Protein-Coupled Receptors in Living Cells. *Nat. Biotechnol.* **2003**, *21*, 807–812.

[71] Bhattacharya, S.; Vaidehi, N. Computational Mapping of the Conformational Transitions in Agonist Selective Pathways of a G-Protein Coupled Receptor. *J. Am. Chem. Soc.* **2010**, *132*, 5205–5214.

[72] Bhattacharya, S.; Lam, A. R.; Li, H.; Balaraman, G.; Niesen, M. J. M.; Vaidehi, N. Critical Analysis of the Successes and Failures of Homology Models of G Protein-Coupled Receptors. *Proteins* **2013**, *81*, 729-39.

[73] Provasi, D.; Filizola, M. Putative Active States of a Prototypic G-Protein-Coupled Receptor from Biased Molecular Dynamics. *Biophys. J.* **2010**, *98*, 2347–2355.

[74] Nygaard, R.; Zou, Y.; Dror, R. O.; Mildorf, T. J.; Arlow, D. H.; Manglik, A.; Pan, A. C.; Liu, C. W.; Fung, J. J.; Bokoch, M. P.; Thian, F. S.; Kobilka, T. S.; Shaw, D. E.; Mueller, L.; Prosser, R. S.; Kobilka, B. K. The Dynamic Process of B2-Adrenergic Receptor Activation. *Cell* **2013**, *152*, 532–542.

[75] Miao, Y.; Nichols, S. E.; Gasper, P. M.; Metzger, V. T.; Mccammon, J. A. Activation and Dynamic Network of the M2 Muscarinic Receptor. *Proc. Natl. Acad. Sci. U. S. A.* **2013**, *110*, 10982-72013.

[76] Kohlhoff, K. J.; Shukla, D.; Lawrenz, M.; Bowman, G. R.; Konerding, D. E.; Belov, D.; Altman, R. B.; Pande, V. S. Cloud-Based Simulations on Google Exacycle Reveal Ligand Modulation of GPCR Activation Pathways. *Nat. Chem.* **2013**, *6*, 15–21.

[77] Goetz, A.; Lanig, H.; Gmeiner, P.; Clark, T. Molecular Dynamics Simulations of the Effect of the G-Protein and Diffusible Ligands on the B2-Adrenergic Receptor. *J. Mol. Biol.* **2011**, *414*, 611–623.

[78] Feng, Z.; Hou, T.; Li, Y. Studies on the Interactions Between Beta2 Adrenergic Receptor and Gs Protein by Molecular Dynamics Simulations. *J. Chem. Inf. Model.* **2012**, *52*, 1005–1014.

[79] Kling, R. C.; Lanig, H.; Clark, T.; Gmeiner, P. Active-State Models of Ternary GPCR Complexes: Determinants of Selective Receptor-G-Protein Coupling. *PLoS One* **2013**, *8*, e67244.

[80] Dror, R. O.; Pan, A. C.; Arlow, D. H.; Borhani, D. W.; Maragakis, P.; Shan, Y.; Xu, H.; Shaw, D. E. Pathway and Mechanism of Drug Binding to G-Protein-Coupled Receptors. *Proc. Natl. Acad. Sci. U. S. A.* **2011**, *108*, 13118–13123.

[81] Kruse, A. C.; Hu, J.; Pan, A. C.; Arlow, D. H.; Rosenbaum, D. M.; Rosemond, E.; Green, H. F.; Liu, T.; Chae, P. S.; Dror, R. O.; Shaw, D. E.; Weis, W. I.; Wess, J.; Kobilka, B. K. Structure and Dynamics of the M3 Muscarinic Acetylcholine Receptor. *Nature* **2012**, *482*, 552–556.

[82] Dror, R. O.; Green, H. F.; Valant, C.; Borhani, D. W.; Valcourt, J. R.; Pan, A. C.; Arlow, D. H.; Canals, M.; Lane, J. R.; Rahmani, R.; Baell, J. B.; Sexton, P. M.; Christopoulos, A.; Shaw, D. E. Structural Basis for Modulation of a G-Protein-Coupled Receptor by Allosteric Drugs. *Nature* **2013**, *503*, 295-299.

[83] Selent, J.; Sanz, F.; Pastor, M.; De Fabritiis, G. Induced Effects of Sodium Ions on Dopaminergic G-Protein Coupled Receptors. *PLoS Comput. Biol.* **2010**, *6*.

[84] Liu, W.; Chun, E.; Thompson, a. a.; Chubukov, P.; Xu, F.; Katritch, V.; Han, G. W.; Roth, C. B.; Heitman, L. H.; IJzerman, a. P.; Cherezov, V.; Stevens, R. C. Structural Basis for Allosteric Regulation of GPCRs by Sodium Ions. *Science* **2012**, *337*, 232–236.

[85] Gutiérrez-de-Terán, H.; Massink, A.; Rodríguez, D.; Liu, W.; Han, G. W.; Joseph, J. S.; Katritch, I.; Heitman, L. H.; Xia, L.; IJzerman, A. P.; *et al.* The Role of a Sodium Ion Binding Site in the Allosteric Modulation of the A2A Adenosine G Protein-Coupled Receptor. *Structure* **2013**, *2*, 1–11.

[86] Whorton, M. R.; Bokoch, M. P.; Rasmussen, S. G. F.; Huang, B.; Zare, R. N.; Kobilka, B.; Sunahara, R. K. A Monomeric G Protein-Coupled Receptor Isolated in a High-Density Lipoprotein Particle Efficiently Activates Its G Protein. *Proc. Natl. Acad. Sci. U. S. A.* **2007**, *104*, 7682–7687.

[87] Ernst, O. P.; Gramse, V.; Kolbe, M.; Hofmann, K. P.; Heck, M. Monomeric G Protein-Coupled Receptor Rhodopsin in Solution Activates Its G Protein Transducin at the Diffusion Limit. *Proc. Natl. Acad. Sci. U. S. A.* **2007**, *104*, 10859–10864.

[88] Fotiadis, D.; Liang, Y.; Filipek, S.; Saperstein, D. A.; Engel, A.; Palczewski, K. Atomic-Force Microscopy: Rhodopsin Dimers in Native Disc Membranes. *Nature* **2003**, *421*, 127–128.

[89] Romano, C.; Yang, W. L.; O'Malley, K. L. Metabotropic Glutamate Receptor 5 Is a Disulfide-Linked Dimer. *J. Biol. Chem.* **1996**, *271*, 28612–28616.

[90] Dorsch, S.; Klotz, K.-N.; Engelhardt, S.; Lohse, M. J.; Bünemann, M. Analysis of Receptor Oligomerization by FRAP Microscopy. *Nat. Methods* **2009**, *6*, 225–230.

[91] Lambert, N. A. GPCR Dimers Fall Apart. *Sci. Signal.* **2010**, *3*, pe12.

[92] Hu, J.; Hu, K.; Liu, T.; Stern, M. K.; Mistry, R.; Challiss, R. A. J.; Costanzi, S.; Wess, J. Novel Structural and Functional Insights into M3 Muscarinic Receptor Dimer/oligomer Formation. *J. Biol. Chem.* **2013**, *288*, 34777–34790.

[93] Manglik, A.; Kruse, A. C.; Kobilka, T. S.; Thian, F. S.; Mathiesen, J. M.; Sunahara, R. K.; Pardo, L.; Weis, W. I.; Kobilka, B. K.; Granier, S. Crystal Structure of the μ-Opioid Receptor Bound to a Morphinan Antagonist. *Nature* **2012**, *485*, 321–326.

[94] Huang, J.; Chen, S.; Zhang, J. J.; Huang, X.-Y. Crystal Structure of Oligomeric B(1)-Adrenergic G Protein-Coupled Receptors in Ligand-Free Basal State. *Nat. Struct. Mol. Biol.* **2013**.

[95] Murakami, M.; Kouyama, T. Crystal Structure of Squid Rhodopsin. *Nature* **2008**, *453*, 363–367.

[96] Shimamura, T.; Shiroishi, M.; Weyand, S.; Tsujimoto, H.; Winter, G.; Katritch, V.; Abagyan, R.; Cherezov, V.; Liu, W.; Han, G. W.; *et al.* Structure of the Human Histamine H1 Receptor Complex with Doxepin. *Nature* **2011**, *475*, 65–70.

[97] Selent, J.; Kaczor, A. A. Oligomerization of G Protein-Coupled Receptors: Computational Methods. *Curr. Med. Chem.* **2011**, *18*, 4588–4605.

[98] Rodríguez, D.; Gutiérrez-de-Terán, H. Characterization of the Homodimerization Interface and Functional Hotspots of the CXCR4 Chemokine Receptor. *Proteins* **2012**, *80*, 1919–1928.

[99] Ingólfsson, H. I.; Lopez, C. A.; Uusitalo, J. J.; de Jong, D. H.; Gopal, S. M.; Periole, X.; Marrink, S. J. The Power of Coarse Graining in Biomolecular Simulations. *Wiley Interdiscip. Rev. Comput. Mol. Sci.* **2013**, n/a–n/a.

[100] Provasi, D.; Johnston, J. M.; Filizola, M. Lessons from Free Energy Simulations of Delta-Opioid Receptor Homodimers Involving the Fourth Transmembrane Helix. *Biochemistry* **2010**, *49*, 6771–6776.

[101] Johnston, J. M.; Aburi, M.; Provasi, D.; Bortolato, A.; Urizar, E.; Lambert, N. A.; Javitch, J. A.; Filizola, M. Making Structural Sense of Dimerization Interfaces of Delta Opioid Receptor Homodimers. *Biochemistry* **2011**, *50*, 1682–1690.

[102] Johnston, J. M.; Wang, H.; Provasi, D.; Filizola, M. Assessing the Relative Stability of Dimer Interfaces in G Protein-Coupled Receptors. *PLoS Comput. Biol.* **2012**, *8*, e1002649.

[103] Periole, X.; Knepp, A. M.; Sakmar, T. P.; Marrink, S. J.; Huber, T. Structural Determinants of the Supramolecular Organization of G Protein-Coupled Receptors in Bilayers. *J. Am. Chem. Soc.* **2012**, *134*, 10959–10965.

[104] Guixà-González, R.; Ramírez-Anguita, J. M.; Kaczor, A. A.; Selent, J. Simulating G Protein-Coupled Receptors in Native-Like Membranes: From Monomers to Oligomers. *Methods Cell Biol.* **2013**, *117*, 63–90.

[105] Franco, R.; Casado, V.; Cortes, A.; Mallol, J.; Ciruela, F.; Ferre, S.; Lluis, C.; Canela, E. I. G-Protein-Coupled Receptor Heteromers: Function and Ligand Pharmacology. *Br. J. Pharmacol.* **2008**, *153*, S90–S98.

[106] Kühhorn, J.; Hübner, H.; Gmeiner, P. Bivalent Dopamine D2 Receptor Ligands: Synthesis and Binding Properties. *J. Med. Chem.* **2011**, *54*, 4896–4903.

[107] Soriano, A.; Ventura, R.; Molero, A.; Hoen, R.; Casadó, V.; Cortés, A.; Fanelli, F.; Albericio, F.; Lluís, C.; Franco, R.; *et al.* Adenosine A2A Receptor-Antagonist/dopamine D2 Receptor-Agonist Bivalent Ligands as Pharmacological Tools to Detect A2A-D2 Receptor Heteromers. *J. Med. Chem.* **2009**, *52*, 5590–5602.

[108] Hiller, C.; Kühhorn, J.; Gmeiner, P. Class A G-Protein-Coupled Receptor (GPCR) Dimers and Bivalent Ligands. *J. Med. Chem.* **2013**, *56*, 6542–6559.

[109] Guixa-Gonzalez, R.; Bruno, a.; Marti-Solano, M.; Selent, J. Crosstalk Within GPCR Heteromers in Schizophrenia and Parkinson's Disease: Physical or Just Functional? *Curr. Med. Chem.* **2012**, *19*, 1119–1134.

[110] Garland, S. L. Are GPCRs Still a Source of New Targets? *J. Biomol. Screen.* **2013**, *18*, 947–966.

CHAPTER 2

Knowledge-Based Drug Repurposing: A Rational Approach Towards the Identification of Novel Medical Applications of Known Drugs

Carolina L. Bellera[1], Mauricio E. Di Ianni[1], María L. Sbaraglini[1], Eduardo A. Castro[2], Luis E. Bruno-Blanch[1] and Alan Talevi[1,*]

[1]Medicinal Chemistry, Department of Biological Sciences, Faculty of Exact Sciences, National University of La Plata and [2]Institute of Physicochemical Theoretical and Applied Research (INIFTA), National Council of Scientific and Technical Research (CONICET) CCT La Plata, Buenos Aires, Argentina

Abstract: Drug repurposing/reprofiling has attracted considerable attention during the last decade. The object of such approach is to discover second or further medical uses of known chemicals, i. e. targeting existing, withdrawn or abandoned drugs, or yet to be pursued clinical candidates to new disease areas. Recently (2011-2012), the US and UK governments launched public-private joint initiatives towards finding new uses of previously shelved compounds (drug rescue). While in the past repurposing emerged from serendipitous findings and/or from rational exploitation of drug side-effects (*e.g.* sildenafil, aspirin), the current tendency in the drug development field focuses on knowledge-based drug repurposing, particularly, computer-aided repositioning approaches. The present chapter reviews different cheminformatic and bioinformatic applications, as well as high-throughput literature analysis, oriented to the discovery of new medical uses of known drugs. Applications of such strategies to the discovery of innovative medications for neglected or rare diseases are discussed. Finally, we also review publicly available resources (*e.g.* chemical libraries) valuable for reprofiling.

Keywords: Bioinformatics, cheminformatics, drug reprofiling, drug repurposing, indication expansion, indication switching, literature-based drug repositioning, neglected diseases, network-based drug repositioning.

INTRODUCTION

Drug repositioning (also known as *drug repurposing* or *drug reprofiling* or

***Corresponding author Alan Talevi:** Medicinal Chemistry, Department of Biological Sciences, Faculty of Exact Sciences, National University of La Plata, Buenos Aires, Argentina; Tel/Fax: 542214235333; Ext. 41; E-mail: atalevi@biol.unlp.edu.ar

Zaheer Ul-Haq and Jeffry D. Madura (Eds). **10.1016/B978-1-60805-865-5.50002-2**

indication expansion or *indication switching*) refers to finding new therapeutic uses for already existing drugs including marketed, discontinued and shelved drugs, and yet-to-be-pursued clinical candidates (recently, the research on new indications for abandoned drugs has been described as *drug rescue*). There are many explanations for the growing attention to drug repositioning within the international drug development community over the last few years (which includes public programs launched by national health authorities in developed countries such as the US and UK) [1-6]. Drug repositioning is intrinsically linked to off-label use, that is, the prescription of a drug by a physician (based upon emerging science or clinical evidence) for indications (or in doses or through routes) not yet evaluated and approved by the health authorities [7]. Off-label use frequently implies the use of a given drug within the medical community for an unapproved therapeutic indication, and it is a very frequent practice in certain branches of Medicine (*e.g.* Psychiatry and Pediatric practices); drug repositioning, especially when sponsored by a pharmaceutical company, aims to the approval of a second medical use (or a medical use different from the originally intended in the case of abandoned and investigational drugs).

Repositioned drugs represent unique translational opportunities, including substantially higher probability of success to market than *de novo* drugs and a reduced development timeline to potentially 3-12 years [8, 9]. Repurposed candidates have (at least) survived preclinical toxicological testing; they have proved tolerable safety and possess adequate, already characterized pharmacokinetic profiles. When the repurposed drug has already been used in clinical practice, manufacturing and stability issues have already been solved; what is more, off-patent repurposed drugs may provide relatively inexpensive solutions for new problems [10]. Successful drug repurposing stories have probably contributed to the interest in indication expansion. E.g. sildenafil was originally investigated for the treatment of hypertension and ischemic heart disease but acquired blockbuster status as a treatment for erectile dysfunction. Aspirin itself has expanded its therapeutic indications and it is at present widely used to prevent heart attacks and strokes in patients with existing cardiovascular disease. More examples are presented in Table **1**.

Table 1: Examples of successful repurposing

Drug	Original Indication	New Indication
Aspirin	Inflammation, pain	Antiplatelet
Amphotericin B	Fungal infections	Leishmaniasis
Bromocriptine	Parkinson's disease, hyperprolactinaemia and galactorrhoea	Diabetes mellitus
Bupropion	Depression	Smoking cessation
Celecoxib	Osteoarthritis and adult rheumatoid arthritis	Familial adenomatous polyposis, colon and breast cancer
Chlorpromazine	Anti-emetic/antihistamine	Non-sedating tranquilizer
Duloxetine	Major depressive disorder	Stress urinary incontinence
Eflornithine	Anti-infective	Reduction of unwanted facial hair in women
Finasteride	Benign Prostatic Hyperplasia	Hair loss
Fluoxetine	Depression	Premenstrual dysphoria
Galantamine	Polio, paralysis and anesthesia	Alzheimer's disease
Gemcitabine	Viral infections	Cancer
Methotrexate	Cancer	Psoriasis, rheumatoid arthritis
Minoxidil	Hypertension	Hair loss
Paclitaxel	Cancer chemotherapeutic agent	Prevention of restenosis of coronary stents
Phentolamine	Hypertension	Impaired night vision
Raloxifene	Breast and prostate cancer	Osteoporosis
Ropinirole	Hypertension	Parkinson's disease and idiopathic restless leg syndrome
Sildenafil	Angina	Erectile dysfunction
Tadalafil	Inflammation and cardiovascular disease	Male erectile dysfunction
Tofisopam	Anxiety-related conditions	Irritable bowel syndrome
Topiramate	Epilepsy	Obesity
Warfarin	Thrombosis prevention	Secondary prophylaxis following myocardial infarction
Zidovudine	Cancer	HIV/AIDS

Second uses have frequently been found through serendipitous observations (typically, intelligent exploitation of drug side-effects). Lately, however, rational, knowledge-based repositioning strategies have been explored, including cheminformatic- and bioinformatic-based approaches [11-17] and high-throughput literature analysis [18, 19]. Repositioning has been signaled as a particularly useful strategy for the discovery of new treatments for orphan, rare and neglected diseases [20-22], which often offer limited potential revenue to pharmaceutical companies and are addressed by private-public joint efforts, the academic sector and non-profit organizations. Throughout this chapter we will review recent trends in the field of computer-aided drug repositioning. We will also discuss the particular application of this strategy in the search of new therapeutic solutions for neglected and rare diseases. Finally, we will present a selection of publicly available *in silico* resources that might be of help to assist drug repositioning initiatives.

BIOINFORMATICS AND DRUG REPURPOSING

Computer-aided drug repositioning relies on two general principles [12]: a) drugs which share biologically relevant molecular features may interact with the same molecular target/s (drug-centric approach) and; b) health disorders linked to the same or similar dysregulated or dysfunctional proteins may be treated with the same drugs (disease-centric approach). Computational methods might be useful to reveal hidden drug-protein or protein-protein relationships. The first approach will be covered separately in the *Cheminformatics and drug repurposing* section of this chapter. High-throughput literature analysis constitutes a third, distinctive approach that will also be discussed.

Bioinformatics deals with the challenge of finding structural similarities and functional connections between gene and gene products, and, more recently, similarities and inverse similarities between genome-wide expression patterns linked to disease and drug-effect signatures.

Genome-wide gene expression profiling offers a snapshot of globally measured transcript levels in a given cell, tissue or organism at a specific point of time under a certain experimental condition [23]. The Broad Institute Connectivity Map is a publicly available resource meant to connect disease and small molecules through gene-profiles [24]. This database was the first to compile gene-expression profiles derived from the treatment of human cells cultured with a large number of perturbagens (drugs and other bioactive compounds). Originally, 164 perturbagens were considered. Currently that number has been expanded to more than 1300 FDA-approved molecules and it has been announced that the Connectivity Map will soon contain around 4000 drug-effect signatures. Query expression signatures can be compared to the stored ones through pattern-matching algorithms: those at the top and bottom of the resulting similarity rank are considered related to the query state by common and opposite expression changes, respectively. How can this resource be used to repurpose drugs? If a signature corresponding to a given disease state is used as query, those drugs whose signatures show an inverse similarity to the query are, hypothetically, a potential therapy to restore physiological state. Alternatively, if a drug-effect is used as query, then all those drug-effect stored signatures similar to the query represent drugs with similar effects (Fig. **1**). Although not related to drug reprofiling, it is interest to note that direct similarity (positive correlation)

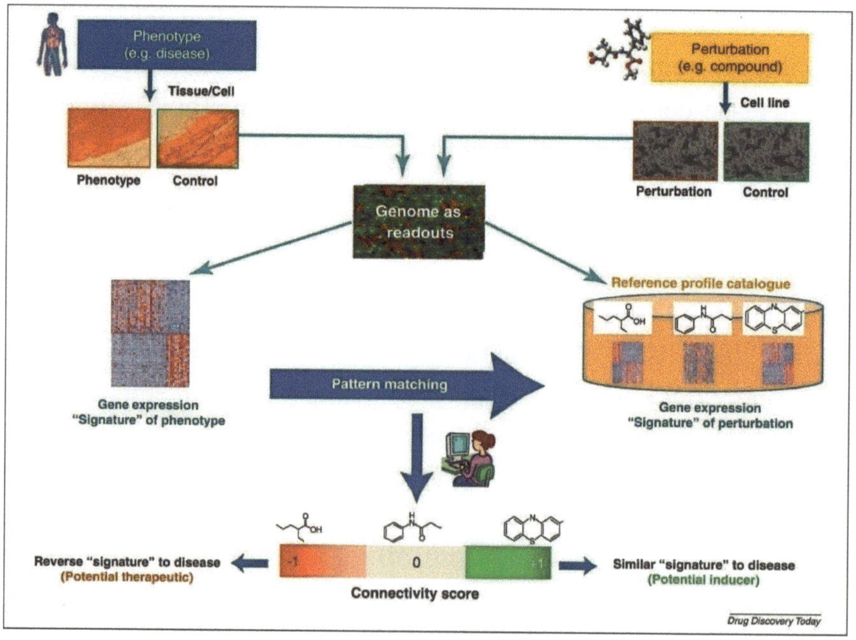

Figure 1: A general scheme showing how comparison of drug-effect and disease signatures can be used to select potential therapeutics. Reproduced under permission of Elsevier from Ku, X. A. *et al*. Applications of the Connectivity Map in drug discovery and development. *Drug. Discov. Today*. 2012, 17(23-24). 1289-1298.

between a given drug-effect signature and a disease signature suggests that the considered perturbagen might exacerbate or induce the disease, thus being contraindicated for those patients suffering such condition (or predisposed to it). In addition, drug-effect signatures provide clues on drug mechanisms of action.

One of the seminal applications of the Connectivity Map to drug repurposing was developed by Sirota *et al.* [25, 26]. These authors produced a large-scale integration of disease signatures with Connectivity Map drug-effect signatures, building a compendium of predicted disease-drug associations. In this way, instead of examining single drug-disease or drug-target pairs or even the potential effect of a large number of drugs on a single target or disease, they considered all the possible drug-disease connections that could be derived from the existing expression data. The method provided significant drug-disease relationships for 53 out of 100 tested conditions. Each of the 164 tested drugs was linked to at least one of the tested conditions. For validation purposes, the therapeutic potential of topiramate (an approved antiepileptic and anti-obesity agent) on inflammatory bowel disease and the effect of cimetidine (an inhibitor of gastric acid secretion) on lung adenocarcinoma were verified through *in vitro* and/or *in vivo* models, with positive results. It was later observed that the associations of cimetidine with cancer and topiramate with inflammatory bowel disease had previously been reported in literature [27]. Nevertheless, many other discoveries have been reported using a similar approach. For instance, Claerhout *et al.* used the top 500 up regulated and the top 500 down regulated genes from microarray data from 65 gastric cancer patients as query signature in the Connectivity Map, finding that vorinostat (a histone deacetylase inhibitor) was a potential candidate to target gastric cancer; the prediction was later validated in gastric cancer cell lines [28]. More recently, the Connectivity Map has been used in combination with Support Vector Machines to optimize drug repurposing for the treatment of hepatocellular carcinoma [29]. Further discussion on some other examples is presented in reference [23]. In the same line are the recent contributions from Sanseau *et al.* and Wang *et al.* [30-32] focusing on disease traits of genetic origin (medical genetics-based drug repositioning). Using the catalog of published Genome-wide Association Studies (GWAS) from the US National Human Genome Research

Institute, Sanseau *et al*. built a list of genes with single nucleotide polymorphisms linked to disease traits [30]. After removing non-replicated data from the list, they analyzed the druggability and biopharmability of the listed genes. 21% of the genes were considered druggable, while 49% were regarded as biopharmable. 15.6% (155) of the listed genes were already associated to launched drugs or ongoing drug projects: 97 matches and 123 mismatches were found between the GWAS traits and the known or pursued therapeutic indication of the drugs. Those 123 mismatches correspond to drug repositioning opportunities. A similar approach was later applied by Wang *et al*. [31, 32], though these authors preferred the OMIM database over GWAS, since OMIM provides more detailed pathogenic information that can help deciding on drug directionality (whether a agonist or antagonist is more adequate to treat a given trait).

A different but very interesting approximation in the field of bioinformatic-based drug repurposing emerges from the very recent work of Haupt *et al.* [33]. These authors demonstrated a connection between ligand promiscuity (a valuable property for drug reprofiling purposes) and global structure similarity and binding site similarity. In order to find the correlation between ligand promiscuity and binding site similarity, 164 ligands co-crystallized with three or more non-redundant targets were extracted from the Protein Data Bank. These ligands were present in 712 non-redundant protein targets (redundancy was defined by 95% sequence identity). All pairs of binding sites for all promiscuous drugs were aligned with the sequence-order-independent profile-profile alignment algorithm implemented in SMAP [34]. Only those sites with consistent binding mode of the ligand (*i.e.* whenever the predicted binding site similarity translated into a similar ligand binding mode) were then kept for subsequent analysis, finding a correlation of $r=0.76$ between the global structure similarity and the degree of promiscuity (drug target count) and a correlation of $r=0.81$ between the square root of the number of similar binding sites and the degree of promiscuity. These findings suggest that one may use the binding site similarity and the global structure similarity as criteria to guide drug repositioning initiatives. Bioinformatics approaches to establish protein-protein connections are briefly discussed in the section covering network-based approaches.

CHEMINFORMATIC-BASED DRUG REPOSITIONING

Cheminformatic-based drug repositioning can be regarded as a very particular type of virtual screening campaign in which the screened library or database only includes approved, discontinued, abandoned and/or investigational drugs. The methods used in cheminformatic-based drug repositioning are thus classified in the same way that for general virtual screening approaches [35]: target-based approaches (prominently, molecular docking) and ligand-based approaches (which roughly include pharmacophore-based, descriptor-based and similarity-based techniques). Lately, parallel and serial combinations of the previously mentioned approximations have been extensively applied [36]. Remarkably, since drug repositioning focuses on an extremely small subset of the known, vast and growing universe of drug-like small molecules, the use of virtual screening for repositioning purposes is particularly efficient, a point that should be taken into consideration especially when target-based approaches are included in the screening protocol. Availability of public repositories of approved, discontinued, abandoned and/or investigational drugs such as DrugBank has smoothed the way for the development of cheminformatic-based drug repositioning campaigns.

Target-based approaches generally involve three stages [37]: i) generation of the molecular model of the target; ii) pre-treatment and conformational sampling of the ligands and; iii) score assignment reflecting the binding energy of the ligand-target complex. Since both proteins and ligands usually possess certain degree of flexibility, a mutual induction of conformational changes favoring the binding event frequently occurs. Multiple approaches have been explored to account for ligand and target flexibility. Methods to tackle the ligand flexibility issue include ligand incremental construction, generation of multiple conformers previous to docking and stochastic methodologies. Treatment of protein flexibility includes using multiple rigid receptor conformations, either computed probable conformations or conformations obtained from experimental (x-ray or RMN) structures. Regarding force-field scoring functions, since originally scoring functions tended to neglect the entropic contribution to binding and the water-mediated ligand-binding, considerable effort is being invested in improving the performance of scoring algorithms, *e.g.* including additional terms to better estimate the solvation effect or the entropy-related change in free energy during binding, and combinations of the output from several scoring functions (*consensus scoring*).

Drug repositioning by molecular docking can be operated *via* a single target approach which aims to identify potential interactions between the drug candidates and a particular target of interest, or inverse docking might otherwise be used to investigate the binding of an existing drug against a panel of known therapeutic targets [37]. Dakshanamurthy *et al.* have developed a proteochemometric method to map the drug-target interaction space and predict new uses for FDA-approved and investigational drugs [38]. They combined shape, topology and chemical signatures (including docking score and functional contact points of the ligand) to predict potential drug-target interactions between 3,671 drugs and 2,335 human proteins. This application uncovered that the antiparasitic mebendazole can inhibit VEGFR2 kinase activity and angiogenesis at doses comparable with the ones used to elicit its known effects on hookworms. They also predicted that the anti-inflammatory drug agent celecoxib binds to cadherin-22, an adhesion molecule relevant in rheumatoid arthritis and poor prognosis malignancies. Regarding the single target approach, Lejal *et al.* recently reported the antiviral effect of the anti-inflammatory naproxen against Influenza A virus [39]. This unknown effect was discovered through application of molecular docking and molecular dynamics simulations on a Sigma-Aldrich catalogue, using the nucleoprotein as molecular target.

Regarding ligand-based approaches, systematic comparisons indicate that, while relatively simpler and efficient approaches (*e.g.* similarity methods based on 2D fingerprints) tend to present good enrichment factors with low computational demand, more elaborated conformation-dependent approaches such as superposition to pharmacophoric hypotheses (which requires conformational analysis of both the reference molecules and the database structures) generally show better scaffold hopping [40, 41]. The prominent role of ensemble learning has been highlighted in the essential article on prospective virtual screening applications from Ripphausen *et al.* [42]. By systematically exploring a variety of targets, Holliday *et al.* have demonstrated that the active enrichment increases when using different reference molecules and different fingerprinting schemes in similarity-based virtual screening campaigns [43].

Recently, Wu *et al.* developed the very attractive idea that two therapeutic indications would be correlated if they share the same or similar drugs (and thus, a network indicating which therapeutic categories have similar drugs would be a valuable framework to systematic drug repositioning). To test their hypothesis, they conceived the indication similarity ensemble approach (iSEA) [44]. Briefly, 1,574 pairs of drug-anatomic therapeutic chemical classes were collected from DrugBank. 1,151 FDA-approved drugs were used to train the network, while 54 experimental or withdrawn drugs involved in 65 drug-anatomic therapeutic chemical classes were used for validation purposes. For each drug pair involved in two anatomic therapeutic chemical classes under comparison, Tanimoto scores were computed using three different fingerprinting schemes that were subsequently averaged. Afterwards, an Evaluating score was computed simply by summing all the average similarity scores for all the possible pairs for the two classes being compared, which was later transformed into a Z score to assess statistical significance. Previous work related to iSEA can be found in the reports from Keiser *et al.* [45, 46]

LITERATURE-BASED DRUG REPOSITIONING

The process of generating novel hypotheses by bridging seemingly unrelated scientific facts (or detecting indirect associations between them) is known as *literature-based discovery* (LBD). LBD is based on the hypothesis that two islands of knowledge or concepts A and C might be related to each other if they share a link to an intermediate concept B [47] (and in fact, the bigger the number of shared concepts between A and C, the more probable the relationship between them). This model is commonly known as Swanson's ABC model after its postulation and first fruitful applications by Swanson during the 1980s and afterwards [48, 49]. The first successful application of the ABC model uncovered the therapeutic potential of oil fish for the treatment of Raynaud's syndrome, following the previously reported observations that a) Raynaud's syndrome is linked to increased blood-viscosity and; b) fish oil reduces blood viscosity. This prediction was inferred through the use of the semi-automated method *Arrowsmith*, which used a "closed" framework in which the user provides the hypothesis (Fig. **2**) [50]. The rapid increase in the volume of the biomedical literature spawns a combinatorial explosion in the number of implicit connections between entities described in it; the possibility of such connections to remain hidden/unnoticed is substantially increased by the progressively more disjointed nature of knowledge as a consequence of specialization [51]; is no longer possible for a researcher to keep up to date with all the relevant literature manually, even on specialized topics [50]. Therefore, the development of automated, high throughput methods for information retrieval and information extraction is becoming progressively essential to researchers; in the context of drug repositioning, an open discovery approach (finding a relationship/hypothesis starting from a disease A and arriving to a drug C, or *vice versa*) (Fig. **2**) seems to be the best approach to find second medical uses [52].

Co-occurrence methods are the simplest approaches to link biomedical terms of interest. Implicit connections between terms that do not co-occur are discovered by finding a third linking term that occurs directly with each of them. This approach is, though, prone to generating false positives (given the large number of possible combinations of bridging terms and potential discoveries) and it does not provide information on the nature of the predicted relationship [47, 51, 53]. Recent research indicates that co-occurrence approaches can be outperformed by Natural Language Processing-based methods, *e.g.* semantic analysis. For instance, Predication-based Semantic Indexing represents concepts and relationships between them as vectors in the hyperdimensional space, and inference takes place as a function of the geometry of such space, providing scalable search and efficient inference [51]. Wilkowski *et al.* have recently used SemRep [54] to extract semantic predications from MEDLINE and built large graphs (predication graphs) of interconnected nodes which are then analyzed through graph-theoretic constructs to find chains of relationships that might guide the research process [55]. To create the graph the user specifies a seed concept to extract predications from the SemRep predication database. Concepts in the resulting graph are ranked according to degree centrality to select a new seed concept and expand the growing graph with additional predications. Path analysis is also applied to reveal potentially interesting associations (*e.g.* longest paths tend to reveal rare associations). Cameron *et al.* expanded Wilkowski's approach by considering not only associations between concepts but also relevant/expressive subgraphs and background knowledge [53]. The idea of exploiting topological motifs analysis to reveal hidden relationships has recently been applied through the Typed Network Motif Comparison Algorithm developed by Choi *et al.* [56].

There are several recent applications of literature-based drug repositioning. Li *et al.* combined text mining with molecular interaction network mining to search for potential new treatments for Alzheimer disease [57]. They retrieved 49 proteins related to Alzheimer from OMIM, and they expanded such list using quality-ranked protein interaction data. Finally, these authors analyzed PubMed abstracts using the resulting 560 Alzheimer disease-relevant proteins as queries, retrieving more than 220,000 related abstracts outside the explicit context of Alzheimer and examining drug terms appearing in them. As a result, diltiazem and quinidine were proposed as potential therapies for Alzheimer.

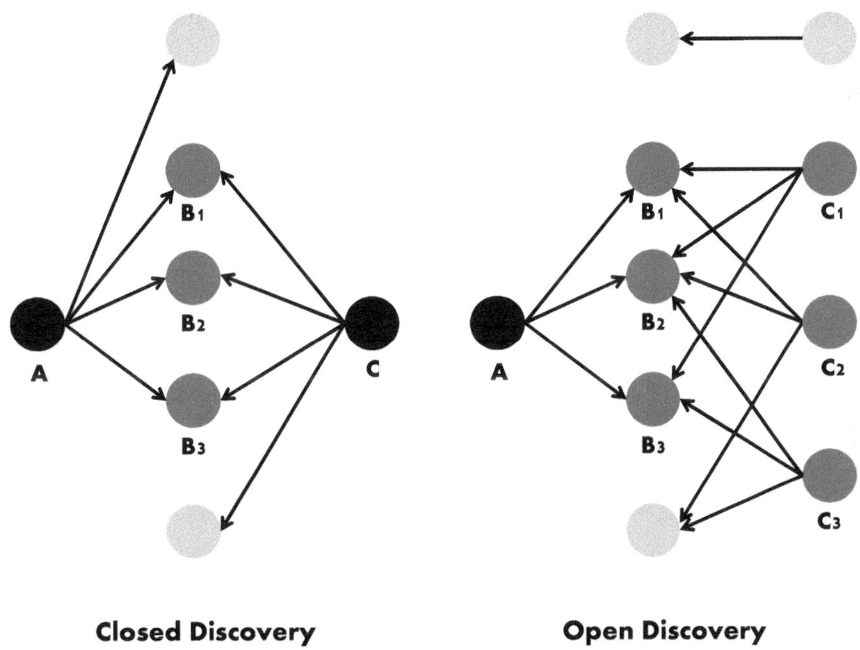

Figure 2: ABC principle of hidden relationships in literature. Closed discovery may be helpful to support previously formulated hypothesis; open discovery is valuable for hypotheses generation and thus for drug repositioning campaigns.

Some years back Fritjers *et al.* developed an ABC-based literature mining tool named CoPub [58]. Using this tool they found interconnections between gene, drugs and diseases. Later, they validated CoPub by finding known and unknown relationships between biomedical concepts [19]. An *R*-scaled score ranging from 1-100, describing the strength of a co-citation between two biological items given their individual frequencies of occurrence, was used to assess the significance of a co-occurrence. A high *R*-scaled score indicates that if two biological concepts occur in literature they are often published together, whereas a low *R*-scaled score indicates that two biological concepts often occur separately in literature [59]. Using this approach the authors predicted the antiproliferative effect of two drugs, damnacanthal and dephostatin; the predictions were later corroborated through *in vitro* assays. Noteworthy, the authors computed the time lag between the average publication date of A-B and B-C intermediates, and compared this date with the date of first appearance of A and C in the literature. They estimated the average lag time in 6.5 years, which indicates to which extent discoveries can be accelerated when this type of relatively simple literature-mining hypothesis generation tools are employed.

NETWORK-BASED DRUG REPOSITIONING

There are many good reasons to resort to large-scale data integration approaches that allow system view on drugs actions to develop drug repositioning campaigns. First, as it has already been pointed-out in the previous section, scientific information is nowadays produced at an unprecedented rate. Manually exploring available literature to find valuable connections is no longer feasible, and computational approaches are needed to digest and bridge such vast amount of data [60]. Elucidating a drug's mechanisms of action is still very time and labor expensive (in fact, new mechanisms are continuously being discovered for drugs that have been clinically used for decades) and on the other hand experimental binding data are incomplete. However, available experimental data on drug-protein interactions may be sufficient to fill the experimental gap by applying computational tools complementarily to high throughput approaches. Integrating multi-dimensional information (chemical, pharmacological and genomic spaces) may help to compensate for intrinsic limitations of isolated approximations/single kinds of information. For instance, cheminformatic structure-based approaches tend to focus on a limited number of proteins such as those with interacting drugs and solved three dimensional structures; structurally similar drugs may bind proteins with no obvious sequence or structural similarity, while structurally dissimilar drugs may bind the same protein (the *activity cliff* issue); in phenotypic effect-based approaches, drugs affecting different targets in the same pathway or biological process may trigger similar responses, while in some cases it may be difficult to distinguish target gene products from downstream regulated genes [61, 62]. Zhao and Li developed three regression models relating "closeness" (on the basis of a protein-protein interaction network derived from the Human Protein Reference Database [63]) to therapeutic similarity, chemical similarity and "multiple" similarity combining the previous two similarities. Remarkably, combining therapeutic and chemical similarities provided the best results in terms of drug-target interactions recognition, leading to outstanding areas under the ROC curve of 0.988 and 0.935 for the training and test sets [61]. Finally, a remarkable paradigm shift has lately taken place in the drug discovery field. Two decades ago, the prevailing paradigm proposed the development of exquisitely selective ligands acting on a single target (the *one drug, one target* paradigm). Selectivity and potency were thus essential aspects to decide whether a drug candidate would progress into further development phases. Such reductionist approach was founded on two notions: a) highly specific drugs would avoid off-target side-effects, thus leading to safer therapeutics and; b) at least some diseases could be adequately treated using a single target intervention. However, recent discoveries have challenged the earlier paradigm in favor of a more holistic approach in line with the philosophy of systems biology. Most of the approved drugs were discovered using "black-box" phenotypic screens and interact with more than one target [64, 65]. Multi-target drugs usually affect their targets only partially, that is, they present low affinity interactions with many of their targets [65]. Contrary to previous beliefs, low-affinity multifunctional drugs may represent and advantage: weak links may stabilize the systems, buffering changes after system perturbations. At last, due to redundant functions and compensatory mechanisms phenotypes are robust, *i.e.* resilient to perturbation [66]. Under this novel perspective, disease can be regarded as a breakdown of the robustness of normal physiological systems and the re-establishment of also robust (and potentially progressive) disease states [64]. The previous discussion explains why multi-target drugs are being pursued today and also sets the logic ground for drug repositioning. However, it should be underlined that selectively non-selective drugs and promiscuous drugs are not exactly equivalent concepts in a drug repositioning scenario: while a certain, convenient degree of promiscuity may be desirable, an excess would certainly represent serious safety issues [67, 68]. Network-based approaches focused on drug repositioning may prove helpful to select candidates with an adequate degree of polypharmacology depending on the pursued new indication.

So what are exactly networks? Networks deal with complexity by simplifying complex systems: *concepts* or *entities* are represented as *nodes* while relationships between nodes are depicted as *edges* [69]. In such representation -naturally connected to Graph Theory- functional and dynamic features of the elements depicted as nodes are often (though not always) lost and emphasis is given to the connectivity between the nodes, *i.e.* the topological architecture of the net. Such connectivity is established through known relationships or through predicted associations (*e.g.* chemical similarity, protein similarity, similar expression-profiles, literature-inferred connections, *etc*). Put in other words, all the approaches overviewed in the last three sections of the chapter are combined holistically and new connections are established by studying the topology (and, more recently, the semantics) of the network. What topological aspects of the network are relevant largely depends on what is being pursued by the researcher. We have mentioned in the *Literature-based drug repositioning* section that highly connected concepts are usually useful as intermediate seeds to reveal hidden connections. If we are seeking for drug targets of interest, the hubs (highly connected nodes) frequently correspond to essential proteins whose modulation deeply impacts the system function. Thus, moderately connected nodes might be of more interest as potential new drug targets. A similar statement may be true when searching for new drugs: a drug node of very high degree may represent a promiscuous agent linked to safety issues, while a moderately connected drug might be the selectively non-selective "master key" being sought.

The current trend in network-based drug repositioning points toward the integration of very heterogeneous types of data and it also studies the introduction of semantic edges. E.g. Chen *et al.* developed a semantic network including a variety of physical and abstract node types: compounds, proteins, side effects, diseases, pathways, tissues, gene ontology terms, and others [70]. Semantic linked data encodes explicit meaning of nodes and edges, allowing traversing from node to node through specific kinds of relationships. The authors connected more than 290,000 nodes through more than 720,000 edges. Every node and edge was semantically annotated using a previously developed ontology [71]. Such annotation allowed the definition of path patterns (paths of nodes and edges that share the same semantics) and the study of path patterns that are particularly valuable to identify relevant links. They also developed the Semantic Link Association Prediction model, which computes an association score from the topology and semantics of the neighborhood. It was demonstrated that this model can identify known drug-target pairs and even indirect drug-target associations such as the change of gene expression level (a type of association undetectable through cheminformatic and non-semantic approaches). The authors claimed that the association scores of a drug against a set of targets constitute a biological signature (which reminds us of the opportunely discussed gene-expression signatures; in this case, however, the signature allows a wider spectrum of relationships besides gene-expression regulation). Although any path between two nodes may support the relation between them, the degree of the contribution depends on the path distance and the weight of the involved edges (*e.g.* a gene ontology molecular function term is considered less informative that a binding term). The area under the ROC curve for the model was 0.92. The authors remarked that their model is capable of clustering biologically similar drugs even if they are not chemically similar.

Another very interesting weighting scheme was presented by Lee *et al.* in their tripartite (drug-protein-disease) knowledge platform PharmDB [72]. These authors measured the importance of a predicted relationship between two nodes through an algorithm that combines the share node count and the share node weight. This share node weight is computed, in turn, as the product of the weights of links bridging the considered nodes. While the weight of directly connected pairs is considered 1, the weight of unconnected pairs is assigned from the connection probability, that is, the fraction of directly connected pairs among the total number of pairs having the given shared nodes count (Fig. **3**). According to ROC curves analysis, the inclusion of such share node weight term in the algorithm clearly empowers the identification of relationships compared to the bare share node count criteria.

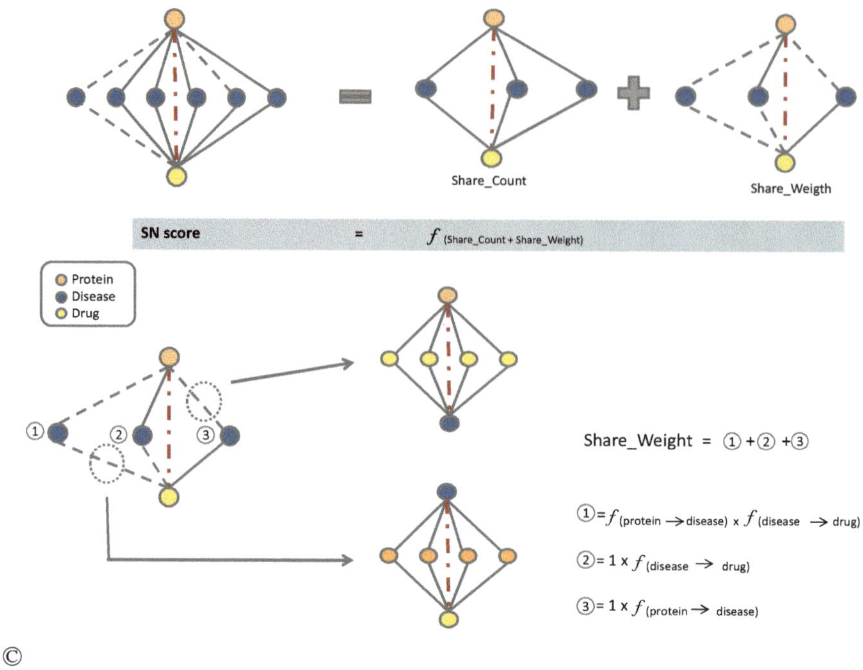

Figure 3: Scheme of the shared neighborhood scoring algorithm proposed by Lee *et al. Reproduced by Creative Commons License from Lee, H. S. et al.* Rational drug repositioning guided by an integrated pharmacological network of protein, disease and drug. *BMC Syst Biol*, 2012, 6(80).

An example of how a network approach can be used to reposition drugs has recently been presented in the 2012 report from Talevi *et al.* on the anticonvulsant effect of non-nutritive sweeteners [73]. A previously reported descriptor-based QSAR model [74] predicted that a number of artificial sweeteners (cyclamate, acesulfame, saccharin) might have anticonvulsant effect in the Maximal Electroshock Seizure (MES) test. Subsequent bibliographic revision showed that one of them, saccharin, had already been evaluated in MES test in 1979 with positive results [75]. The predictions of the model were later validated experimentally, and both acesulfame and cyclamate showed anticonvulsant effects. The results made us wonder whether a link could exist between the sweet taste receptor and any known molecular target of antiepileptic drugs. Bibliographic search indicated that a family of proteins named T1R is the major mediator of the sweet and umami responses in mammals [76, 77]. In humans, sweet sensation is elicited by a heterodimer formed by T1R2 and T1R3, while umami flavor is detected by the combination of T1R1 and T1R3. Noteworthy, one of the main stimuli sensed by the umami receptor is no other that glutamate, and response to glutamate flavor is totally abolished in T1r3 KO mice [78]. These findings led to search for similar sequences to T1R3 in the NCBI non-redundant protein database, in order to investigate the possible link between T1R3 and molecular targets of antiepileptic drugs. BLAST revealed that several of the significantly aligned sequences corresponded to metabotropic glutamate receptors (mGlu) from different species, among them human, rat and mouse. This was of most interest since subtypes of mGlu that were retrieved in the BLAST search (mGluR1 and mGluR5) are upregulated in epileptogenesis and kindling models of epilepsy, and in patients with complex partial seizures [79-81]. They are also associated to increase of NMDA receptors activity, arachidonic acid release, excitotoxicity and neuronal injury [82, 83]. A synthesis of the previous analysis is depicted in the network of Fig. **4**. The relationships that had been experimentally established previously to the report from Talevi *et al*. are presented as thick edges; associations that were predicted with computational (bioinformatic, cheminformatic) tools are presented as thin edges. Further examples of network-based approaches are presented in the following sections.

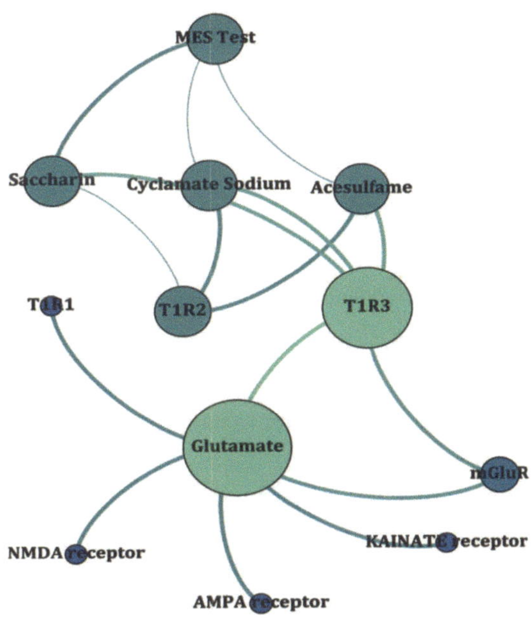

Figure 4: Thick lines in the network illustrate experimentally corroborated links; the thin lines correspond to cheminformatic (cyclamate and acesulfame anticonvulsant effect) and bioinformatic (T1R3-mGluR) predictions. The first two predictions were later validated experimentally.

COMPUTER AIDED DRUG REPURPOSING FOR RARE/ORPHAN AND NEGLECTED DISEASES

Tropical diseases such as Chagas disease, African sleeping sickness, leishmaniasis, lymphatic filariasis, dengue and schistosomiasis are still among the main causes of mortality and morbidity in the world. They belong to a group of diseases collectively known as neglected (tropical) diseases. Although there is little consensus on what constitutes a neglected disease, is it generally accepted that they disproportionally affect people in the developing world and that there exists a need of improved diagnosis and/or treatment products [84]. Even though according to the World Health Organization (WHO) neglected tropical diseases affect more than 750 million people throughout the world, only 21 (1.3%) out of 1556 medications registered between 1975 and 2004 were specifically developed for these conditions, which reflects market flaws and failure of public policies [85].

On the other hand, the expression *rare diseases* denotes a group of health conditions which affect relatively small patient populations. In the US, for example, a disease is considered rare if it affects less than 200,000 people, or it affects more than 200,000 people in the US but is not expected to recover the cost of development and marketing (Orphan Drug Act and implementing regulations). Though very different in nature, rare and neglected diseases share the reluctance within the private pharmaceutical sector to invest in R&D of new treatments, owing to the perceived limited commercial revenue. Thus, the public sector, the academy and non-profit organizations play a prominent role in the development of new solutions to these diseases [84, 86, 87]. In fact, the G-Finder study reveals that around 90% of the funding for R&D on neglected diseases comes from the public sector and non-profit organizations [84].

Recently, the use of drug repurposing as a key strategy within academia and public research institutes has been extensively discussed [87-89]. Public institutions, including public research laboratories and universities, have contributed to the development of nearly 90% of new indications for previously approved drugs [88]. Interestingly, the possibility of repurposing drugs without commercialization (by direct incorporation of the research output to the clinical practice after examination of the data by regulatory authorities) has been suggested as an option unique to academic discoveries [87]. The importance of computer-aided drug repositioning in the developing countries, where most of the limited R&D investment comes from the government and where the private sector seems reluctant to invest in R&D has also been underlined [35].

The previous discussion explains why drug repositioning constitutes a key strategy in the field of drug discovery and development for orphan diseases, where there is an obvious need of collaborative public-private partnerships [20, 22, 90, 91]. Several initiatives such as WHO Special Programme for Research and Training in Tropical Disease, the Medicines for Malaria Venture, the Global Alliance for TB Drug Development, Drugs for Neglected Diseases and the Open Source Drug Discovery initiative have recognized drug repositioning as an attractive option to provide low-cost access to medications in developing countries [92]. The potential of computer-aided drug repurposing focused on neglected/rare diseases has recently been reinforced by the ongoing replication/transference of the Open Source model (which proposes collective knowledge production and dissemination) within the drug discovery field [93-95]. This model facilitates the entry of firms/players from emerging markets/countries. Open Source initiatives promote the exchange of chemical and biological data, chemical libraries, software and computational resources; in the field of neglected diseases, resources related to the Open Source philosophy include, among many others, the open access publication PloS Neglected Tropical Diseases, the ChEMBL - Neglected Tropical Disease (ChEMBL - NTD) archive (an open repository for primary screening and medicinal chemistry data directed at neglected diseases, which to the moment compiles contributions from GlaxoSmithKline, Novartis, the Drugs for Neglected Diseases Initiative - DNDi, St. Jude Children's Research Hospital and the University of California) and the Indian Open Source Drug Discovery initiative (OSDD, a global platform for collaborative research on tropical diseases such as malaria, tuberculosis and leishmaniasis).

There exist several examples of drug repositioning focused on neglected and rare diseases (see Tables **2** and **3**, respectively).

Table 2: Examples of repositioned drugs for neglected tropical diseases currently in development

Drug	Original use	New Use	Refs.
Amiodarone	Anti-arrhythmic	Chagas disease	[96, 97]
Bromocriptine	Parkinson's disease	Chagas disease	[97]
Tamoxifen	Antiestrogen	Leishmania amazonensis	[98]
Amphotericin B	Fungal infections	Leishmaniasis	[99]
Ivermectin	Antiparasitic (river blindness)	Malaria	[100]
Eflornitine	Anticancer	African sleeping sickness	[101]
Astemizole	Antihistamine	Malaria	[102]
Cycloserine	Infections caused by *Giardia*	Tuberculosis	[103]

Table 3: Examples of currently approved repositioned drugs for rare diseases

Drug	Original use	New Use
Azathioprine	Rheumatoid arthritis	Renal transplant
Colchicine	Gout	Mediterranean fever
Cyclosporine	Rheumatoid arthritis. Psoriasis	Transplant rejection
Everolimus	Renal cancer	Renal transplant
Histrelin	Prostate cancer	Precocious puberty
Infliximab	Ulcerative colitis. Psoriasis	Crohn's disease
Interferon alfa	Hepatitis B and C	Various cancers

Source: Food & Drug Administration

There are many examples of computer-assisted indication expansion campaigns focused on neglected and rare diseases. Florez *et al.* [104] developed a protein-protein interaction (PPI) network for the pathogenic trypanosomatid *Leishmania major* and identified 142 potential drug targets after homology filtering with the human proteome. In addition to selecting important proteins from PPI networks and analyzing metabolic pathways that link metabolites and reactions on a system level, this network can also shed light on disease mechanisms and assist drug target discovery. The topological analysis of the network of proteins has allowed the identification of a set of candidate proteins that may be both (1) essential for parasite survival and (2) without human orthologs, thus being potentially attractive and safe drug targets (Fig. **5**).

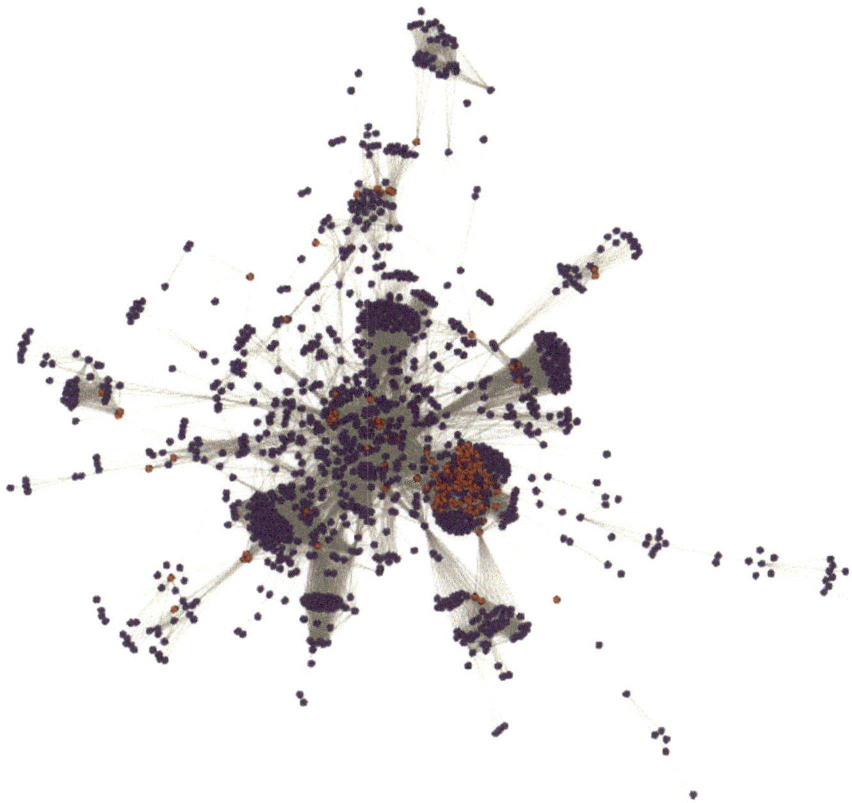

Figure 5: Cytoscape Network for the Leishmania major interactome. The nodes highlighted in red are predicted essential nodes without human orthologs. *Reproduced under Creative Commons License from* Florez, A. F.; Park, D.; Bhak, J. *et al.* Protein network prediction and topological analysis in *Leishmania major* as a tool for drug target selection. *BMC Bioinformatics*, 2010, 11, 484-492.

Raman *et al.* [105] proposed a drug target identification pipeline, namely targetTB, to predict and refine drug targets for the tuberculosis bacteria, combining important proteins/genes from both the interactome and the reactome of *Mycobacterium tuberculosis.* Potential drug targets candidates can be inferred from similar known drug targets. To this purpose, known drug-target relationships and drug similarity and target similarity measures are required. Once a compound has been identified as a ligand for a given target, the related targets and compounds can be predicted using algorithms for similarity comparison. Potential off targets can be identified *via* similarities of their ligand-binding pockets. Using such an approach, it was determined that the enoyl-acyl carrier protein reductase of *M. tuberculosis* has a similar structure to that of rat Catechol-omethyltransferase, the molecular target of the Parkinson's disease drug entacapone. This compound was found to inhibit both the activity of enoyl-acyl carrier protein reductase and the growth of the pathogen [106].

Bellera *et al.* [97] have recently developed and implemented a virtual screening campaign on the DrugBank repository (see *Valuable publicly available resources for in silico repositioning campaigns* section) to find new antichagasic drug candidates acting through reversible inhibition of cruzipain (Cz). The authors generated a conformation-independent computational model (discriminant function) based on Dragon 4.0 molecular descriptors and capable of identifying novel inhibitors of Cz. The 2D classification model was developed from a 163-compound dataset which includes both Cz inhibitors and non-inhibitors. 54 approved drugs (the straightforward candidates for repositioning purposes) belonging to the model's applicability domain were selected from DrugBank 3.0 database. Four candidates were experimentally tested in enzymatic and inhibitory assays. Among them, amiodarone (approved as antiarrhythmic) and bromocriptine (traditionally used against Parkinson and more recently repurposed for the treatment of diabetes) showed a weak but dose-dependent inhibition on Cz activity with clear effects on *T. cruzi* proliferation and morphology. The same authors obtained a second model (this time using Dragon 6.0) and applied it once again in the screening of DrugBank, finding that levothyroxine inhibits Cz in a dose-dependent manner [107]. It is worth noting, however, that the IC_{50}s of the drug candidates selected in these campaigns are far from the steady-state plasma levels obtained when administering the drugs for their original indications. What is more: levothyroxine is contraindicated in cardiac patients, and Chagas patients frequently develop cardiac symptoms. Thus, although the findings are valuable to validate the predictive ability of the models, the results are far from being ideal for reprofiling ends.

The results of these examples illustrate the possibilities of computer-aided drug repositioning in the search of novel medications for neglected diseases.

VALUABLE PUBLICLY AVAILABLE RESOURCES FOR *IN SILICO* REPOSITIONING CAMPAIGNS

There is a wide spectrum of computational resources that may result helpful in drug repurposing campaigns. We will briefly discuss some of them, although the reader should remember that the following list is far from being comprehensive. Some other resources have been compiled by Loging *et al.* [15]. The reader should also take into account that the information compiled by these resources is being continuously expanded; therefore, the figures discussed have no hope of being updated.

BindingDB [108] is a publicly accessible database presently containing above 1 million binding data for 6,589 protein targets including isoforms and mutational variants, and more that 400,000 small molecule ligands. The data are extracted from the scientific literature; the collection focuses on proteins that are drug-targets or candidate drug-targets and for which structural data are present in the Protein Data Bank. The BindingDB website supports a range of query types, including searches by chemical structure, substructure and similarity; protein sequence; ligand and protein names; affinity ranges and; molecular weight. Data sets generated by BindingDB queries can be downloaded in the form of annotated SDF files for further analysis, or used as the basis for virtual screening of a compound database uploaded by the user. The data in BindingDB are linked both to structural data in the PDB *via* PDB IDs and chemical and sequence searches, and to the literature in PubMed *via* PubMed IDs. Interestingly, it provides protein-ligand validation sets (cogeneric series with at least one associated protein-ligand co-crystal structure). Although structural data are available for the protein targets included in BindingDB, the resource collects data for many ligands that are not represented in the PDB. It continuously curates a set of publications not covered by other public databases.

BioLiP is a semi-manually curated database of biologically relevant ligand-protein interactions [109]. Most binding sites prediction tools use the protein structures from the Protein Data Bank (PDB) as templates. However, not all ligands present in the PDB are biologically relevant, as small molecules are often used as additives to solve the protein structures. To facilitate template-based ligand-protein docking, virtual screening and protein function annotations, a hierarchical procedure was developed for assessing the biological relevance of the ligands present in the PDB structures. The entries in BioLiP contain annotations on ligand-binding residues, ligand-binding affinity and catalytic sites. Moreover, a new consensus-based algorithm (COACH) has been developed to predict ligand binding sites from protein sequence or 3D structure. The BioLiP database is updated weekly and the current release contains 204,223 high quality ligand-protein interactions, involving 50,621 proteins from the PDB. The ligand distribution in BioLiP database is shown in Fig. **6**.

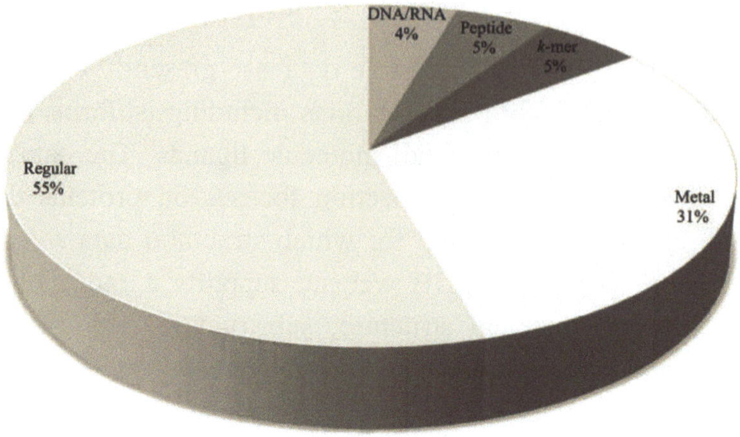

Figure 6: Distribution of ligands in BioLiP. 'Regular' represents the common small-molecule ligands except for the DNA/RNA, peptide, k-mer and metal ligands. *Reproduced under Creative Commons License from Yang, J.; Roy, A.; Zhang, Y.* BioLiP: a semi-manually curated database for biologically relevant ligand-protein interactions. *Nucleic Acids Res.*, 2013, 41(Database issue), D1096-D1103.

The **Connectivity Map** comprises a large public catalogue of gene-expression data from cultured human cells perturbed with many chemicals and genetic reagents, along with pattern-matching tools to detect similarities among them [24]. The Gene Set Enrichment Analysis (GSEA) approach is a non parametric, rank-based pattern-based strategy applied for identifying small molecules with similar effects. GSEA starts with a ''query signature'' and assesses its similarity to each of the reference expression profiles in the data set. A query signature is any list of genes whose expression is correlated with a state of interest. Examples could include genes correlated with a subtype of disease (*e.g.* drug-resistant *versus* drug-sensitive leukemia) or regulated by a biological process of interest (*e.g.* experimental activation of a signaling pathway). Each gene in the query signature carries a sign, indicating whether it is up-regulated or down-regulated. The reference gene-expression profiles in the Connectivity Map data set are also represented in a nonparametric fashion. Each profile is compared to its corresponding vehicle-treated control. The genes on the array are rank-ordered according to their differential expression relative to the control; each treatment instance thus gives rise to a rank-ordered list of ~22,000 genes [24, 110].

Drugbank is a drug-focused database currently containing more than 6,000 entries, among them 1,424 FDA-approved small molecules, 132 biotechnological drugs and 5,210 experimental drugs. More than 4,000 non-redundant protein sequences (*e.g.* drug targets, enzymes and transporters) are linked to those entries. It thus combines detailed drug information (chemical, pharmacological and pharmaceutical data) with comprehensive drug target info [111].

The **NCGC Pharmaceutical Collection** is a comprehensive, non-redundant and freely available electronic resource compiling FDA-approved drugs, plus drug listings from the UK National Health Service Information Authority, Health Canada's Drug Products Database, the European Medicines Agency and the Japanese Pharmacopeia [112]. Veterinary products listed in the Green Book (the FDA-approved animal drug list) and drugs previously approved for human use but subsequently withdrawn from the market are also included. A physical collection of small molecules suitable for high-throughput screening is available through collaborations. The collection has been conceived for drug repurposing with a focus on rare and neglected diseases.

Ondex is an integrated platform currently linking 120,000 concepts through 570,000 relations. Many types of data (multiple concept classes and relation types) can be brought together in the same graph, allowing nodes and edges to be annotated with semantically rich metadata. 15 concept classes (*e.g.* compound, drug, disease, protein, target, pathway) and 29 relation types (*e.g.* family relationships predicted with Pfam, sequence similarities computed through Blast, molecular similarities annotated between compounds, semantic similarity to score protein-protein interactions) are allowed. It includes sequence analysis, text mining and graph analysis tools [113].

Open Source Drug Discovery (OSDD) is a Council of Scientific and Industrial Research (CSIR) Team India consortium that provides a global (though Indian-centric) platform for collaborative discovery work of novel therapies for neglected tropical diseases (originally, it focused on tuberculosis). OSDD utilizes several websites including a publicly available information website (www.osdd.net), online collaboration forums (Sysborg, http://sysborg2.osdd.net, and many others for general, non-task related discussions) and a publicly-available web portal (http://crdd.osdd.net) that provide access to a huge set of computer tools valuable for target identification, virtual screening and drug design (including many of the ones previously discussed in this section) [114]. It also includes open-access biological, chemical and document repositories. Possible contributions include sharing experimental data, submitting well-characterized, pure potential active compounds, providing computing time/bandwidth, providing access to laboratories, and others [115]. There are some points that should be underlined regarding OSDD. OSDD License terms and conditions of use specify that OSDD owns all contents posted to Sysborg. Therefore, OSDD content is not of public domain. What is more, all improvements based upon data within Sysborg must be contributed back to OSDD under a worldwide royalty-free non-exclusive license. OSDD owned data may not be used by other entities without entering into a contract with OSDD. It is, therefore, a proprietary knowledge repository and the License can be considered viral. Patented inventions of OSDD are meant to ensure that drugs are licensed non-exclusively as generic drugs; patented data submitted to OSDD are used by the on-line collaboration system to track individual contributions and assure attribution (micro-attribution). It has been discussed that the viral clauses in the License agreement assure that subsequent innovation following on the existing patents remain openly accessible. Finally, it has been criticized that the decision-making processes are not entirely transparent. A more detailed discussion on these issues can be found in refs [114, 116].

PharmGKB is a pharmacogenomics knowledge resource encompassing clinical information on potentially clinically useful gene-drug associations and genotype-phenotype relationships. It annotates genetic variants and gene-drug-disease relationships and summarizes important pharmacogenomic genes and associations between genetic variants and drugs, and drug pathways [117].

PPI networks represent another domain of genome-wide data for disease-centric repositioning studies. Integration of the numerous available PPI databases that are experimentally generated and manually curated might enhance the accuracy of a

PPI network. **PrePPI** is a database that combines predicted and experimentally determined protein-protein interactions (PPIs) using a Bayesian framework [118]. Probabilities of being correct are assigned to predicted interactions. These probabilities are derived from calculated likelihood ratios (LRs) by combining structural, functional, evolutionary and expression information, with the most important contribution coming from structure. Experimentally determined interactions are compiled from a set of public databases that manually collect PPIs from the literature and are also assigned LRs. A final probability is then computed for every interaction by combining the LRs for both predicted and experimentally determined interactions. The present version of PrePPI contains 2 million PPIs with a probability above 0.1. Among them 60,000 PPIs for yeast and 370,000 PPIs for human are considered of high confidence (probability> 0.5). The PrePPI database differs from others on the following four novel features: (i) PrePPI provides structural information for many more interactions than has previously been possible using structure-enabled approaches and databases [119-121]; (ii) the predicted PPIs in PrePPI are obtained by combining structural and non-structural information merged through a Bayesian algorithm (iii) the PrePPI database contains integrative information of PPIs from major PPI databases and provides a measure of the confidence level of these interactions; and (iv) the PrePPI database assigns a single probability for each interaction using a Bayesian framework that combines quantitative results based on computational predictions with evidence contained in publicly available databases.

Stitch is a searchable database which integrates information about drug-protein interactions derived from: a) repositories of experimental information; b) manually curated sources of drug targets (*e.g.* Matador) and; c) manually curated pathway databases [122]. Additionally, interaction information is predicted through literature mining through co-occurrence and Natural Language Processing. Currently, the number of chemicals covered by Stitch surpasses 300,000, while the number of proteins goes beyond 2.6 million. Interestingly, a confidence score is assigned to each interaction reflecting its level of significance.

SuperTarget is a database containing information on more than 330,000 drug-target relationships. It provides tools for 2D drug screening and sequence comparison of molecular targets. It presently includes more than 6,200 targets and

195,000 chemical compounds [123]. A subset of more extensively annotated and manually curated drugs is provided separately in the **Matador** database, which includes both direct (binding) and indirect interactions between proteins and chemicals (*e.g.* binging a drug metabolite instead of the parent compound and drug-induced changes in gene expression). Whether only direct or both direct and indirect interactions are considered is left to the user to decide [123, 124].

The **SWEETLEAD** database is a recently launched, highly curated and publicly available database compiling chemical structures of globally approved drugs (drugs approved in the USA, India, China, Australia, Brazil, the EMA, the WHO Essential Medicines List, and those listed in the NGCG Pharmaceutical Collection have been included), as well as chemical isolated from traditional medicinal herbs and other regulated chemicals. In other words, all the natural candidates for drug repurposing campaigns are compiled there. It currently holds more than 4,400 chemical entities [125].

The **Therapeutic Target Database (TTD)** is a drug database that provides information on known therapeutic proteins and nucleic acid targets, the targeted disease, pathway information and the corresponding approved, clinical trial and investigative drugs directed at those targets. Two relevant features of the last version are the compilation of 2D and 3D QSAR models directed at different targets and chemical families, and the inclusion of the structure and potency of more than 3,600 multi-target agents [126]. It currently includes 2,025 targets and 17,816 drugs.

CONCLUSION

Drug repositioning has lately led to an explosion of activity in the public and private sectors (usually prompting public-private or academy-industry joint collaborations). From the numerous applications and resources overviewed throughout this chapter, a clear trend emerges showing a remarkable shift from serendipitous drug repositioning to knowledge-based systematic approaches that often involve computer methods. Summarizing, we have described three basic approaches to computer assisted repositioning campaigns, namely bioinformatic, cheminformatic and literature-based approximations. These approaches might be (and are being) combined in integrative, network-based approximations. Among these, multipartite networks (which contain a wide diversity of concept types/node types) including semantic associations between their nodes seem to be, in our opinion, the most complete and promising methodologies so far. The impressive and exponentially growing volume of information that shall be analyzed to comprehensively envision and understand drug-protein interactions and drug effects on phenotype, and to test the subsequent emerging hypothesis on novel therapies, explains why collaborative efforts are needed to fully develop the potential of indication expansion. Is no surprising, thus, that a diversity of publicly available resources valuable to repositioning ends has been developed, and that efforts to transfer the open source model to the field of drug discovery are being performed. Computer-aided drug repositioning poses an excellent alternative for the development of new therapies for orphan and neglected diseases, which is frequently driven by limited public funding and government incentives. After the computer-generated hypotheses are experimentally validated, it can also provide an interesting framework to guide off-label prescriptions. Although exceeding the scope of the chapter, it should be noted that intellectual property considerations are critical to drug repurposing [127].

AKNOWLEDGEMENTS

We thank Elsevier for the permission to reproduce Fig. (**1**). Figs. (**3**), (**5**) and (**6**) reproduced by license Creative Commons. The authors would like to thanks UNLP and CONICET. C. L. Bellera and M. E. Di Ianni are fellowship holders from the National Council of Scientific and Technical Research (CONICET). A. Talevi and E. Castro are members of the Scientific Research Career at CONICET. L. E. Bruno Blanch is a researcher of Faculty of Exact Sciences, National University of La Plata.

CONFLICT OF INTEREST

The authors confirm that this chapter contents have no conflict of interest.

REFERENCES

[1] Allarakhia, M. Open-source approaches for the repurposing of existing or failed candidate drugs: learning from and applying the lessons across diseases. *Drug Des. Devel. Ther.*, **2013**, **2013**(7), 753-766.

[2] Allison, M. NCATS launches drug repurposing program. *Nat. Biotechnol.*, **2012**, 30, 571-572.

[3] Marusina, K.; Welsch, D. J.; Rose, L.; Brock, D.; Bahr, N. The CTSA pharmaceutical assets portal - a public-private partnership model for drug repositioning. *Drug Discov. Today Ther. Strat.*, **2011**, 8(3-4), 77-83.

[4] Novac, N. Challenges and opportunities of drug repositioning. *Trends Pharmacol. Sci.*, **2013**, 34(5), 267-272.

[5] Murteira, S.; Ghezaiel, Z.; Karray, S.; Lamure, M. Drug reformulations and repositioning in pharmaceutical industry and its impact on market access: reassessment of nomenclature. *J. Mark Access Health Pol.*, **2013**, 1(21131).

[6] Wadman, M. NIH gambles on recycled drugs. *Nature*, **2013**, 499(7458), 263-264.

[7] Meadows, W. A.; Hollowell, B. D. 'Off-label' drug use: and FDA regulatory term, not a negative implication of its medical use. *Int. J. Impot. Res.*, **2008**, 20(2), 135-144.

[8] Arrowsmith, J.; Harrison R. Drug repositioning: the business case and current strategies to repurpose shelved candidates and marketed drugs. In: *Drug Repositioning: Bringing New Life to Shelved Assets and Existing Drugs,* Barrat, M. J. and Frail, D. E., Eds.; John Wiley & Sons, NJ, **2012**; 9-32.

[9] Ashburn, T.T.; Thor, K. B. Drug repositioning: identifying and developing new uses for existing drugs. *Nat. Rev. Drug Discov.*, **2004**, 3(8), 673-683.

[10] Aubé, J. Drug repurposing and the medicinal chemist. *ACS Med. Chem. Lett.*, **2012**, 3(6), 442-444.

[11] Napolitano, F.; Zhao, Y.; Moreira, V. M.;Tagliaferri, R.; Kere, J.; D'amato, M.; Greco, D. Drug repositioning: a machine-learning approach through data integration. *J. Cheminform.*, **2013**, 5(30).

[12] Liu, Z.; Fang, H.; Reagan, K.; Xu, X.; Mendrick, D. L.; Slikker, W. Jr.; Tong, W. *In silico* drug repositioning: what we need to know. *Drug Discov. Today,* **2013**, 18(3-4), 110-115.

[13] Issa, N. T.; Kruger, J.; Byers, S. W.; Dakshanamurthy, S. Drug repurposing a reality: from computers to the clinic. *Expert Rev. Clin. Pharmacol.*, **2013**, 6(2), 95-97.

[14] Lussier, Y. A.; Chen, J. L. The emergence of genome-based drug repositioning. *Sci. Transl. Med.*, **2011**, 3(96), 96-131.

[15] Loging, W.; Rodriguez-Esteban, R.; Hill, J.; Freeman, T.; Miglietta, J. Cheminformatic/bioinformatic analysis of large corporate databases: application to drug repurposing. *Drug Discov. Today Ther. Strat.*, **2011**, 8(3-4), 109-116.

[16] Wu, Z.; Wang, Y; Chen, L. Network-based drug repositioning. *Mol. BioSyst.*, **2013**, 9(6), 1268-1281.

[17] Haupt, V. J.; Schroeder, M. Old friends in new guise: repositioning of known drugs with structural bioinformatics. *Brief. Bioinform.*, **2011**, 12(4), 312-326.

[18] Deftereos, S. N.; Andronis, Ch.; Friedla, E. J.; Persidis, A.; Persidis, A. Drug repurposing and adverse event prediction using high-throughput literature analysis. *Wiley Interdiscip. Rev. Syst. Biol. Med.*, **2011**, 3(3), 323-334.

[19] Frijters, R.; van Vugt, M.; Smeets, R.; van Schaik, R.; de Vlieg, J.; Alkema, W. Literature mining for the discovery of hidden connections between drugs, genes and diseases. *PLoS Comput. Biol.*, **2010**, 6(9), e1000943.

[20] Sardana, D.; Zhu, C.; Zhang, M.; Gudivada, R. C.; Yang, L.; Jegga, A. G. Drug repositioning for orphan diseases. *Brief. Bioinform.*, **2011**, 12(4), 346-356.

[21] Muthyala, R. Orphan/rare drug discovery through drug repositioning. *Drug Discov. Today Ther. Strat.*, **2011**, 8(3-4), 71-76.

[22] Ekins, S.; Williams, A. J.; Krasowski, M. D.; Freundlich, J. S. *In silico* repositioning of approved drugs for rare and neglected diseases. *Drug Discov. Today*, **2011**, 16(7-8), 298-310.

[23] Qu, X. A.; Rajpal, D. K. Applications of Connectivity Map in drug discovery and development. *Drug Discov. Today,* **2012**, 17, (23-24), **1289**-1298.

[24] Lamb, J.; Crawford, E. D.; Peck, D.; Modell, J. W.; Blat, I. C.; Wrobel, M. J.; Lerner, J.; Brunet, J. P.; Subramanian, A.; Ross, K. N.; Reich, M.; Hieronymus, H.; Wei, G.; Armstrong, S. A.; Haggarty, S. J.; Clemons, P. A.; Wei, R.; Carr, S. A.; Lander E. S.; Golub, T. R. The Connectivity Map: using gene-expression signatures to connect small molecules, genes, and, disease. *Science,* **2006**, 313(5795). 1929-1935.

[25] Sirota, M.; Dudley, J. T.; Kim, J.; Chiang, A. P.; Morgan, A. A.; Sweet-Cordero, A.; Sage, J.; Butte, A. J. Discovery and preclinical validation of drug indications using compendia of public gene expression data. *Sci. Transl. Med.,* **2011**, 3(96). 96ra77.

[26] Dudley, J. T.; Sirota, M. Shenoy, M.; Pai, R. K.; Roedder, S.; Chiang, A. P.; Morgan, A. A.; Sarwal, M. M.; Pasricha, P. J; Butte, A. J. Computational repositioning of the anticonvulsant topiramate for inflammatory bowel disease. *Sci. Transl. Med.,* **2011**, 3(96). 96ra76.

[27] Cavalla, D. Predictive methods in drug repurposing: gold mine or just a bigger haystack?. *Drug. Discov. Today*, **2013**, 18(11-12), 523-532.

[28] Claerhout, S.; Lim, J. Y.; Choi, W. Park, Y.; Kim, K.; Kim, S.; Lee, J.; Mills, G.; Cho, J. Y. Gene expression signature analysis identifies vorinostat as a candidate therapy for gastric cancer. *Plos One,* **2011**, 6(9), e24662.

[29] Yang, W. R.; Lee, Y.; Chen, M.; Chao, K. M.; Huang, C. Y. *In silico* drug screening and potential target identification for hepatocellular carcinoma using support vector machines based on drug screening result. *Gene,* **2013**, 518(1), 201-208.

[30] Sanseau, P.; Agarwal, P.; Barnes, M. R.; Pastinen, T.; Richards, J. B.; Cardon, L. R.; Mooser, V. Use of genome-wide association studies for drug repositioning. *Nat. Biotechnol.* **2012**, 30(4), 317-320.

[31] Wang, Z. Y.; Zhang, H. Y. Rational drug repositioning by medical genetics. *Nat. Biotechnol.* **2013**, 31(12), 1080-1082.

[32] Wang, Z. Y.; Quang, Y.; Zhang, H. Y. Medical genetic inspirations for anticancer drug repurposing. *Trends Pharmacol. Sci.* **2014**, 35(1), 1-3.

[33] Haupt, V. J.; Daminelli, S.; Schroeder, M. Drug promiscuity in PDB: protein binding site similarity is key. *Plos One,* **2013**, 8(6), e65894.

[34] Xie, L.; Xie, L.; Bourne, P.E. A unified statistical model to support local sequence order independent similarity searching for ligand-binding sites and its application to genome-based drug discovery. *Bioinformatics*, **2009**, 25(12), i305-i312.

[35] Talevi, A.; Castro, E. A.; Bruno-Blanch, L. E. Virtual screening: an emergent, key methodology for drug development in an emergent continent. A bridge towards

patentability. In: *Advanced methods and applications in cheminformatics: research progress and new applications*, Castro, E. A. and Haghi, A. K., Eds.; IGI Global, Hershey, PA, **2011**, 229-245.

[36] Talevi, A.; Gavernet, L.; Bruno-Blanch, L. E. Combined virtual screening strategies. *Curr. Comput.-Aid. Drug.,* **2009**, 5(1), 23-37.

[37] Ma, D. L.; Chan, D. S.; Leung, Ch. Drug repositioning by structure-based virtual screening. *Chem. Soc. Rev.*, **2013**, 42(5), 2130-2141.

[38] Dakshanamurthy, S.; Issa, N. T.; Assefnia, S.; Seshasayee, A.; Peters, O. J.; Madhavan, S.; Uren, A.; Brown, M. L.; Byers, S. W. Predicting new indications for approved drugs using a proteochemometric method. *J. Med. Chem.,* **2012**, 55(15), 6832-6848.

[39] Lejal, N.; Tarus, B.; Chenavas, S.; Bertho, N.; Delmas, B.; Ruigrok, R. W.; Di Primo, C.; Slama-Schwok, A. Structure-based discovery of the novel antiviral properties of naproxen against the nucleoprotein of influenza A virus. *Antimicrob. Agents Chemother.*, **2013**, 57(5), **2231**-2242.

[40] Zhang, Q; Muegge, L. Scaffold hopping through virtual screening using 2D and 3D similarity descriptors: ranking, voting, and consensus scoring. *J. Med. Chem.*, **2006**, 49(5), **1536**-1548.

[41] Good, A. A.; Hermsmeier, M. A.; Hindle, S. A. Measuring CAMD technique performance: a virtual screening case study in the design of validation experiments. *J Comput Aided Mol Des.*, **2004**, 18(7-9), 529-536.

[42] Ripphausen, P.; Nisius, B.; Peltason, L.; Bajorath, J. Quo vadis, virtual screening? A comprehensive survey of prospective applications. *J. Med. Chem.*, **2010**, 53(24), 8461-8467.

[43] Holliday, J. D.; Kanoulas, E.; Malim, N.; Willett, P. Multiple search methods for similarity-based virtual screening: analysis of search overlap and precision. *J. Cheminform.,* **2011**, 3(29).

[44] Wu, L.; Ai, N; Liu, Y.; Wang, Y.; Fan, X. Relating anatomical therapeutic indications by the ensemble similarity of drug sets. *J. Chem. Inf. Model.*, **2013**, 53(8), 2154-2160.

[45] Keiser, K. L.; Roth, L. B.; Armbruster, B. N.; Ernsberger, P.; Irwin, J. J.; Shoichet, B. K. Relating protein pharmacology by ligand chemistry. *Nature Biotechnology,* **2007**, 25(2), 197-206.

[46] Keiser, M. J.; Irwin, J. J.; Laggner, Ch.; Abbas, A. I.; Hufeisen, S. J.; Jensen, N. H.; Kuijer, M. B.; Matos, R. C.; Tran, T. B.; Whaley, R.; Glennon, R. A.; Hert, J.; Thomas, K. L.; Edwards, D. D.; Shoichet, B. K.; Roth, B. L. Predicting new molecular targets for known drugs. *Nature*, **2009**, 462(7270), 175-181.

[47] Lekka, E.; Deftereos, S.; Persidis, A.; Persidis, A.; Andronis, Ch. Literature analysis for systematic drug repurposing: a case study from Biovista. *Drug Discovery Today*, **2011**, 8(3-4), 103-108.

[48] Swanson, D. R. Fish oil, Raynaud's syndrome, and undiscovered public knowledge. *Perspect Biol Med.*, **1986**, 30(1), 7-18.

[49] Swanson, D.R. Migraine and magnesium: eleven neglected connections. *Perspect. Biol. Med.*, **1988**, 31(4), 526-557.

[50] Jensen, L. J.; Saric, J.; Bork, P. Literature mining for the biologist: from information retrieval to biological discovery. *Nat. Rev. Genetics,* **2006**, 7, 119-129.

[51] Cohen, T.; Widdows, D.; Schvaneveldt, R. W.; Davies, P.; Rindflesch, T. Discovering discovery patterns with predication-based Semantic Indexing. *J. Biomed. Informatics*, **2012**, 45(6), **1049**-1965.

[52] Weeber, M.; Kors, J. A.; Mons, B. Online tools to support literature-based discovery in the life sciences. *Brief Bioinform.*, **2005**, 6(3), 277-286.

[53] Cameron, D.; Bodenreider, O.; Yalamanchili, H.; Dahn, T.; Vallabhaneni, S.; Thirunarayan, K.; Sneth, A. P.; Rindflesch, T. C. A graph-based recovery and decomposition of swanson's hypothesis using semantic predications. *J. Biomed. Inform.*, **2013**, 46(2), 238-251.

[54] Rindflesch, T. C.; Fiszman. The interaction of domain knowledge and linguistic structure in natural language processing: interpreting hypernymic propositions in biomedical text. *J. Biomed. Inform.*, **2003**, 36(6), 462-477.

[55] Wilkowski, B.; Fiszman, M.; Miller, Ch.; Hristovski, D.; Arabandi, S.; Rosemblat, G.; Rindflesch, T. Graph-based methods for discovery browsing with semantic predications. *AMIA Annu Symp Proc.*, **2011**, 1514-1523.

[56] Choi, J.; Kim, K.; Song, M.; Lee, D. Generation and application of drug indication inference models using typed network motif comparison analysis. *BMC. Med. Inform. Decis. Mak.,* **2013**, 13(1), s2.

[57] Li, J.; Zhu, X.; Chen, J. Y. Building disease-specific drug-protein connectivity maps from molecular interaction networks and PubMed abstracts. *PLoS Comput Biol.*, **2009**, 5(7), e1000450.

[58] Frijters, R.; Heupers, B.; van Beek, P.; Bouwhuis, M.; van Schaik, R.; de Vlieg, J.; Polman, J.; Alkema, W. CoPub: a literature-based keyword enrichment tool for microarray data analysis. *Nucleic Acids Res.*, **2008**, 36(web server issue), W406-W410.

[59] Alako, B.T.; Veldhoven, A.; van Baal, S.; van Baal, S.; Jelier, R.; Verhoeven, S.; Rullmann, T.; Polman, J.; Jenster, G. CoPub mapper: mining MEDLINE based on search term co-publication. *BMC Bioinformatics*, **2005**, 6(51), 1-15.

[60] Iskar, M.; Zeller, G.; Zhao, X.; van Noort, V.; Bork, P. Drug discovery in the age of systems biology: the rise of computational approaches for data integration. *Curr. Opin. Biotech.*, **2012**, 23(4), 609-616.

[61] Zhao, S.; Li, S. Network-based relating pharmacological and genomic spaces for drug target identification. *Plos One*, **2010**, 5(7), e11764.

[62] Hu, Y.; Stumpfe, D.; Bajorath, J. Advancing the activity cliff concept. *F1000Research*, **2013**, 2(199).

[63] Keshava Prasad, T.S.; Goel, R.; Kandasamy, K.; Keerthikumar, S.; Kumar, S.; Mathivanan, S.; Telikicherla, D.; Raju, R.; Shafreen, B.; Venugopal, A.; Balakrishnan, L.; Marimuthu, A.; Banerjee, S.; Somanathan, D. S.; Sebastian, A.; Rani, S.; Ray, S.; Harrys Kishore, C. J.; Kanth, S.; Ahmed, M.; Kashyap, M. K.; Mohmood, R.; Ramachandra, Y. L.; Krishna, V.; Rahiman, B. A.; Mohan, S.; Ranganathan, P.; Ramabadran, S.; Chaerkady, R.; Pandey, A. human protein reference database—2009 update. *Nucleic Acids Res.*, **2009**, 37(database issue), D767-D772.

[64] Yildirim, M. A.; Goh, K.I.; Cusick, M.E.; Barabasi, A. L.; Vidal, M. Drug-target network. *Nat Biotechnol.* **2007**, 25(10), **1119**-1126.

[65] Mencher, S. K.; Wang, L. G. Promiscuous drugs compared to selective drugs (promiscuity can be a virtue). *BMC Clin Pharmacol.*, **2005**, 5(3).

[66] Hopkins, A. L. Network pharmacology: the next paradigm in drug discovery. *Nat Chem Biol.,* **2008**, 4(11), 682-690.

[67] Merino, A.; Bronowska, A. K.; Jackson, D. B.; Cahill, D. J. Drug profiling: knowing where it hits. *Drug Discov. Today,* **2010**, 15(17-18), 749-756.

[68] Medina-Franco, J. L.; Giulianotti, M. A.; Welmaker, G. S.; Houghten, R. A. Shifting from the single to the multitarget paradigm in drug discovery. *Drug Discov Today*, **2013**, 18(9-10), 495-501.

[69] Vidal, M.; Cusick, M. E.; Barabási, A. L. Interactome network and human disease. *Cell,* **2011**, 144(6), 987-998.

[70] Chen, B.; Ding, Y.; Wild, D. J. Assessing drug target association using semantic linked data. *Plos. Comput. Biol.,* **2012**, 8(7), e1002574.

[71] Chen, B.; Ding, Y.; Wild, D. J. Improving integrative searching of systems chemical biology data using semantic annotation. *J Cheminform.*, **2012**, 4(6).

[72] Lee, H. S.; Bae, T.; Lee, J. H.; Kim, D. G.; Oh, Y. S.; Jang, Y.; Kim, J. T.; Lee, J. J.; Innocenti, A.; Supuran, C. T.; Chen, L.; Rho, K.; Kim, S. Rational drug repositioning guided by an integrated pharmacological network of protein, disease and drug. *BMC Syst. Biol.*, **2012**, 6(80).

[73] Talevi, A.; Enrique, A. V.; Bruno-Blanch, L. E. Anticonvulsant activity of artificial sweeteners: a structural link between sweet-taste receptor T1R3 and brain glutamate receptors. *Bioorg. Med. Chem. Lett.,* **2012**, 22(12), 4072-4074.

[74] Talevi, A.; Bellera, C. L.; Castro, E. A.; Bruno-Blanch, L. E. A successful virtual screening application: prediction of anticonvulsant activity in the MES test of widely use pharmaceutical and food preservatives methylparaben and propylparaben. *J Comput Aided Mol Des.,* **2007**, 21(9), 527-538.

[75] Shkulev, V. A.; Aboyan, L. S.; Dzhagatspanyan, I. A.; Akopyan, N. E.; Mndzhoyan, O. L. Saccharin derivatives. *Pharm. Chem. J.,* **1979**, 13(2), 144-147.

[76] Nelson, G.; Hoon, M. A.; Chandrashekar, J.; Zhang, Y.; Ryba, N. J.; Zuker, C. S. Mammalian sweet taste receptors. *Cell,* **2001**, 106(3), 381-390.

[77] Chen, Q. Y.; Alarcon, S.; Tharp, A.; Ahmed, O. M.; Estrella, N. L.; Greene, T. A.; Rucker, J.; Breslin, P. A. Perceptual variation in umami taste and polymorphism in TAS1R taste receptor genes. *Am J Clin Nutr.,* **2009**, 90(3), 770S-779S.

[78] Lemon, C. H.; Margolskee, R. F. Contribution of the T1r3 taste receptor to the response properties of central gustatory neurons. *J Neurophysiol.,* **2009**, 101(5), 2459-2471.

[79] Akiyama, K.; Daigen, A.; Yamada, N.; Itoh, Y.; Kohira, I.; Ujike, H.; Otsuki, S. Long-lasting enhancement of metabotropic excitatory amino acid receptor-mediated polyphosphoinoisitide hydrolysis in the amygdale/pyriform cortex of deep prepiriform cortical kindled rats. *Brain Res.,* **1992**, 569(1), 71-77.

[80] Keele, N. B.; Zinebi, F.; Neugebauer, V.; Shinnick-Gallaher, P. Epileptogenesis up-regulates metabotropic glutamate receptor activation of sodium-calcium exchange current in the amiygdala. *J Neurophysiol.,* **2000**, 83(4), 2458-2462.

[81] Notenboon, R. G.; Hampson, D. R.; Jansen, G. H.; van Rijen, P. C.; van Veelen, C. W.; van Nieuwenhuizen, O.; de Graan, P. N. Up-regulation of hippocampal metabotropic glutamate receptor 5 in temporal lobe epilepsy patients. *Brain,* **2006**, 129(1), 96-107.

[82] Skeberdis, V. A.; Lan, J.; Opitz, T.; Zheng, X.; Bennett, M. V.; Zukin, R. S. mGluR1-mediated potentiation of NMDA receptors involves a rise in intracellular calcium and activation of protein kinase C. *Neuropharmacology,* **2001**, 40(7), 856-865.

[83] Lea, P. M.; Custer, S. J.; Vicini, S.; Faden, A. Neuronal and glial mGluR5 modulation prevents stretch-induced enhancement of NMDA receptor current. *Pharmacol Biochem Behav.,* **2002**, 73(2), 287-298.

[84] Moran, M.; Guzman, J.; Ropars, A.; McDonald, A.; Jameson, N.; Omune, B.; Ryan, S.; Wu, L. Neglected disease research and development: how much are we really spending? *PLoS Medicine*, **2009**, 6(2), 137-146.

[85] Chirac, P. and Torreele, E. Global framework on essential health R&D. *Lancet*, **2006**, 367(9522), **1560**-1561.

[86] Griggs, R.C.; Batshaw, M.; Dunkle, M.; Gopal-Srivastava, R.; Kaye, E.; Krischer, J.; Nguyen, T.; Pulus, K.; Merkel, P. A. Clinical research for rare disease: opportunities, challenges, and solutions. *Mol. Genet. Metab.,* **2009**, 96(1), 20-26.

[87] Coles, L.D.; Cloyd, J. C. The role of academic institutions in the development of drugs for rare and neglected diseases. *Clin Pharmacol Ther.*, **2012**, 92(2), 193-202.

[88] Stevens, A.J.; Jensen, J. J.; Wyller, K.; Kilgore, P. C.; Chatterjee, S.; Rohrbaugh, M. L. The role of public-sector research in the discovery of drugs and vaccines. *N. Engl. J. Med.*, **2011**, 364(6), 535-541.

[89] Oprea, T.I.; Bauman, J. E.; Bologa, C. G.; Buranda, T.; Chigaev, A.; Edwards, B. S.; Jarvik, J. W.;, Gresham, H. D.;, Haynes, M. K.; Hjelle, B.; Hromas, R.; Hudson, L.; Mackenzie, D. A.; Muller, C.Y.; Reed, J. C.;, Simons, P. C.; Smagley, Y.; Strouse, J.; Surviladze, Z.; Thompson, T.; Ursu, O.; Waller, A.; Wandinger-Ness, A.; Winter, S. S.; Wu, Y.; Young, S. M., Larson, R. S.; Willman, C.; Sklar, L. A. Drug repurposing from an academic perspective. *Drug Discov Today*, **2011**, 8(3-4), 61-69.

[90] Pollastri, M.P.; Campbell, R.K. Target repurposing for neglected diseases. *Future Med. Chem.*, **2012**, 3(10), 1307-1315.

[91] Radish, J. More medicines for neglected and emerging infectious diseases. *Bull World Health Org.*, **2007**, 85(8), 569-648.

[92] Bost, F.; Jacobs, R. T.; Kowalczk, P. Informatics for neglected diseases collaboration. *Current Opinion in Drug Discovery & Development*, **2010**, 13(3), 286-296.

[93] Allarakhia, M. Open source biopharmaceutical innovation- A mode of entry for firms in emerging markets. *J. Bus. Chem.*, **2009**, 6 (1), 11-30

[94] Allarakhia, M.; Ajuwon, L. Understanding and creating value from open source drug discovery for neglected tropical diseases. *Expert Opin Drug Discov.*, **2012**, 7(8), 643-657.

[95] Allarakhia M. Developing a framework for understanding and enabling open source drug discovery. *Expert Opin Drug Discov.*, **2010**, 5(8), 709-714.

[96] Oldfield, E. Targeting isoprenoid biosynthesis for drug discovery: bench to bedside. *Acc. Chem. Res.*, **2010**, 43(9), 1216-1226.

[97] Bellera, C.; Balcazar, D.; Alberca, L.; Labrilola, C.; Talevi, A.; Carrillo, C. Application of computer-aided drug repurposing in the search of new cruzipain inhibitors: discovery of amiodarone and bromocriptine inhibitory effects. *J. Chem. Inf. Model.*, **2013**, 53(9), 2402-2408.

[98] Miguel, D.C.; Jenicer, K. U.; Uliana, S. R. B. Tamoxifen is effective in the treatment of *Leishmania amazonensis* infections in mice. *PLoS Negl. Trop. Dis.*, **2008**, 2(6), e249.

[99] Freitas-Junior, L. H.; Chatelain, E.; Andrade Kim, H.; Siqueira-Neto, J. L. Visceral leishmaniasis treatment: What do we have, what do we need and how to deliver it? *Int J Parasitol*, **2012**, 2, 11-19.

[100] Kobylinski, K.C.; Sylla, M.; Chapman, P. L.; Sarr, M. D.; Foy, B. D. Ivermectin mass drug administration to humans disrupts malaria parasite transmission in snegalese villages. *Am J Trop Med Hyg.,* **2011**, 85(1), 3-5.

[101] [No authors listed]. One world problem, one world solution? *Nature Rev Drug Discov.*, **2005**, 4, 701.

[102] Chong, C.R.; Chen, X.; Shi, L.; Liu, J. O.; Sullivan, D. J. A clinical drug library screen identifies astemizole as an antimalarial agent. *Nat. Chem. Biol.*, **2006**, 2(8), 415-416.

[103] de Carvalho, L. P.; Lin, G.; Jiang, X.; Nathan, C. Nitazoxanide kills replicating and non replicating Mycobacterium tuberculosis and evades resistance. *J. Med. Chem.*, **2009**, 52(19), 5789-5792.

[104] Florez, A. F.; Park, D.; Bhak, J.; Kim, B.; Kuchinsky, A.; Morris, J.; Espinosa, J.; Muskus, C. Protein network prediction and topological analysis in *Leishmania major* as a tool for drug target selection. *BMC Bioinformatics*, **2010**, 11(484).

[105] Raman, K.; Yeturu, K.; Chandra, N. targetTB: a target identification pipeline for *Mycobacterium tuberculosis* through an interactome, reactome and genome-scale structural analysis. *BMC Syst Biol.*, **2008**, 2(109).

[106] Kinnings, S.L.; Liu, N.; Buchmeier, N.; Tonge, P.; Xie, L.; Bourne, P. E. Drug discovery using chemical systems biology: repositioning the safe medicine comtan to treat multi-drug and extensively drug resistant tuberculosis. *PLoS Comput. Biol.*, **2009**, 5(7), e1000423.

[107] Bellera, C.; Balcazar, D.; Alberca, L.; Labriola, C.; Talevi, A.; Carrillo, C. Identification of levothyroxine antichagasic activity through computer-aided drug repurposing. *The Scientific World Journal,* **2013**, *in press.*

[108] Liu, T.; Lin, Y.; Wen, X.; Jorissen, R. N.; Gilson, M. K. BindingDB: a web-accessible database of experimentally determined protein-ligand binding affinities. *Nucleic Acids Research*. **2007**, 35(Data base issue), D198-D201.

[109] Yang, J.; Roy, A.; Zhang, Y. BioLiP: a semi-manually curated database for biologically relevant ligand-protein interactions. *Nucleic Acids Res.*, **2013**, 41(Database issue), D1096-D1103.

[110] Subramanian, A.; Tamayo, P.; Mootha, V. K.; Mukherjee, S.; Ebert, B. L.; Gillette, M. A.; Paulovich, A.; Pomeroy, S. L.; Golub, T. R.; Lander, E. S.; Mesirov, J. P. Gene set enrichment analysis: A knowledge-based approach for interpreting genome-wide expression profiles. *Proc Natl Acad Sci USA*., **2005**, 102(43), 15545-15550.

[111] Knox, C.; Law, V.; Jewison, T.; Liu, P.; Ly, S.; Frolkis, A.; Pon, A.; Banco, K.; Mak, C.; Neveu, V.; Djoumbou, Y.; Eisner, R.; Guo, A. C.; Wishart, D. S. DrugBank 3.0: a comprehensive resource for 'omics' research on drugs. *Nucleic Acids Res.,* **2011**, 39 (Database issue), D1035-D1041.

[112] Huang, R.; Southall, N.; Wang, Y.; Yasgar, A.; Shinn, P.; Jadhav, A.; Nguyen, D.; Austin, Ch. The NCGC pharmaceutical collection: a comprehensive resource of clinically approved drugs enabling repurposing and chemical genomics. *Sci. Transl. Med.*, **2011**, 3(80), 80-96.

[113] Cockell, S.J.; Weile, J.; Lord, P.; Wipat, C.; Andriychenko, D.; Pocock, M.; Wilkinson, D.; Young, M.; Wipat, A. An integrate database for *in silico* discovery. *JIB*, **2010**, 7(3), 116.

[114] Ardal, Ch.; Rottingen, J. A. Open source drug discovery in practice: a case study. *PLOS Negl. Trop. Dis.*, **2012**, 6(9), e1827.

[115] Bhardwaj, A.; Scaria, V.; Raghava, G. P.; Lynn, A. M.; Chandra, N.; Banerjee, S.; Raghunandanan, M. V.; Pandey, V.; Taneja, B.; Yadav, J.; Dash, D.; Bhattacharya, J.; Misra, A.; Kumar, A.; Ramachandran, S.; Thomas, Z.; Open Source Drug Discovery

Consortium; Brahmachari, S. K. Open source drug discovery - a new paradigm of collaborative research in tuberculosis drug development. *Tuberculosis (Edinb)*, **2011**, 91(5), 479-486.

[116] Sagumaran, G. Open Source Drug Discovery - redefining IPR through open source innovations. *Curr Sci*, **2012**, 102(12), **1637**-1639.

[117] Thorn, C. F.; Klein, T. E.; Altman, R.B. Pharmacogenomics and bioinformatics: PharmGKB. *Pharmacogenomics,* **2010**, 11(4), 501-505.

[118] Zhang, Q. C.; Petrey, D.; Garzón, J. I; Deng, L.; Honig, B. PrePPI: a structure-informed database of protein-protein interactions. *Nucleic Acid Res.,* **2013**, 41(Database issue), D828-D833.

[119] Stein, A.; Céol, A.; Aloy, P. 3did: identification and classification of domain-based interactions of known three-dimensional structure. *Nucleic Acids Res.*, **2011**, 39(Database issue), D718-D723.

[120] Lo, Y. S.; Chen, Y.; Yang, J. M. 3D-interologs: an evolution database of physical protein-protein interactions across multiple genomes. *BMC Genomics*, **2010**, 11(3), S7.

[121] Davis, F.P.; Sali, A. PIBASE: a comprehensive database of structurally defined protein interfaces. *Bioinformatics,* **2005**, 21(9), 1901-1907.

[122] Kuhn, M.; Szklarczyk, D.; Franceschini, A.; van Mering, C.; Jensen, L. J.; Bork, P. STITCH 3: zooming in on protein-chemical interactions. *Nucleic Acid Res.*, **2012**, 40 (Database issue), D876-D880.

[123] Hecker, N.; Ahmed, J.; von Eichborn, J.Dunkel, M.; Macha, K.; Eckert, A.; Gilson, M.; Bourne, P. SuperTarget goes quantitative: Update on drug-target interactions. *Nucleic Acids Res.,* **2012**, 40 (D1), D1113-D1117.

[124] Günther, S.; Kuhn, M.; Dunkel, M.; Campillos, M.; Senger, C.; Petsalaki, E.; Ahmed, J.; Urdiales, E. G.; Gewiess, A.; Jenses, L. J.; Schneider, R.; Skoblo, R.; Russell, R. B.; Bourne, P. E.; Bork, P.; Preissner, R. SuperTarget and Matador: Resources for exploring drug-target relationships. *Nucleic Acids Res.,* **2008**, 36 (Database issue), D919-D922.

[125] Novick, P.A.; Ortiz, O.; Poelman, J.; Abdulhay, A. Y.; Pande, V.S. SWEETLEAD: an *in silico* database of approved drugs, regulated chemicals, and herbal isolates for computer-aided drug discovery. *PloS One*, **2013**, 8 (11), e79568.

[126] Zhu, F.; Shi, Z.; Tao, L.; Liu, X.; Xu, F.; Zhang, L.; Song, Y.; Liu, X.; Zhang, J.; Han, B.; Zhang, P; Chen, Y. Therapeutic target database update **2012**: a resource for facilitating target-oriented drug discovery. *Nucleic Acids Res.*, **2012**, 40(D1), D1128-D1136.

[127] Witkowski, T. X. Intellectual property and other legal aspects of drug repurposing. *Drug Discov. Today Therap. Strat.* **2011**, 8(3-4), 131-137.

CHAPTER 3

Tuning the Solvation Term in the MM-PBSA/GBSA Binding Affinity Predictions

Irene Maffucci and Alessandro Contini*

Dipartimento di Scienze Farmaceutiche - Sezione di Chimica Generale e Organica "Alessandro Marchesini", Università degli Studi di Milano, Via Venezian, 21 20133 Milano, Italy

Abstract: Molecular Mechanics Poisson-Boltzmann Surface Area (MM-PBSA) and Molecular Mechanics Generalized Born Surface Area (MM-GBSA) are widely used methods for the prediction of binding free energies in drug design/discovery. Indeed, their computational efficiency makes them applicable also within virtual screening protocols. Thus, in order to be useful for drug design/discovery purposes, MM-PBSA and MM-GBSA binding energy predictions have to correlate well with experimental activities. Nowadays the global effort to find a way to improve the predictivity of MM-PBSA/GBSA calculations is also focused on the solvation term by using various approaches. This chapter reports on the application of MM-PBSA/GBSA methods within the process of drug discovery and, in particular, on strategies that can be applied to improve the correlation between MM-PBSA/GBSA predicted binding affinities and experimental pharmacological activities by acting on the way the solvent is treated in such calculations. Indeed, in PB and GB models, the solvent is described as a continuous medium with a fixed dielectric constant (*i.e.* $\varepsilon = 80$ for water), while a low internal dielectric constant is assigned to the solute (generally $\varepsilon_{in} = 1$ or 2 for proteins). However, the default approach could in some cases lead to a weak correlation between predicted binding free energies and experimental data. The aim of this chapter is to present and exemplify the ways to improve the prediction of ligand binding affinity by acting on the solvation term. Different methods are observed in the literature, *e.g.* tuning the ε_{in} value depending on the features of the binding site, including a selection of explicit water molecules in order to better describe the solute-solvent interactions, tuning the grid size in PB calculations and/or using different PB solvers, or modifying the non-polar term of the solvation free energy. The pros and cons of the above mentioned methods will be critically discussed in order to help the reader in choosing the most performing protocol in terms of both calculation time and prediction quality, depending on the molecular system under evaluation.

Keywords: Binding energy, MM-PBSA, MM-GBSA, molecular dynamics, solvation.

***Corresponding author Alessandro Contini:** Dipartimento di Scienze Farmaceutiche - Sezione di Chimica Generale e Organica "Alessandro Marchesini", Università degli Studi di Milano, *Via* Venezian, 21 20133 Milano, Italy; E-mail: alessandro.contini@unimi.it

INTRODUCTION

Nowadays, the accurate prediction of binding energies is one of the most attractive goals in drug design [1-3]. Many computational methods developed for this purpose are based on Molecular Dynamics (MD) simulations, which provide a statistically meaningful conformational ensemble for thermodynamic calculations [4, 5]. The MD based approaches can be divided into pathway or end-point approaches [6]. The former methods require the interconversion of the system from the initial state to the final state through finite or infinitesimal alchemical changes of the system energy function. The most common pathway methods are the Free Energy Perturbation (FEP) [7] and the Thermodynamic Integration (TI) [1], which are very rigorous, but computationally expensive. Furthermore, their application for drug design/discovery purposes can often be non-trivial.

Computational costs can be reduced by considering only the end-point states during the binding energy prediction. One of the most popular end-point methods is Molecular Mechanics Poisson-Boltzmann/Generalized Born Surface Area (MM-PBSA/GBSA) [4, 8], which represents a good trade-off between calculation efficiency and accuracy in binding energy calculations [9, 10]. Thus, MM-PBSA/GBSA methods are getting more and more used in this field and their application within virtual screening protocols has also been reported [11, 12].

In MM-PBSA and MM-GBSA, the free energy of a ligand (L) binding to a protein (R) in order to create the complex (RL) is calculated by eq. (1):

$$\Delta G_{bind} = G_{RL} - G_R - G_L \tag{1}$$

Each term is considered as the sum of a gas-phase energy (E_{MM}), a solvation free energy (G_{solv}), and an entropy term (TS) as reported in eq. (2).

$$G = E_{MM} + G_{solv} - TS \tag{2}$$

Thus, the binding free energy is computed by eq. (3):

$$\Delta G_{bind} = \Delta E_{MM} + (\Delta G_{solv,RL} - \Delta G_{solv,R} - \Delta G_{solv,L}) - T\Delta S \tag{3}$$

ΔE_{MM} is approximated by the molecular mechanics energies of the complex and it is determined from the MD force field, which contains terms for bond, angle and torsion energies and for van der Waals and electrostatic interactions [13].

The entropy change can be divided into translational, rotational and vibrational terms; the first two of these are calculated with a standard statistical-mechanical expression, while the latter is generally calculated with normal mode or quasi-harmonic analyses and it is estimated with a rigid-rotor harmonic oscillator approximation [14].

The solvation energy is one of the most relevant terms, since the solvent is strongly involved in ligand-receptor and protein-protein interactions, stabilization of protein tertiary structure, and consequently, protein function. Thus, an accurate treatment of the solvation term is fundamental when computing binding free energies [15].

Accordingly to eq. (4), in MM-PBSA/GBSA the solvation term is decomposed into a polar and an apolar term.

$$\Delta G_{solv} = \Delta G_{pol,solv} + \Delta G_{nonpol,solv} \quad (4)$$

$\Delta G_{nonpol,solv}$ is commonly considered to be proportional to the solvent accessible surface area (SASA) [4, 16], although it can be estimated by different approaches which will be discussed later in the chapter.

$\Delta G_{pol,solv}$ is calculated by solving, for each state, the linearized Poisson-Boltzmann (PB) equation [17] or the Generalized Born (GB) equation [18, 19], an approximation of the PB equation. Thus, $\Delta G_{pol,solv}$ represents the contribution of charge and electrostatic interactions between the solute and the solvent to ΔG_{solv}.

Both PB and GB methods assume that the solvent can be macroscopically described as a continuous dielectric medium [20]. Thus, the physical system for calculating the polar contribution to the solvation free energy of a molecule can be simplified to a distribution of charges in a solvent-inaccessible low dielectric cavity surrounded by a homogeneous high dielectric medium. Commonly, the charge distribution coincides with the partial charges located at the atomic centers and the molecular surface can be considered as the dielectric boundary. The external dielectric constant is specific for the considered solvent (*i.e.* $\varepsilon = 80$, for water), while the internal dielectric constant should be set to reproduce the solute dielectric constant and is thus set to a value close to that of vacuum, although this choice is quite debated [21].

The solvation free energy for a molecule X can be calculated by eq. (5) [22]:

$$G_{PB(GB)}(X) = \frac{1}{2}\sum_{i,j\in X} q_i q_j\, g_{ij}^{PB(GB)} \tag{5}$$

Where q_i and q_j are the atomic charges and g_{ij}^{PB} represents the solution of PB equation eq. (6) [23]:

$$\vec{\nabla}\left[\varepsilon(\vec{r})\vec{\nabla}\psi(\vec{r})\right] = -4\pi\rho(\vec{r}) - 4\pi\sum_i c_i^{\infty} q_i^{ion}\lambda(\vec{r}) \cdot e^{\frac{-q_i^{ion}\psi(\vec{r})}{k_B T}} \tag{6}$$

where $(\vec{r})$ represents the position dependence, $\vec{\nabla}\psi$ is the gradient of the electrostatic potential, ρ is the solute charge distribution, c_i^{∞} is the bulk charge density of ion q_i^{ion} and λ is the accessibility of position $(\vec{r})$ to the ions in solution; k_B and T are the Boltzmann constant and absolute temperature, respectively.

The classic solvers need to solve eq. (6) twice, once in vacuum and once in the solvent environment, accordingly to eq. (7):

$$\Delta G_{solv,polar} = G_{polar}^{\varepsilon=80} - G_{polar}^{\varepsilon=1} \tag{7}$$

Since the numerical solution of the PB equation is computationally expensive and parameters have to be accurately tuned, many approaches have been developed for its approximation with a minor loss in accuracy. Among all of them, the most popular is based on the GB formalism [24].

Within the GB model, the g_{ij} term is calculated by eq. (8):

$$g_{ij}^{GB} = \left(\frac{1}{\varepsilon} - 1\right)\left[r_{ij}^{n} + \alpha_i\alpha_j exp\left(-\frac{r_{ij}^{n}}{A\alpha_i\alpha_j}\right)\right]^{-1/n} \tag{8}$$

Where A and n are constants (4 and 2, respectively, in the original formulation [19]) ε is the solvent dielectric constant and r_{ij} is the distance between atoms i and j.

In the GB method, if ε_{in} is not equal to 1, the term in the first brackets becomes $\left(\frac{1}{\varepsilon} - \frac{1}{\varepsilon_{in}}\right)$ [24]. Moreover, it is necessary to include the distance from each atom to the dielectric boundary (α_i, α_j), that is the generalized Born radius.

In theory, the Born radius for a certain atom can be calculated from the PB equation by assigning a unit charge to the atom itself, while keeping the rest of the molecule uncharged, but present so that it can be used to define the dielectric boundary [21]. Nevertheless, the Born radii can also be calculated by applying the so-called Coulomb field approximation (CFA) [24], which assumes that the dielectric displacement follows a Coulombic form and doesn't depend on the external dielectric. Thanks to the CFA, the GB method is able to reproduce quite well the results obtained by the PB model, but at a fraction of its computational cost.

In principle, in order to solve eq. (3), separate MD trajectories would be needed for the complex, the unbounded receptor and the ligand. However, an advantage of the MM-PBSA/GBSA method is given by the possibility to estimate the binding free energy of a given ligand from a single trajectory, the one obtained for the complex. The average of the interaction energies between the receptor and the ligand is then obtained by analyzing a pre-established number of snapshots for the complex, receptor and ligand, all taken from the trajectory of the solvated complex. As reported by Page and Bates [25], who applied MM-PBSA/GBSA for the prediction of binding free energies of six protein kinase inhibitors, and by Gohlke and Case [26], who studied the performance of MM-PBSA/GBSA on the H-Ras/C-Raf1 complexes, the use of single trajectories reduces both the required computational effort and the uncertainties typical of a multiple trajectory approach. However, it should be noted that separated MD simulations for the ligand, receptor and complex should be preferred in those cases where the receptor undergoes to relevant conformational changes after ligand binding [25].

Although MM-PBSA/GBSA is considered a reliable and efficient method for the estimate of binding free energies, it also presents some weaknesses that should be taken in account. In particular, a source of error can be represented by the entropy contribution, which is often neglected when relative binding free energies of similar molecules are computed. Furthermore, the quality of results strictly depends on how accurately the whole conformational space is sampled, as well as on parameters used for the description of the molecular system, such as the force field, the internal dielectric constant and the set of atomic radii [13].

Moreover, MM-PBSA shows some limits in the estimation of binding free energies of highly polar or charged molecules, since the uncertainty in the calculation of the solvation free energy is proportional to the polarity of the considered molecules [27]. Those limits are particularly relevant for buried ligands, because of the inhomogeneity of the interior of biomacromolecules [28] which might not always be correctly represented by a unique internal dielectric constant.

In addition, the implicit solvation model cannot describe the explicit solute-solvent interactions that might contribute to the binding free energy [29], such as those observed when a water molecules bridges the interaction between the ligand and the receptor [30].

As it will be explained later, the accuracy and/or the predictivity of MM-PBSA/GBSA methods can be improved by using some expedients, most of them acting on the solvation term. Indeed, in the present chapter, we will describe some of the approaches, reported so far in the literature, aiming to ameliorate the treatment of both the polar and non-polar term of the solvation free energy. Each method herein discussed is summarized in Table **1**, where references about its original application are also reported. For each method, specific applications will be discussed.

Table 1: Summary of methods described in this chapter and corresponding references

		References
Methods affecting PB calculations		
	Choice of different PB solvers	[21]
	Tuning the grid mesh	[31]
Methods affecting GB calculations		
	Choice of different GB models	[32]
Methods affecting either PB and GB calculations		
	Tuning the internal dielectric constant ε_{in}	[28, 33-39]
	Inclusion of crystallographic waters	[3, 40-43]
	Inclusion of specific water molecules selected from MD trajectory analysis (H-bonds, B factors)	[30, 40, 43]
	Inclusion of hydration shells comprising a fixed number of water molecules	[44, 45]
	Chimera methods	[34, 46, 47]
Methods affecting the non-electrostatic contribution		
	CD	[48, 49]
	PCM	[48, 49]
	LIE (β)[50]	[34]

1. TUNING THE PARAMETERS SPECIFIC TO MM-PBSA

1.1. The PB Solver

As previously described, in MM-PBSA the polar term of solvated free energy is estimated by solving the PB equation, commonly with a finite difference method.

This calculation can be made using different solvers, some of them available as "stand alone" packages (*e.g.* DelPhi [51], Adaptive Poisson Boltzmann Solver (APBS) [52, 53] and ZAP [54]), while other are available as built modules in MD simulations software (*e.g.* CHARMM PBEQ [55], and Amber *pbsa* [56]).

Most solvers use the Finite Difference Poisson-Boltzmann (FDPB) method [57, 58], which implies that charges and dielectric constants are discretized over a grid. The system is defined as a molecular surface (MS) and mapped onto a three dimensional grid with a user-defined density, which is used to obtain the finite difference solutions of the PB equation. Accuracy and precision of the FDPB method depend on the quality of the MS mapping over the grid as well as on the grid density.

Traditional PB solvers compute the electrostatic free energies by calculating the product of the electrical potential and charge at each grid point where a real charge has been mapped. This implies two FDPB calculations: in the first, half of the sum of the product of the charge at each grid point by the corresponding grid potential is calculated for the molecule in solution, generally a medium having a large ε (*e.g.* $\varepsilon = 80$, for water). In the second, the same calculation is done in vacuum, or in a medium having the same ε as the solute. The two results are than combined to obtain the reaction field energies. This method uses the potential at charged points, where it is infinitely large, and consequently inaccuracies are introduced in the calculation of solvation energies. Modern solvers are generally able to provide higher accuracy in comparable or even better computation time. For example, DelPhi uses an algorithm based on induced charges, the Scaled Solvation Energy method [51]. This approach relies on the fact that reaction field effects due to a dielectric boundary can be properly reproduced by an appropriate distribution of induced polarization charges placed at the dielectric boundary, which, in FDPB, are obtained through the numerical solution of the Gauss's law. The reaction field energy is then obtained by solving the Coulomb's law between induced polarization charges and real charges, just as in vacuum. Therefore, only one PB run is needed instead of two and this could lead to an appreciable saving in computation time. Moreover, within this method, the potential used for deriving induced charges is positioned at the MS, where there aren't fixed charges, so the calculation is also more accurate [51].

Other commonly used PB equation solver are APBS [52] and ZAP [54]. The former is a multigrid FDPB solver and performs the calculation by using initially a coarser grid and then a finer one for the refinement. ZAP is a very fast PB solver, although it is less accurate since it uses a Gaussian-based molecular volume method to build MSs.

The choice of a particular solver might then influence both the quality of results and the speed of calculations. Feig and coworkers [21] compared commonly used PB solvers (CHARMM PBEQ [55], MEAD [59], DelPhi [51], APBS [52-53],

Impact PBF [60, 61], REBEL [62] and ZAP [54]), although without exhaustively considering parameters such as the grid size and the time needed to reach convergence. Results obtained with CHARMM PBEQ were used as a reference. The authors observed that the FDPB solvers such as CHARMM PBEQ, MEAD, DelPhi and APBS were equivalent in term of accuracy (0.2% relative error), although this equivalence was obtained by using different grid resolutions. Indeed, DelPhi and APBS calculations were conducted with a grid spacing of 0.4 or 0.5 Å, while PBEQ and MEAD calculations required a grid spacing of 0.25 Å to perform as the former two software, at the expenses of computation time.

1.2. Acting on the Grid Resolution and on the MS

When predicting binding free energies by MM-PBSA calculations, the grid density might be critical: properly setting the grid resolution in order to gain the best ratio between speed and accuracy might not be trivial. Harris *et al.* [31] reported on how the grid resolution can affect the accuracy of the prediction of the electrostatic term ($\Delta\Delta G_{el}$) and the time required to reach convergence.

A grid spacing of 0.5 Å, commonly used as the default setting in many PB solvers, can lead to unacceptably large errors in the estimate of $\Delta\Delta G_{el}$. Indeed, while with a grid spacing of 0.5 Å the electrostatic contribution of each component of the complex is consistent with the experimental data [63], $\Delta\Delta G_{el}$ are much more influenced by errors due to their typically smaller magnitudes.

The authors observed that, by using van der Waals and Solvent Exposed (SE) surfaces, the average error in the estimate of $\Delta\Delta G_{el}$ can exceed 30 kcal/mol with a grid spacing of 0.5 Å and up to 100 kcal/mol with a grid spacing of 1.0 Å.

Nevertheless, a universally accepted grid spacing cannot be defined, because it strictly depends on the system under investigation. For instance, in some of the reported examples [31] the use of a grid spacing of 0.3 Å did not allow the PB calculations of achieving convergence. In those cases, the authors attempted the use of a Gaussian surface, which does not present crevices, edges and cusps, as the MS for PB calculations. Calculations readily converged, but the use of this kind of MS still need to be thoroughly validated [31].

2. TUNING THE PARAMETERS SPECIFIC TO MM-GBSA

In MM-GBSA calculations, the polar contribution to the solvation free energy is calculated by using the GB formalism, an efficient approximation of the PB equation (see eqs. (6) and (8)).

Save for the salt concentration and the external and internal dielectric constants, which are common parameters among PB and GB, the latter model is somehow less customizable. The main, and often not easy, choice the user needs to do while approaching GB calculations regards the kind of GB model to be used. Available models generally differ on the set of atomic radii (usually based on the Born atomic radii) and on the algorithm used to generate the MS. The choice of the GB model might severely affect the accuracy of results and their agreement with MM-PBSA calculations [64].

At the moment, the most used GB models are GB-HCT [65], GB-OBC(I) and GB-OBC(II) [66], GB-Neck [67] and GB-Neck2 [32].

GB-HCT and GB-OBC use the van der Waals surface for the definition of the solute-solvent boundary. The GB-OBC (I) and (II) models introduce tunable empirical parameters for scaling up buried atoms radii, which are underestimated in GB-HCT. Nevertheless, GB-OBC models are now considered obsolete, while GB-HCT, which corresponds to the standard pairwise generalized Born model (*i.e.* igb = 1, in the *sander* and *pmemd* modules of the Amber package), is still used for calculations involving nucleic acids in combination with the default set of atomic radii (bondi) [68].

The GB-Neck implementation includes a correction term to better describe the MS, however it shows no improvement in solvation energy accuracy [67] and sometimes it tends to destabilize the native protein structures, because of intramolecular hydrogen bonds and implicit solvent interactions are not properly balanced [69, 70].

Recently, the GB-Neck2 model has been introduced for MM-GBSA analysis of protein systems and it was shown to be able to lead to a significant improvement compared to GB-OBC(II) or GB-Neck models [32]. Indeed, it reproduced well the

PB solvation energies for many protein systems; moreover, it was able to overcome the secondary structure and salt bridge biases observed in GB-HCT and GB-OBC models. Concerning the Amber implementation of the GB-Neck2 model (*i.e.* igb = 8), the developers suggest to use the mbondi2 [66] or mbondi3 [32] intrinsic Born radii set, depending on the objectives of the analysis. In the mbondi3 set, modifications were introduced to the carboxylate oxygen radii and to arginine hydrogens at the guanidine nitrogen atoms, and its use is particularly recommended to get rid of the salt bridge bias typical of the GB model [32]. It should also be noted that all the above mentioned models are based on the CFA approximation, which has been shown to overestimate the effective radii [71], although GB-Neck2 was designed to empirically overcome this overestimation.

3. MM-PBSA *VS* MM-GBSA

One of the main needs in the field of drug design/discovery is represented by reliable, simple and as fast as possible computational approaches able to predict free energy of binding. However, speed and accuracy are rarely combined in a computational protocol, so the choice between the fastest, but approximate, GB and the most accurate, but computationally demanding, PB may not be trivial.

Many works comparing results obtained with the two methods have been published [2, 21, 33-34, 72, 73] and in several cases it has been observed that the choice between PB and GB may depend on the specific goal of the researcher.

Feig *et al.* [21] compared several GB and PB protocols applied to five differently populated test sets, noticing that the requirements of accuracy and speed are the critical factors for this choice. For instance, GB is preferred for scoring large sets of ligands, *e.g.* the binding conformations obtained by previous docking experiments. Otherwise, if higher levels of accuracy are desired, especially when absolute energies are going to be predicted, PB can be the first choice. However, the superior accuracy of PB *versus* GB in reproducing the experimental results is quite controversial. Indeed, Feig and coworkers reported that the GB-OBC(II) implementation was able to reduce to 1% the errors in the calculation of solvation free energies [21].

Furthermore, Wang and coworkers [73] correlated the computed binding free energies, obtained by MD simulations in explicit solvent followed by MM-PBSA and MM-GBSA analyses of eleven proteins bounded to fourteen small molecules, with the experimental free energy of binding obtained by isothermal titration calorimetry (ITC). The authors observed that, although MM-PBSA has reproduced better the experimental absolute free energies, MM-GBSA leaded to a higher correlation with ITC experiments, with a Pearson's correlation coefficient of 0.75 against 0.37 obtained by MM-PBSA. Moreover, they noticed that MM-PBSA was more sensitive than MM-GBSA to the conformation used during the analysis, while MM-GBSA performed well using random snapshots taken from the MD simulation. Indeed, by filtering the MD trajectory with scoring functions favoring the native-like poses, MM-PBSA showed the correct rank [73].

The same conclusions were stated in another recent work [72] where the HIV protease in complex with six inhibitors was used as a test set for MD based MM-PBSA/GBSA calculations. The authors observed that both MM-PB and GBSA analyses generally overestimated binding free energies; however, MM-GBSA results showed a better correlation with the experiments, whichever method was used to derive atomic partial charges during the system setup.

Hou *et al.* [33, 39] confirmed these results in a study about the accuracy of binding free energy predictions by MD, where the performance of MM-PBSA/GBSA was evaluated on fifty-nine ligands interacting with six different proteins (α-thrombin, penicillopepsin, neuraminidase, avidin, cytochrome C peroxidase and P450cam). In this case also, the MM-PBSA predicted absolute binding free energies were closer to experimental value than those obtained by MM-GBSA, but this latter method showed better performances in predicting relative binding affinities for most systems. In particular, MM-GBSA led to better correlation with experiments for α-thrombin, penicillopepsin and neuraminidase; comparable results between the two methods were obtained for avidin, while for cytochrome C peroxidase and P450cam better correlations were observed by using MM-PBSA. Probably, for the latter two systems, the lowest performance of MM-GBSA was due to the absence of optimized GB parameters for the iron ions in the binding sites.

In a recent study, where the effect of explicit solvation shells around the ligands was evaluated, we also observed for MM-GBSA a better correlation between the predicted binding affinities and the experimental data in topoisomerase, α-thrombin and avidin systems [44]. Better correlation was instead obtained by MM-PBSA for penicillopepsin, although positive binding affinities were predicted.

Genheden and Ryde [34] used avidin, with seven ligands, and fXa protein, with nine inhibitors, to compare binding affinity predictions obtained by several methods based on end-point MD simulations. In this case, the comparison between MM-GBSA and MM-PBSA showed diverging results for avidin and fXa systems. In particular, MM-PBSA gave a better correlation with the experimental data for the avidin complexes, while MM-GBSA leaded to a better correlation and lower standard errors for fXa. However, it should be noticed that the MD protocols for the simulations on the two systems were slightly different, so the results might not be safely compared due to the aforementioned dependence of the PB method on the sampling of conformations [73].

The strict influence of MD simulation protocols on the quality of MM-PBSA binding free energy predictions was also observed by Srivastava and Sastry [2], who studied the inhibition of HIV protease by fourteen selected ligands. The authors submitted each complex to a 10 ns run of MD simulation and ΔG_{bind} were computed by MM-PBSA/GBSA at different time intervals, spanning from 0-1 to 0-10 ns. The results were then correlated with the experimental IC_{50} and the authors observed for both methods an excellent correlation between the predicted and the experimental binding free energies or biological activities. By analyzing the results both qualitatively and quantitatively, they noticed that MM-GBSA computed energies had a constant number of incorrect trends during the whole MD simulation and the Pearson correlation coefficient (r^2) reached its maximum (about 0.86) after only 3 ns, then it remained constant. On the other hand, MM-PBSA showed a number of incorrect trends that decreased from 3 to 0 when rising the MD simulation time; the quantitative correlation became meaningful only after 9 ns of simulation and reached its maximum at 10 ns (r^2 = 0.91). This behavior probably depends from the accuracy of the electrostatic contribution estimations, which raises with simulation time for the PB method, while it remains constant in GB.

In conclusion, MM-PBSA might lead to a better accuracy, but a large MD simulation time is needed to reach convergence. On the other hand, MM-GBSA should be preferred for quicker calculations on large data sets. To make a comparison with the shooting sports, PB and GB could be represented as a rifle and a shotgun, respectively. While the former, in the hands of a skilled shooter, might perform better, with the latter is generally easier to hit the target.

4. TUNING PARAMETERS COMMON TO PB AND GB

4.1. Tuning the Internal Dielectric Constant ε_{in}

The most common user-modifiable parameters that are present in both PB and GB models are the ionic strength and the external and internal dielectric constants. Among these, the internal dielectric constant (ε_{in}) is particularly critical in computing the polar term of the solvation energy and, as consequence, in the binding free energy predictions.

The external dielectric constant (ε_{ext}) represents a well-defined property and depends only by the solvent used for the simulation (ε_{ext} = 80 for water); conversely, ε_{in} is not well defined since complex molecules seldom are uniform electrostatic media. Actually, in MM-PBSA/GBSA calculations ε_{in} does not represent a real physical constant, but a parameter which depends on the used method [74]. A value of $\varepsilon_{in} = 1$ is usually assigned by default [4], but the choice of the solute dielectric constant is object of debate and many works have been published discussing on the dependency of the performance of the binding free energy predictions from ε_{in} [33-37, 39].

The treatment of ε_{in} is controversial, especially when ranking the ligand-receptor binding free energies, because it was observed that the use of $\varepsilon_{in} = 1$ can lead to an overestimation of the ligand-receptor electrostatic interactions [35, 75-77].

In literature, two approaches for the modification of ε_{in} have been described: one is based on a systematic scanning of ε_{in} from 1 to 25 [33-34, 37, 39], while the other implies the use of a variable ε_{in}, depending on the physicochemical properties of the interacting residues [35-36].

This latter approach was tested by Ravindranathan and coworkers on six pharmaceutically relevant targets [35], namely CDK2, fXa, p38_u, PDE10A, human carbonic anhydrase and a second p38 chemical series (p38_pp), in complex with several ligands. They assigned five different ε_{in} values (1, 2, 4, 8 and 20) to each polar or ionizable residue (Ser, Thr, Asn, Gln, His, Lys, Arg, Asp, Glu) and to all the other residues, which were considered altogether. Then, for each system the best set of dielectric constants, evaluated in terms of r^2 and predictive index (PI), was selected and discussed. However, this approach leaded to minimal improvements in r^2 and PI values in comparison with the standard electrostatic treatment. This was especially observed for those systems with binding sites prevalently made by non-polar residues, such as PDE10A and p38_pp, where the ligand-receptor electrostatic interactions are not appreciably large.

A similar approach was used to rank the inhibitory activity against HIV-1 gp41 fusion peptide of mutants of the virus inhibitory peptide (VIRIP) [36] having known IC_{50}s. The authors initially assigned $\varepsilon_{in} = 2$ to the wild type VIRIP-gp41 complex, and a variable ε_{in} to the mutated complexes. Those latter were assigned by treating each mutated peptide accordingly to the following rule: $\varepsilon_{in} = 2$ was assigned to the non-polar residues, $\varepsilon_{in} = 3$ to the polar residues and $\varepsilon_{in} = 4$ to the charged residues, accordingly to a previously reported protocol [38]. Final ε_{in}s used in MM-PBSA analyses, ranging from 2 to 6, were then obtained by averaging the contributions of each residue. With this approach the authors obtained an improvement in the correlation between the experimental activities and the MM-PBSA binding energies of about 30%, if compared to the standard approach were ε_{in} was set to 2 for all complexes. It should be noted that the best results were obtained by analyzing the MD trajectories performed with weak restraints on the backbone atoms. Moreover, although the use of multiple internal dielectric constants leaded to a clear separation between strong and weak ligands, it also produced quite large standard deviations.

From the studies described above, it appears that setting ε_{in} on the bases of the different dielectric constants assigned to each different residue can be non-trivial and computationally quite expensive, in face of a generally modest improvement in correlation between theoretical and experimental data.

An easier approach, although less rigorous, can be the simple increase of the ε_{in} value in order to scale down the overestimated electrostatic interactions [28], however a universal dielectric constant suitable for every protein has not been found so far [33-34, 37]. Indeed, the choice of ε_{in} is strictly system-dependent and the binding site need to be accurately investigated to gather the most appropriate ε_{in} [33]. Hou *et al.* evaluated the correlation between the predicted and the experimental binding free energies, in terms of Spearman correlation coefficient (r_s), of six systems (α-thrombin, avidin, cytochrome C peroxidase, neuraminidase, P450cam and penicillopepsin) to which ε_{in} = 1, 2 and 4 was assigned in both PB and GB calculations. For the neuraminidase and α-thrombin systems, characterized by highly charged binding sites and consequent ability to form ion-ion interactions with negatively charged ligands [78], the best correlation for PB calculations was obtained by using ε_{in} = 4 (r_s = 0.68 and 0.81, respectively). Similar results, although slightly better, were obtained by using the GB-HCT model with ε_{in} = 4 (r_s = 0.78 and 0.90, respectively, for GB), even if for α-thrombin good results were also obtained for ε_{in} = 2 (r_s = 0.88 and 0.91 for GB-HCT and GB-OBC models, respectively). Coherent results were obtained for α-thrombin by Yang and coworkers [37] who correlated with experiments the MM-PBSA/GBSA binding free energies computed for twenty-eight ligands by setting ε_{in} = 1 and ε_{in} = 4. In this case also, the best correlation was obtained for ε_{in} = 4 (r^2 = 0.74 and 0.73 for PB and GB calculations, respectively).

Conversely, for penicillopepsin, where only one charged residue, able to interact with ligands, is present in the active site, the best correlation in PB calculations was obtained with ε_{in} = 2 (r_s = 0.41), while GB-OBC provided comparable results for ε_{in} = 2 and 4 (r_s = 0.73 and 0.73, respectively) [33]. For avidin, which does not have charged residue in the binding pocket, the optimal ε_{in} value was 1 for both PB and GB models (r_s = 0.92 and 0.93 for PB and GB-OBC models, respectively). Coherent results were reported by Genheden and Ryde [34] who noticed, for the same system, a decrease in the r^2 value from 0.60 to 0.13 when setting ε_{in} to 1, 2, 4, 10 and 25 in MM-GBSA calculations.

Thus, the reported works underscore that the choice of the optimum ε_{in} strictly depends on the features of both the binding site and the ligand. Moreover, it has been suggested that this approach can be safely applied only for calculations of relative binding energies of ligands having a similar total charge, because in this way the limits of continuum solvation might be reduced due to the cancellation of errors [79, 80]. It should also be taken in account that the reported method might be useful to improve the correlation between the predicted and the experimental binding energies, an important objective in drug design or discovery [37], but might not be similarly effective for the prediction of absolute binding energies.

Although better predictions were sometimes obtained by setting $\varepsilon_{in} > 1$, this does not mean that a more accurate description of the solute-solvent interactions is also obtained. Indeed, the effect of increasing ε_{in} is that the contribution of the non-bonded electrostatic energy term, proportional to $1/\varepsilon_{in}$, and the total electrostatic term, proportional to $1/\varepsilon_{in}^2$ are reduced. As a result, the estimated free energy is dominated by the non-polar and entropy terms [73].

4.2. Inclusion of Explicit Solvent Molecules

It is well known that water molecules play a relevant role in ligand-receptor and protein-protein interactions, since they can take part in stable water-mediated hydrogen bonds or stabilize the complex by creating transient hydrogen bonds bridging the ligand and the receptor [45, 81-84].

Therefore, water is usually explicitly included in the MD simulations and its effects on the binding free energy estimate have been deeply studied [3, 30, 40-44, 73].

In the MM-PBSA/GBSA calculations, water molecules are usually stripped before the analysis, however this could lead to erroneous results for those systems where water is known to mediate hydrogen bonds between the ligand and the receptor.

Thus, in these cases, some explicit water molecules should be included in the calculation in order to take into account those solvent effects not adequately managed by an implicit solvation model.

In theory, this approach appears an obvious expedient, but in practice its application is non-trivial and, in some cases, useless or even detrimental [3, 43]. However, many examples in which the explicit consideration of water molecules in the MM-PBSA/GBSA calculations led to an improvement in the binding energy predictions, in particular by increasing the correlation between the computed and the experimental binding free energies, have been reported [30, 40-42, 44-45].

When one wants to consider the inclusion of an explicit water residue, the most critical question is how the selection of the solvent residues to be considered in the MM-PBSA/GBSA calculation should be made. The most intuitive approach is to include those molecules which are known from crystallography to bridge the receptor with one or more ligands [3, 41-43].

Otherwise, it is possible to consider a certain number of water molecules selected from a MD simulation conducted with explicit solvent. In this case, the selection could be made on the basis of a water occupancy analysis of the trajectory [30, 40]. Another possibility is to choose those solvent residues which are placed within an established distance from the ligand [3]. It has also been reported that the inclusion of a pre-determined number of water molecules (generally a number from 20 to 70), which are frame by frame the closest to the ligand during the whole simulation time can lead to an improvement of correlation between the predicted and the experimental binding affinities [44].

When explicit water molecules are going to be considered in the MM-PBSA/GBSA calculations, it can be advisable to include them in the receptor mask, otherwise higher standard deviations could be obtained without improving the binding free energy estimation. For instance, Treesuwan and Hannongbua [40] compared the performance of two different approaches, one consisting in the calculation of a term for the contribution of water alone, the other including the water effect in the receptor term. They observed that the binding energies obtained with this latter approach were more reliable and led to lower residuals.

The pros and cons of considering explicit water molecules in the binding free energy predictions will be discussed in details in the following paragraphs, where specific examples for each of the above mentioned strategies are discussed.

4.2.1. Inclusion of Crystallographic Water Molecules

As previously mentioned, one of the most common approaches is the selection of these water molecules which are found to mediate ligand-receptor interactions in crystallographic structures.

For example, Nurisso and coworkers [41] reported how the inclusion of a crystallographic water molecule affected the ranking of three isomeric disaccharides in binding the *Pseudomonas aeruginosa* Lectin I (PA-IL) protein. Indeed, the crystallographic structures showed that one water molecule, bridging ligand and receptor, was conserved in the binding site of the three complexes; therefore, it was retained during the preparation of the structures to be subjected to the MD simulation. MM-PBSA analyses were then performed either by including this residue in the receptor mask or not. The authors observed that, only by including the bridging water molecule, the experimental ranking was respected and the contribution of each monosaccharide was correctly determined. Moreover, when analyzing the single contributions to the binding free energy, it was observed how the electrostatic interactions played the most relevant role in determining the binding affinity, coherently with the experimental results.

Furthermore, other authors achieved a correct ranking of the experimental binding poses of eight protein-ligand complexes only by both including structural bridging waters during ensemble-averaged MM-PBSA analyses and using the polarized protein specific charge model (PPC) [42].

Also, in a study on the comparison of nevirapine affinity for the wild type and a mutant type HIV-1 reverse transcriptase, it was observed that the inclusion of a bridging crystallographic water improved the estimation of the binding free energies [40].

It is important to underline that, in the examples mentioned above, selected water residues remained stable during the whole simulation time. This behavior is not obvious as structural water can be replaced by another one even in a crystal structure where each water residue is identified as a mean electron density which might also result from a fast switching between neighboring waters, even if H-bounded [40, 44, 85].

The inclusion of a water molecule, selected from a crystal structure, in MM-PBSA/GBSA calculations in some cases resulted detrimental on the correlation between computed and experimental binding free energies [3, 43], an effect attributed by Checa *et al.* to an incomplete treatment of the interactions between the solvent and the solute [43]. Indeed, while using MM-PBSA to compute the binding energies of trypsin complexes with seven similar flavonoids [43], they observed that the inclusion of four crystallographic water molecules in the calculations worsened the correlation between the computed and the experimental activities; a correlation comparable to that obtained by using the default solvent model was instead observed when including a cap of 530 water molecules surrounding the active site. Although better correlations were not observed, the authors suggested that the water cap might properly balance the solute-solvent interactions in MM-PBSA calculations.

Greenidge and coworkers assessed the performance of MM-GBSA on a dataset of 855 complexes taken from the Protein Data Bank [3]. Although water-mediated H-bonds between ligand and receptor were observed in several examples, the inclusion of all the crystallographic waters within 3.5 Å from the ligand leaded to any improvement in correlation between the predicted and the experimental binding energies, if compared with implicit solvation only.

4.2.2. Inclusion of Water Molecules Identified from MD Trajectory Analysis

Relevant water molecules to be considered in MM-PBSA/GBSA binding energy predictions can also be selected by a careful analysis of the solvated MD trajectories. Different methods can be used to identify the critical water molecules, thus affecting the efficacy and/or the computational cost of the approach.

Commonly, the selection can be made by post-processing the MD trajectories through H-bond analysis [40], B-factor analysis [30], water density or water occupancy analyses [86], or by selecting those water molecules which are the closest to the ligand or to other relevant residues during the whole simulation [44-45].

Unlike the bare inclusion of a few crystallographic water molecules, in most cases this approach showed to benefit the correlation between computed binding free energies and experimental data.

For instance, Wallnoefer and coworkers [30] used MM-PBSA/GBSA for the prediction of binding free energies of six ligands in complex with factor Xa and noted that the binding affinities were severely affected by the presence of some relevant water molecules. Indeed, by including any explicit water an inverse correlation was obtained by both PB ($r_s = -0.76$) and GB ($r_s = -0.48$) methods. On the other hand, a direct correlation between predicted and experimental binding energies ($r_s = 0.85$ and 0.93 for PB and GB methods, respectively) was obtained by including a single crystallographic water only for those systems where binding energies were overestimated. Finally, the MD trajectories were submitted to B-factor analysis and those waters showing a B-factor for the oxygen atom below 100 (about 20 residues) were explicitly included in the MM-PBSA/GBSA calculations. This approach provided a further improvement in correlation, leading to $r_s = 0.93$ and 0.97 for PB and GB methods, respectively.

It should be noted that a not negligible benefit of this approach consists in its generalizability and reproducibility. Relevant water residues can be indeed selected also for those complexes not available as X-ray structures. Moreover, the choice is less affected by subjectivity, being indeed based on selecting those residues which pass a given numerical threshold.

The selection of a certain number of solvent residues that are frame by frame the closest to an established position (*i.e.* the ligand or a protein-protein interaction surface) during the whole simulation has been also successful [44-45].

This method was initially applied by Wong and coworkers [45] while studying the binding affinity of the wild type and two mutant T-cell receptors (TCR) in complex with the staphylococcal enterotoxin 3 (SEC3). In a first instance, MM-PBSA calculations were performed by excluding explicit water molecules. Predicted binding affinities were correctly ranked, but their uncertainties overlapped, thus reducing the statistical significance of results. Therefore, the authors decided to include some explicit water molecules during the analysis, since the key role of the solvent in the TCR-SEC3 system had been previously shown [87]. Two different approaches had been followed: in one case, the 200 water molecules closest to the protein-protein interaction interface were included in the calculation, in the other one only two interfacial solvent residues were considered. The first approach was not able to correctly rank the binding affinities and the absolute values and statistical errors were far too high, possibly because of the larger contribution of the electrostatic and van der Waals interactions due to the explicit solvation [45]. From our experience, the incapacity of the method to yield a correct energetic trend could also depend on the excessively high number of explicit water molecules considered. Indeed, it is known that extended protein-protein interface are often rich in hydrophobic residues [88, 89], thus the inclusion of a large number of water molecules should not be useful and can introduce background noises probably responsible of the erroneous ranking in binding affinities [44]. This hypothesis is supported by the excellent results obtained by the inclusion of only two explicit water molecules, selected among those closest to hydrophilic TCR residues Asn54 and Glu56 and the SEC3 residue Phe206 which, in the crystal structure [87], presents its carbonyl group pointing toward the protein-protein interaction interface. Indeed, in the crystal structure, two interfacial water residues were found to interact with these three protein residues. By applying this approach, the authors obtained the correct energetic ranking with a statistical significance.

Starting from the above findings, we decided to systematically investigate the effect of the inclusion, as a part of the receptor, of explicit water shells populated by a defined number of water molecules (N_{wat}) which were selected to be the closest to the ligand atoms during an MD simulation [44]. This approach is easily applicable by processing an MD trajectory by the *ptraj* software included in the Amber Tool package [56] using the "closest" keyword. We initially evaluated this protocol on the complexes between DNA-topoisomerase I and nine recently published [90] camptothecin derivatives. MM-PBSA/GBSA analyses conducted by only using implicit solvation yielded an incorrect ranking of the ligand binding affinities and no correlation between predicted and experimental activities (r^2 = 0.19 and 0.02 for GB and PB methods, respectively). No better results were obtained by including in the calculation the two crystallographic water molecules which were known to bridge hydrogen bonds between topoisomerase and its ligand topotecan [91]. Indeed, we noticed that during the MD simulation the selected waters were frequently replaced by neighboring ones. In this way, the water mediated bridge observed in the crystal structure was always maintained in the MD simulation, but it was originated by a cluster of fast-switching water molecules. This was also confirmed by a water density analysis, which showed that several areas with high water density were present around the ligand (Fig. **1 A**). We thus investigated the effect of the explicit water shells by systematically varying N_{wat} from 10 to 50, in order to find the condition able to maximize the correlation with experimental data (Table **2**). This was obtained for N_{wat} = 20 (r^2 = 0.87 and 0.51 for GB and PB methods, respectively), while higher N_{wat} values slightly reduced the correlation, probably due to the background noises caused by the unnecessary water residues, coherently with the results obtained by Wong [45].

In order to assess the performance of this approach, we tested it on three other ligand-receptor complexes previously used for benchmarking MM-PBSA/GBSA calculations: α-thrombin, avidin and penicillopepsin [33].

Table 2: Correlations between the Experimental $-\log_{10}(IC_{50})$ (Topoisomerase) or ΔG_{bind} (α-Thrombin, Penicillopepsin, and Avidin) and MM-GBSA Binding Energies Obtained for N_{wat} = 0-70. Adapted with permission from *J. Chem. Theory and Comput.* 2013, *9* (6), 2706-2717. Copyright 2013American Chemical Society

	Topoisomerase	**α-thrombin**	**Penicillopepsin**	**Avidin**
N_{wat}=0	0.19	0.67	0.46	0.72
N_{wat}=10	0.45	0.58	0.56	0.83
N_{wat}=20	0.87	0.61	0.60	0.84
N_{wat}=30	0.79	0.69	0.69	0.84
N_{wat}=40	0.75	0.76	0.73	0.85
N_{wat}=50	0.72	0.80	0.77	0.85
N_{wat}=60	NC	0.82	0.78	0.85
N_{wat}=70	NC	0.83	0.79	0.85

NC = not calculated

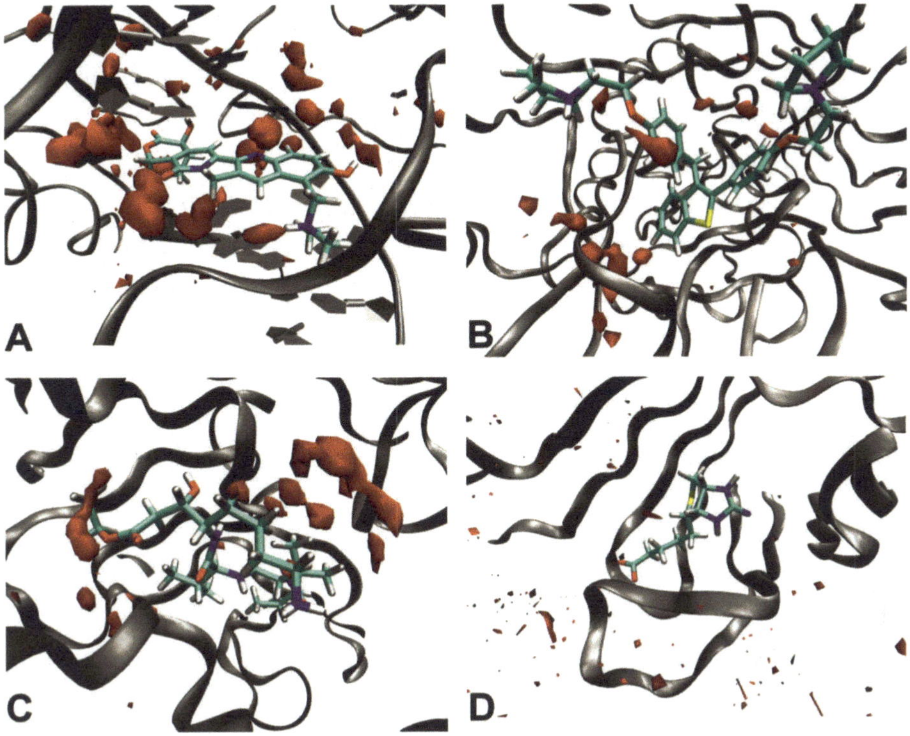

Figure 1: Water density plots obtained by grid analysis of topoisomerase-DNA-TTC (A), α-thrombin-BT2 (B), penicillopepsin-APT (C) and avidin-BTN2 complexes. (ptraj; grid box = 50x50x50 Å, mesh = 0.5 Å; visualization with VMD specifying an isovalue = 45 (A, B and C) or 25 (D)) Adapted with permission from *J. Chem. Theory and Comput.* 2013, *9* (6), 2706-2717. Copyright 2013 American Chemical Society.

MM-GBSA results (Table **2**) for α-thrombin and avidin complexes well correlated with experiments also without considering explicit waters (r^2 = 0.67 and 0.72, respectively). This was not surprising, as the water density analyses suggested for these systems a weaker involvement of explicit water in mediating ligand-receptor interactions (Fig. **1B** and **D**, respectively), if compared to the topoisomerase complexes (Fig. **1A**).

Nevertheless, starting from a value of N_{wat} = 30 we observed in both cases an improvement in the correlation, although the best was observed for N_{wat} = 70 (r^2 = 0.83 and 0.85 for α-thrombin and penicillopepsin, respectively), probably due to the contribution of several, but transient, hydrogen bonds between the water and the solute.

The behavior of the penicillopepsin complexes was instead similar to that observed for topoisomerase, as water was shown to play a relevant role in the ligand-receptor interaction (Fig. **1C**). By only using implicit solvation, a low correlation was obtained between the predicted and the experimental binding energies (r^2 = 0.46), but an improvement up to an r^2 = 0.69 was obtained for N_{wat} = 30 and a plateau was reached between N_{wat} = 60 and 70 (r^2 = 0.78 and 0.79, respectively).

In conclusion, this method, as well as the use of B-factors [30], appears to be applicable to different kind of ligand-receptor complexes without the need of an accurate analysis of the binding site features and seems to be of easier application and less affected by subjectivity if compared to the tuning of ε_{in} (see section 1.4.1). Even if the value of N_{wat} should be tuned for optimal performances, by setting N_{wat} = 30 an improvement over implicit solvation only was always obtained in the above-mentioned examples as well as in some other cases not yet published. However, it should be noted that an increase in the correlation between predicted binding affinities and the corresponding experimental data does not mean that a better prediction of absolute binding free energies can also be obtained.

4.3 "Chimera" Methods

The prediction of the binding free energy can also be achieved by applying methods obtained from the combination of MM-PBSA/GBSA with other approaches, which are used for the solution of one or more terms of eq. (3).

Commonly, these alternative methods are employed for the modulation of the solvation free energy polar term [46, 47], but other "chimera" methods, which modify both the electrostatic and the non-electrostatic contributions have been proposed and tested [34].

These approaches aim to improve the accuracy in the estimation of the binding free energy at about the same computational cost of MM-PBSA/GBSA, by working around some weaknesses of the original method.

One of this weaknesses can be considered the insufficient description of the solute-solvent interaction due to the default implicit solvation model [47].

To overcome this limitation, Freedman *et al.* developed a new approach called Molecular Mechanics Poisson Boltzmann/Linear Response Approximation Surface Area (MM-PB/LRA-SA) [47]. This method computes the ΔG_{solv} upon binding for the receptor with the PB or GBSA approaches, while the contribution of the ligand is obtained by using the LRA-SA approach (eq. 9):

$$\Delta G_{solv} = [G_{sol,R(complex)} - G_{sol,R(free)}]^{\text{PB/GB-SA}} + [G_{sol,L(complex)} - G_{sol,L(free)}]^{\text{LRA-SA}} \quad (9)$$

The LRA assumes that a free energy change can be approximately considered to be in a linear dependence to the ligand charge [34]. The LRA method calculates the solvation free energy as a function of the solute-solvent radial distribution functions, which are determined as the factors multiplying the ideal solvent density to give the real density for each one of the solvent atoms, as a function of its distance from an established solute atom. These functions have to be determined at the two end-points in order to calculate the ligand solvation free energy.

As a result, the MM-PB/LRA-SA approach should be able to determine the electrostatic and the attractive van der Waals terms for the ligand as a function of the interactions between the solute and the solvent, evaluated during an explicit solvent MD simulation. In the meanwhile, interactions with the bulk water are discarded beyond a fixed radius from each one of the solute atom.

The authors tested the MM-PB/LRA-SA method on an RNA aptamer bounded to theophylline and four of its derivatives. The results were then compared with the experiments and with those obtained by the standard MM-PB and GBSA methods. While the latter methods were not able to correctly rank the ligands binding free energies, with the MM-PB/LRA-SA approach all but one binding affinities were ranked coherently with the experimental data.

However, the computational cost of this approach is about twice that of MM-PBSA, because two separate MD simulations are required for each complex: one with the unmodified force field and one with a modified force field where the Coulomb and the attractive van der Waals interactions between the ligand and the solvent atoms are zeroed. Moreover, two MD simulations are also necessary for the solvated ligand [47].

A combination of the PB and LRA approaches, obtained by replacing the Langevin dipole solvation model in the PDLD/s-LRA/β method [92] by PB or GB, for the calculation of the polar term of solvation free energy was also applied by Genheden and Ryde on the avidin and fXa systems [34]. Nevertheless, although such an approach should be theoretically more rigorous than both GB and PB, the latter methods provided better results, in terms of correlation between the predicted and the experimental binding free energies, at a fraction of the computational cost, because the LRA approximation requires two additional simulations, the free ligand and the complex with zeroed receptor charges.

Taken together, the above reported examples confirm that, although it is obvious that the less approximated methods are more computationally demanding, there is no guarantee that their application in the prediction of the free energies of binding actually lead to a better correlation between the calculated and the experimental data.

5. ACTING ON THE NON-POLAR SOLVATION TERM

The tuning of the calculation methods for polar solvation free energy has been deeply studied, but some examples have also been reported on the methods aiming to improve the non-polar solvation term ($\Delta G_{nonpolsolv}$).

Commonly, $\Delta G_{nonpolsolv}$ is considered to be linearly proportional to the solvent-accessible surface area (SASA) (eq. 10):

$$\Delta G^{SASA}{}_{nonpolsolv} = \gamma_{\mathrm{SASA}}\mathrm{SASA} + \mathrm{b}_{\mathrm{SASA}} \qquad (10)$$

where the surface tension coefficient γ_{SASA} is the contribution to the $\Delta G^{SASA}{}_{nonpolsolv}$ per unit of surface area and b_{SASA} can be obtained from a linear regression analysis of the solvation free energies of a set of small apolar molecules in water [93-95].

However, this model showed a quite low accuracy in correlating the solvation free energies computed with the SA model and those computed with an explicit solvent model [26, 46, 96, 97].

Thus, different models have been developed in order to improve the accuracy of the estimation of $\Delta G_{nonpolsolv}$. One of the most applied approaches is the cavity-dispersion (CD) model [98], which relies on the observation that $\Delta G_{nonpolsolv}$ is the result of two different contributes: a repulsive (ΔG_{rep}) and an attractive free energy (ΔG_{att}), modeled and computed separately (eq. 11):

$$\Delta G^{CD}{}_{nonpolsolv} = \Delta G_{rep} + \Delta G_{att} \tag{11}$$

ΔG_{rep} is the solvation free energy raising from the repulsive interactions between the solute and the solvent and from the formation of the solute cavity, while ΔG_{att} corresponds to the free energy for the formation of the attractive solute-solvent interactions and for the reorganization of the bulk solvent.

The repulsive contribution was found to well correlate with the SASA, independently from the kind of MS (SA or SE) [98, 99]. The attractive term can be considered as equal to the van der Waals attractive interaction potential energy between the solute and the solvent [100].

A different approach has been attempted by using the polarized continuum model (PCM) which, in addition, implies the inclusion of a term for the exchange repulsion (eq. 12):

$$\Delta G^{PCM}{}_{nonpolsolv} = \Delta G_{cavity} + \Delta G_{rep} + \Delta G_{att} \tag{12}$$

The ΔG_{cavity} term is obtained by the expressions of the radius of each atom to the power of 0 to 3, which consists in the consideration of an area and a volume term [50].

SASA, CD and PCM approaches have been compared in two studies by Genheden *et al.*, who considered TI as a reference [48-49]. The first study focused on the prediction of the free energy of binding of benzene to the T4 lysozyme Leu99Ala mutant, for which the TI results were concordant with the experiments. The predicted free energy value was decomposed accordingly to the MM-GBSA formalism, with $\Delta G_{nonpol} = \Delta G_{vdW}^{free} - \Delta G_{vdW}^{bound}$. As regards to the MM-GBSA and CD approaches, ΔG_{solv} was computed by using the GB method and the $\Delta G_{nonpolsolv}$ by using the SASA and CD methods, respectively. For the PCM based approach, the whole solvation free energy was computed by PCM. In a first instance, the SASA approach gave the most accurate estimation of $\Delta G_{nonpolsolv}$. However, the failure of the other two approaches was attributed to the assumption that the binding site in the free protein is filled with water, although this is in contrast with the calculations and the experiments showing that the cavity is empty at the free state [101]. The same assumption is made by the SASA method, but SASA terms are smaller and so they did not lead to an incorrect estimation of the apolar term. This misbehavior has been worked around by filling the cavity of the free protein with a non-interacting ligand and consequently the results of the CD and PCM approaches were decidedly improved, with the latter providing the best estimation of each term.

It should be noticed that PCM uses a van der Waals surface for the estimation of ΔG_{cavity}; this kind of MS generates crevices and cavities inside the proteins, which are too small for being occupied by the solvent molecules. In addition, it leads to an incorrect trend in the protein surface area change after the ligand binding [48].

The authors applied SASA, CD and PCM methods to some proteins whose binding sites were variably accessible to the solvent: galectin-3, which binds its galactoside ligand on the surface, trypsin, which binds 2-aminobenzimidazole in a cleft partly exposed to the solvent, avidin and ferritin, which bind a biotin analogue and phenol, respectively, in a buried and an hidden cavity [49].

The results from the calculations made on these latter complexes confirmed those obtained for the benzene-T4 lysozyme mutant complex [48]. However, the improvement in accuracy of the $\Delta G_{nonpolsolv}$ estimation, obtained with the inclusion of a dummy ligand in the binding cavity of the free protein, was considered to be fortuitous, because it did not considered water molecules to be displaced by the ligand.

As regards to galectin-3 and trypsin, the authors obtained poor results in any case, with errors of 22-73 kJ/mol for the $\Delta G_{nonpolsolv}$ estimate. The authors observed that $\Delta G_{nonpolsolv}$ obtained by TI was somehow between the values obtained by considering the cavity filled with continuum water and those obtained by including a dummy ligand. Thus, a combined method which worked well for any complex, save for trypsin, was developed. Within this approach (eq. 13), the cavity term was derived by performing the calculation with the binding site filled with a non-interacting ligand (method P0), while the dispersion and repulsion terms were obtained from the calculations with the cavity filled by a continuum solvent (P method) [49]:

$$\Delta G_{np}{}^{bound} = \xi G_{np}{}^{bound}(\mathrm{P}) + (1 - \xi)\, G_{np}{}^{bound}(\mathrm{P0}) \tag{13}$$

where ξ is related to the solvent exposure (SE) of the bounded ligand.

As it can be noticed, the treatment of the apolar term of the solvation energy is controversial. Indeed, the commonly used SASA approach can be considered accurate only for those proteins with buried binding sites, while the combined method developed by Genheden and coworkers is preferable for the cavities which are more exposed to the solvent. Nevertheless, the use of P0 approach also results in a change of the polar term, because of the binding site being filled with a non-interacting ligand instead of the solvent molecules.

Moreover, although it is fundamental to know the hydration state of the binding cavity, unexpected results might be obtained: for example, the ferritin crystal structure shows four water molecules in the free binding site, but the results obtained by using the P0 approach are decidedly better than those obtained with the standard P approach.

Furthermore, it has to be observed that all the evaluated continuum solvation method failed in predicting the $\Delta G_{nonpolsolv}$ term for the solvent-exposed binding sites, such as galectin-3, because in these cases the water molecules have not bulk-like properties [49].

Another interesting approach, alternative to SASA, had been also reported by Genheden and Ryde [34], who took the non-electrostatic part from the LIE method (eq. 14):

$$\Delta G_{nonpol}^{LIE} = \beta\left(\langle E_{vdw}^{L-(R+S)} \rangle_{RL} - \langle E_{vdw}^{L-(R+S)} \rangle_{L}\right) \tag{14}$$

where β is 0.18 and $E_{vdw}^{L-(R+S)}$ are the van der Waals interaction energies between the ligand and the surroundings (receptor + solvent) in the simulation of the complex (RL) and the free ligand (L).

The authors tested this method (referred as MM-PB/GB-β) on avidin in complex with seven biotin-like ligands and on fXa binding nine derivatives of 3-amidinobenzyl-1*H*-indol-2-carboxamide. They observed that, for the latter system, standard MM-PBSA/GBSA gave the best results in term of correlation between the predicted and the experimental binding free energies, while MM-PB/GB-β leaded to the highest correlation for the avidin complexes, with MM-PB-β being the most precise method [34].

CONCLUDING REMARKS

When applying the MM-PBSA/GBSA methods for the binding energy predictions, several parameters, affecting either the quality of the prediction or the computation time, should be properly set to obtain the maximum performance. This task is anything but easy; however, specific examples on how one or the other parameter is able to affect the method reliability have been reported in the literature in the last few years. The majority of them are related to the calculation of the electrostatic contribution to the solvation free energy, although some studies on the optimization of the non-electrostatic contribution were also reported.

The principal aim of this chapter has been to describe such examples, collected in specific paragraphs accordingly to the parameter being investigated, in order to make the reader aware about the pros and the cons of each strategy.

It has to be underscored that the effectiveness of each of the methods herein described strictly depends on the scope of the analysis. Therefore, the choice of the approach should be carefully evaluated and validated, if possible, by a comparison with the experimental data.

ACKNOWLEDGEMENTS

The authors kindly acknowledge the Italian Ministry of Education, University and Research for financial support.

CONFLICT OF INTEREST

The authors confirm that this chapter contents have no conflict of interest.

REFERENCES

[1] Pearlman, D. A.; Charifson, P. S., Are Free Energy Calculations Useful in Practice? A Comparison with Rapid Scoring Functions for the p38 MAP Kinase Protein System†. *J. Med. Chem.* **2001,** *44* (21), 3417-3423.

[2] Srivastava, H. K.; Sastry, G. N., Molecular Dynamics Investigation on a Series of HIV Protease Inhibitors: Assessing the Performance of MM-PBSA and MM-GBSA Approaches. *J. Chem. Inf. Model.* **2012,** *52* (11), 3088-3098.

[3] Greenidge, P. A.; Kramer, C.; Mozziconacci, J.-C.; Wolf, R. M., MM/GBSA Binding Energy Prediction on the PDBbind Data Set: Successes, Failures, and Directions for Further Improvement. *J. Chem. Inf. Model.* **2012,** *53* (1), 201-209.

[4] Massova, I.; Kollman, P., Combined molecular mechanical and continuum solvent approach (MM-PBSA/GBSA) to predict ligand binding. *Perspect. Drug Discovery Des.* **2000,** *18* (1), 113-135.

[5] Wang, W.; Wang, J.; Kollman, P. A., What determines the van der Waals coefficient β in the LIE (linear interaction energy) method to estimate binding free energies using molecular dynamics simulations? *Proteins: Struct., Funct., Bioinf.* **1999,** *34* (3), 395-402.

[6] Deng, Y.; Roux, B. t., Computations of Standard Binding Free Energies with Molecular Dynamics Simulations. *J. Phys. Chem. B* **2009,** *113* (8), 2234-2246.

[7] Price, D.; Jorgensen, W., Improved convergence of binding affinities with free energy perturbation: Application to nonpeptide ligands with pp60src SH2 domain. *J. Comput.-Aided Mol. Des.* **2001,** *15* (8), 681-695.

[8] Kollman, P. A.; Massova, I.; Reyes, C.; Kuhn, B.; Huo, S.; Chong, L.; Lee, M.; Lee, T.; Duan, Y.; Wang, W.; Donini, O.; Cieplak, P.; Srinivasan, J.; Case, D. A.; Cheatham, T. E., 3rd, Calculating structures and free energies of complex molecules: combining molecular mechanics and continuum models. *Acc. Chem. Res.* **2000,** *33* (12), 889-897.

[9] Srinivasan, J.; Cheatham, T. E.; Cieplak, P.; Kollman, P. A.; Case, D. A., Continuum Solvent Studies of the Stability of DNA, RNA, and Phosphoramidate−DNA Helices. *J. Am. Chem. Soc.* **1998,** *120* (37), 9401-9409.

[10] Kuhn, B.; Kollman, P. A., Binding of a Diverse Set of Ligands to Avidin and Streptavidin: An Accurate Quantitative Prediction of Their Relative Affinities by a Combination of Molecular Mechanics and Continuum Solvent Models. *J. Med. Chem.* **2000,** *43* (20), 3786-3791.

[11] Xu, D. In *Autodock2MMGBSA, A multi-level virtual screening rescoring and refinement scheme that combines consensus scoring, simulated annealing and MM-GBSA binding free energy methods*, American Chemical Society: 2012; pp NORM-153.

[12] Xu, D.; Sawaya, N.; McCammon, J. A.; Li, W. W. In *Autodock2MMGBSA, A multi-level virtual screening rescoring and refinement scheme that combines consensus scoring and MM-GBSA binding free energy methods*, American Chemical Society: 2010; pp COMP-78.

[13] Weis, A.; Katebzadeh, K.; Söderhjelm, P.; Nilsson, I.; Ryde, U., Ligand Affinities Predicted with the MM/PBSA Method: Dependence on the Simulation Method and the Force Field. *J. Med. Chem.* **2006,** *49* (22), 6596-6606.

[14] Case, D. A., Normal mode analysis of protein dynamics. *Curr. Opin. Struct. Biol.* **1994,** *4* (2), 285-290.

[15] Hayes, J. M.; Skamnaki, V. T.; Archontis, G.; Lamprakis, C.; Sarrou, J.; Bischler, N.; Skaltsounis, A.-L.; Zographos, S. E.; Oikonomakos, N. G., Kinetics, *in silico* docking, molecular dynamics, and MM-GBSA binding studies on prototype indirubins, KT5720, and staurosporine as phosphorylase kinase ATP-binding site inhibitors: the role of water molecules examined. *Proteins* **2011,** *79*, 703-719.

[16] Sanner, M. F.; Olson, A. J.; Spehner, J.-C., Reduced surface: An efficient way to compute molecular surfaces. *Biopolymers* **1996,** *38* (3), 305-320.

[17] Jackson, J. D., *Classical Electrodynamics*. Wiley: New York, 1999.

[18] Constanciel, R.; Contreras, R., Self consistent field theory of solvent effects representation by continuum models: Introduction of desolvation contribution. *Theor. Chim. Acta* **1984,** *65* (1), 1-11.

[19] Still, W. C.; Tempczyk, A.; Hawley, R. C.; Hendrickson, T., Semianalytical treatment of solvation for molecular mechanics and dynamics. *J. Am. Chem. Soc.* **1990,** *112* (16), 6127-6129.

[20] Simonson, T., Macromolecular electrostatics: continuum models and their growing pains. *Curr. Opin. Struct. Biol.* **2001,** *11* (2), 243-252.

[21] Feig, M.; Onufriev, A.; Lee, M. S.; Im, W.; Case, D. A.; Brooks, C. L., Performance comparison of generalized born and Poisson methods in the calculation of electrostatic solvation energies for protein structures. *J. Comput. Chem.* **2004,** *25* (2), 265-284.

[22] Simonson, T., Electrostatics and dynamics of proteins. *Rep. Prog. Phys.* **2003,** *66* (5), 737-787.

[23] Fogolari, F.; Brigo, A.; Molinari, H., The Poisson-Boltzmann equation for biomolecular electrostatics: a tool for structural biology. *J. Mol. Recognit.* **2002,** *15* (6), 377-392.

[24] Bashford, D.; Case, D. A., GENERALIZED BORN MODELS OF MACROMOLECULAR SOLVATION EFFECTS. *Annu. Rev. Phys. Chem.* **2000,** *51* (1), 129-152.

[25] Page, C. S.; Bates, P. A., Can MM-PBSA calculations predict the specificities of protein kinase inhibitors? *J. Comput. Chem.* **2006,** *27* (16), 1990-2007.

[26] Gohlke, H.; Case, D. A., Converging free energy estimates: MM-PB(GB)SA studies on the protein-protein complex Ras-Raf. *J. Comput. Chem.* **2004,** *25* (2), 238-250.

[27] Kongsted, J.; Söderhjelm, P.; Ryde, U., How accurate are continuum solvation models for drug-like molecules? *J. Comput.-Aided Mol. Des.* **2009,** *23* (7), 395-409.

[28] Singh, N.; Warshel, A., Absolute binding free energy calculations: On the accuracy of computational scoring of protein-ligand interactions. *Proteins: Struct., Funct., Bioinf.* **2010,** *78* (7), 1705-1723.

[29] Homeyer, N.; Gohlke, H., Free Energy Calculations by the Molecular Mechanics Poisson–Boltzmann Surface Area Method. *Mol. Inform.* **2012,** *31* (2), 114-122.

[30] Wallnoefer, H. G.; Liedl, K. R.; Fox, T., A challenging system: Free energy prediction for factor Xa. *J. Comput. Chem.* **2011,** *32* (8), 1743-1752.

[31] Harris, R. C.; Boschitsch, A. H.; Fenley, M. O., Influence of Grid Spacing in Poisson-Boltzmann Equation Binding Energy Estimation. *J. Chem. Theory Comput.* **2013,** *9* (8), 3677-3685.

[32] Nguyen, H.; Roe, D. R.; Simmerling, C., Improved Generalized Born Solvent Model Parameters for Protein Simulations. *J. Chem. Theory Comput.* **2013,** *9* (4), 2020-2034.

[33] Hou, T.; Wang, J.; Li, Y.; Wang, W., Assessing the Performance of the MM/PBSA and MM/GBSA Methods. 1. The Accuracy of Binding Free Energy Calculations Based on Molecular Dynamics Simulations. *J. Chem. Inf. Model.* **2011,** *51* (1), 69-82.

[34] Genheden, S.; Ryde, U., Comparison of end-point continuum-solvation methods for the calculation of protein-ligand binding free energies. *Proteins: Struct., Funct., Bioinf.* **2012,** *80* (5), 1326-1342.

[35] Ravindranathan, K.; Tirado-Rives, J.; Jorgensen, W. L.; Guimarães, C. R. W., Improving MM-GB/SA Scoring through the Application of the Variable Dielectric Model. *J. Chem. Theory Comput.* **2011,** *7* (12), 3859-3865.

[36] Venken, T.; Krnavek, D.; Munch, J.; Kirchhoff, F.; Henklein, P.; De Maeyer, M.; Voet, A., An optimized MM/PBSA virtual screening approach applied to an HIV-1 gp41 fusion peptide inhibitor. *Proteins* **2011,** *79* (11), 3221-3235.

[37] Yang, T.; Wu, J. C.; Yan, C.; Wang, Y.; Luo, R.; Gonzales, M. B.; Dalby, K. N.; Ren, P., Virtual screening using molecular simulations. *Proteins: Struct., Funct., Bioinf.* **2011,** *79* (6), 1940-1951.

[38] Moreira, I. S.; Fernandes, P. A.; Ramos, M. J., Computational alanine scanning mutagenesis—An improved methodological approach. *J. Comput. Chem.* **2007,** *28* (3), 644-654.

[39] Hou, T.-J.; Wang, J.-M.; Li, Y.-Y.; Wang, W., Assessing the performance of the molecular mechanics/Poisson Boltzmann surface area and molecular mechanics/generalized Born surface area methods. II: The accuracy of ranking poses generated from docking. *J. Comput. Chem.* **2011,** *32*, 866-877.

[40] Treesuwan, W.; Hannongbua, S., Bridge water mediates nevirapine binding to wild type and Y181C HIV-1 reverse transcriptase—Evidence from molecular dynamics simulations and MM-PBSA calculations. *J. Mol. Graphics Modell.* **2009,** *27* (8), 921-929.

[41] Nurisso, A.; Blanchard, B.; Audfray, A.; Rydner, L.; Oscarson, S.; Varrot, A.; Imberty, A., Role of Water Molecules in Structure and Energetics of Pseudomonas aeruginosa Lectin I Interacting with Disaccharides. *J. Biol. Chem.* **2010,** *285* (26), 20316-20327.

[42] Liu, J.; He, X.; Zhang, J. Z. H., Improving the Scoring of Protein-Ligand Binding Affinity by Including the Effects of Structural Water and Electronic Polarization. *J. Chem. Inf. Model.* **2013,** *53* (6), 1306-1314.

[43] Checa, A.; Ortiz, A. R.; de Pascual-Teresa, B.; Gago, F., Assessment of solvation effects on calculated binding affinity differences: trypsin inhibition by flavonoids as a model system for congeneric series. *J. Med. Chem.* **1997,** *40* (25), 4136-4145.

[44] Maffucci, I.; Contini, A., Explicit Ligand Hydration Shells Improve the Correlation between MM-PB/GBSA Binding Energies and Experimental Activities. *J. Chem. Theory Comput.* **2013,** *9* (6), 2706-2717.

[45] Wong, S.; Amaro, R. E.; McCammon, J. A., MM-PBSA Captures Key Role of Intercalating Water Molecules at a Protein−Protein Interface. *J. Chem. Theory Comput.* **2009,** *5* (2), 422-429.

[46] Genheden, S.; Luchko, T.; Gusarov, S.; Kovalenko, A.; Ryde, U., An MM/3D-RISM Approach for Ligand Binding Affinities. *J. Phys. Chem. B* **2010,** *114* (25), 8505-8516.

[47] Freedman, H.; Huynh, L. P.; Le, L.; Cheatham, T. E.; Tuszynski, J. A.; Truong, T. N., Explicitly Solvated Ligand Contribution to Continuum Solvation Models for Binding Free Energies: Selectivity of Theophylline Binding to an RNA Aptamer. *J. Phys. Chem. B* **2010,** *114* (6), 2227-2237.

[48] Genheden, S.; Kongsted, J.; Söderhjelm, P.; Ryde, U., Nonpolar Solvation Free Energies of Protein−Ligand Complexes. *J. Chem. Theory Comput.* **2010,** *6* (11), 3558-3568.

[49] Genheden, S.; Mikulskis, P.; Hu, L.; Kongsted, J.; Söderhjelm, P.; Ryde, U., Accurate Predictions of Nonpolar Solvation Free Energies Require Explicit Consideration of Binding-Site Hydration. *J. Am. Chem. Soc.* **2011,** *133* (33), 13081-13092.

[50] Cossi, M.; Tomasi, J.; Cammi, R., Analytical expressions of the free energy derivatives for molecules in solution. Application to the geometry optimization. *Int. J. Quantum Chem.* **1995,** *56* (S29), 695-702.

[51] Rocchia, W.; Sridharan, S.; Nicholls, A.; Alexov, E.; Chiabrera, A.; Honig, B., Rapid grid-based construction of the molecular surface and the use of induced surface charge to calculate reaction field energies: Applications to the molecular systems and geometric objects. *J. Comput. Chem.* **2002,** *23* (1), 128-137.

[52] Baker, N.; Holst, M.; Wang, F., Adaptive multilevel finite element solution of the Poisson-Boltzmann equation II. Refinement at solvent-accessible surfaces in biomolecular systems. *J. Comput. Chem.* **2000,** *21* (15), 1343-1352.

[53] Holst, M.; Baker, N.; Wang, F., Adaptive multilevel finite element solution of the Poisson-Boltzmann equation I. Algorithms and examples. *J. Comput. Chem.* **2000,** *21* (15), 1319-1342.

[54] Grant, J. A.; Pickup, B. T.; Nicholls, A., A smooth permittivity function for Poisson-Boltzmann solvation methods. *J. Comput. Chem.* **2001,** *22* (6), 608-640.

[55] Im, W.; Beglov, D.; Roux, B., Continuum solvation model: Computation of electrostatic forces from numerical solutions to the Poisson-Boltzmann equation. *Comput. Phys. Commun.* **1998,** *111* (1-3), 59-75.

[56] D.A. Case, T. A. D., T.E. Cheatham, III, C.L. Simmerling, J. Wang, R.E. Duke, R. Luo, R.C. Walker, W. Zhang, K.M. Merz, B. Roberts, S. Hayik, A. Roitberg, G. Seabra, J. Swails, A.W. Goetz, I. Kolossváry, K.F. Wong, F. Paesani, J. Vanicek, R.M. Wolf, J. Liu, X. Wu, S.R. Brozell, T. Steinbrecher, H. Gohlke, Q. Cai, X. Ye, J. Wang, M.-J. Hsieh, G. Cui, D.R. Roe, D.H. Mathews, M.G. Seetin, R. Salomon-Ferrer, C. Sagui, V. Babin, T. Luchko, S. Gusarov, A. Kovalenko, and P.A. Kollman *AMBER 12*, 2012.

[57] Honig, B.; Nicholls, A., Classical electrostatics in biology and chemistry. *Science* **1995,** *268* (5214), 1144-1149.

[58] Gilson, M. K.; Sharp, K. A.; Honig, B. H., Calculating the electrostatic potential of molecules in solution: Method and error assessment. *J. Comput. Chem.* **1988,** *9* (4), 327-335.

[59] Bashford, D., An object-oriented programming suite for electrostatic effects in biological molecules An experience report on the MEAD project. In *Scientific Computing in Object-Oriented Parallel Environments*, Ishikawa, Y.; Oldehoeft, R.; Reynders, J. W.; Tholburn, M., Eds. Springer Berlin Heidelberg: 1997; Vol. 1343, pp 233-240.

[60] Cortis, C. M.; Friesner, R. A., An automatic three-dimensional finite element mesh generation system for the Poisson-Boltzmann equation. *J. Comput. Chem.* **1997,** *18* (13), 1570-1590.

[61] Cortis, C. M.; Friesner, R. A., Numerical solution of the Poisson-Boltzmann equation using tetrahedral finite-element meshes. *J. Comput. Chem.* **1997,** *18* (13), 1591-1608.

[62] Totrov, M.; Abagyan, R., Rapid boundary element solvation electrostatics calculations in folding simulations: Successful folding of a 23-residue peptide. *Biopolymers* **2001,** *60* (2), 124-133.

[63] Nicholls, A.; Mobley, D. L.; Guthrie, J. P.; Chodera, J. D.; Bayly, C. I.; Cooper, M. D.; Pande, V. S., Predicting Small-Molecule Solvation Free Energies: An Informal Blind Test for Computational Chemistry. *J. Med. Chem.* **2008,** *51* (4), 769-779.

[64] Onufriev, A.; Case, D. A.; Bashford, D., Effective Born radii in the generalized Born approximation: The importance of being perfect. *J. Comput. Chem.* **2002,** *23* (14), 1297-1304.

[65] Hawkins, G. D.; Cramer, C. J.; Truhlar, D. G., Pairwise solute descreening of solute charges from a dielectric medium. *Chem. Phys. Lett.* **1995,** *246* (1-2), 122-129.

[66] Onufriev, A.; Bashford, D.; Case, D. A., Exploring protein native states and large-scale conformational changes with a modified generalized born model. *Proteins: Struct., Funct., Bioinf.* **2004,** *55* (2), 383-394.

[67] Mongan, J.; Simmerling, C.; McCammon, J. A.; Case, D. A.; Onufriev, A., Generalized Born Model with a Simple, Robust Molecular Volume Correction. *J. Chem. Theory Comput.* **2006,** *3* (1), 156-169.

[68] Bondi, A., van der Waals Volumes and Radii. *J. Phys. Chem.* **1964,** *68* (3), 441-451.

[69] Roe, D. R.; Okur, A.; Wickstrom, L.; Hornak, V.; Simmerling, C., Secondary Structure Bias in Generalized Born Solvent Models: Comparison of Conformational Ensembles and Free Energy of Solvent Polarization from Explicit and Implicit Solvation. *J. Phys. Chem. B* **2007,** *111* (7), 1846-1857.

[70] Shell, M. S.; Ritterson, R.; Dill, K. A., A Test on Peptide Stability of AMBER Force Fields with Implicit Solvation. *J. Phys. Chem. B* **2008,** *112* (22), 6878-6886.

[71] Mongan, J.; Svrcek-Seiler, W. A.; Onufriev, A., Analysis of integral expressions for effective Born radii. *J. Chem. Phys.* **2007,** *127* (185101), 1-10.

[72] Oehme, D. P.; Brownlee, R. T. C.; Wilson, D. J. D., Effect of atomic charge, solvation, entropy, and ligand protonation state on MM-PB(GB)SA binding energies of HIV protease. *J. Comput. Chem.* **2012,** *33* (32), 2566-2580.

[73] Wang, B.; Li, L.; Hurley, T. D.; Meroueh, S. O., Molecular Recognition in a Diverse Set of Protein-Ligand Interactions Studied with Molecular Dynamics Simulations and End-Point Free Energy Calculations. *J. Chem. Inf. Model.* **2013,** *53* (10), 2659-2670.

[74] Schutz, C. N.; Warshel, A., What are the dielectric "constants" of proteins and how to validate electrostatic models? *Proteins: Struct., Funct., Bioinf.* **2001,** *44* (4), 400-417.

[75] Mikulskis, P.; Genheden, S.; Rydberg, P.; Sandberg, L.; Olsen, L.; Ryde, U., Binding affinities in the SAMPL3 trypsin and host-guest blind tests estimated with the MM/PBSA and LIE methods. *J. Comput.-Aided Mol. Des.* **2012,** *26* (5), 527-541.

[76] Guimarães, C. R. W., A Direct Comparison of the MM-GB/SA Scoring Procedure and Free-Energy Perturbation Calculations Using Carbonic Anhydrase as a Test Case: Strengths and Pitfalls of Each Approach. *J. Chem. Theory Comput.* **2011,** *7* (7), 2296-2306.

[77] Guimarães, C. R. W.; Mathiowetz, A. M., Addressing Limitations with the MM-GB/SA Scoring Procedure using the WaterMap Method and Free Energy Perturbation Calculations. *J. Chem. Inf. Model.* **2010,** *50* (4), 547-559.

[78] Udommaneethanakit, T.; Rungrotmongkol, T.; Bren, U.; Frecer, V.; Stanislav, M., Dynamic Behavior of Avian Influenza A Virus Neuraminidase Subtype H5N1 in Complex with Oseltamivir, Zanamivir, Peramivir, and Their Phosphonate Analogues. *J. Chem. Inf. Model.* **2009,** *49* (10), 2323-2332.

[79] Kongsted, J.; Ryde, U., An improved method to predict the entropy term with the MM/PBSA approach. *J. Comput.-Aided Mol. Des.* **2009,** *23* (2), 63-71.

[80] Genheden, S.; Ryde, U., How to obtain statistically converged MM/GBSA results. *J. Comput. Chem.* **2010,** *31* (4), 837-846.

[81] Wang, R.; Lu, Y.; Wang, S., Comparative Evaluation of 11 Scoring Functions for Molecular Docking. *J. Med. Chem.* **2003,** *46* (12), 2287-2303.

[82] Abel, R.; Young, T.; Farid, R.; Berne, B. J.; Friesner, R. A., Role of the Active-Site Solvent in the Thermodynamics of Factor Xa Ligand Binding. *J. Am. Chem. Soc.* **2008,** *130* (9), 2817-2831.

[83] Barillari, C.; Taylor, J.; Viner, R.; Essex, J. W., Classification of Water Molecules in Protein Binding Sites. *J. Am. Chem. Soc.* **2007,** *129* (9), 2577-2587.

[84] Poornima, C. S.; Dean, P. M., Hydration in drug design. 1. Multiple hydrogen-bonding features of water molecules in mediating protein-ligand interactions. *J. Comput.-Aided Mol. Des.* **1995,** *9* (6), 500-512.

[85] Schiffer, C.; Hermans, J., Promise of Advances in Simulation Methods for Protein Crystallography: Implicit Solvent Models, Time-Averaging Refinement, and Quantum Mechanical Modeling. In *Methods Enzymol.*, Carter, C. W., Jr.; Sweet, R. M., Eds. Academic Press: New York, 2003; Vol. 374, p 412-461.

[86] Henchman, R. H.; McCammon, J. A., Structural and dynamic properties of water around acetylcholinesterase. *Protein Sci.* **2002,** *11* (9), 2080-2090.

[87] Cho, S.; Swaminathan, C. P.; Yang, J.; Kerzic, M. C.; Guan, R.; Kieke, M. C.; Kranz, D. M.; Mariuzza, R. A.; Sundberg, E. J., Structural Basis of Affinity Maturation and Intramolecular Cooperativity in a Protein-Protein Interaction. *Structure* **2005,** *13* (12), 1775-1787.

[88] Glaser, F.; Steinberg, D. M.; Vakser, I. A.; Ben-Tal, N., Residue frequencies and pairing preferences at protein-protein interfaces. *Proteins: Struct., Funct., Bioinf.* **2001,** *43* (2), 89-102.

[89] Ansari, S.; Helms, V., Statistical analysis of predominantly transient protein-protein interfaces. *Proteins: Struct., Funct., Bioinf.* **2005,** *61* (2), 344-355.

[90] Samorì, C.; Guerrini, A.; Varchi, G.; Fontana, G.; Bombardelli, E.; Tinelli, S.; Beretta, G. L.; Basili, S.; Moro, S.; Zunino, F.; Battaglia, A., Semisynthesis, Biological Activity, and Molecular Modeling Studies of C-Ring-Modified Camptothecins. *J. Med. Chem.* **2009,** *52* (4), 1029-1039.

[91] Staker, B. L.; Hjerrild, K.; Feese, M. D.; Behnke, C. A.; Burgin, A. B.; Stewart, L., The mechanism of topoisomerase I poisoning by a camptothecin analog. *Proc. Natl. Acad. Sci. U. S. A.* **2002,** *99* (24), 15387-15392.

[92] Sham, Y. Y.; Chu, Z. T.; Tao, H.; Warshel, A., Examining methods for calculations of binding free energies: LRA, LIE, PDLD-LRA, and PDLD/S-LRA calculations of ligands binding to an HIV protease. *Proteins: Struct., Funct., Bioinf.* **2000,** *39* (4), 393-407.

[93] Hermann, R. B., Theory of hydrophobic bonding. II. Correlation of hydrocarbon solubility in water with solvent cavity surface area. *J. Phys. Chem.* **1972,** *76* (19), 2754-2759.

[94] Sitkoff, D.; Sharp, K. A.; Honig, B., Accurate Calculation of Hydration Free Energies Using Macroscopic Solvent Models. *J. Phys. Chem.* **1994,** *98* (7), 1978-1988.

[95] Tan, C.; Tan, Y.-H.; Luo, R., Implicit Nonpolar Solvent Models. *J. Phys. Chem. B* **2007,** *111* (42), 12263-12274.

[96] Wagoner, J.; Baker, N. A., Solvation forces on biomolecular structures: A comparison of explicit solvent and Poisson-Boltzmann models. *J. Comput. Chem.* **2004,** *25* (13), 1623-1629.

[97] Levy, R. M.; Zhang, L. Y.; Gallicchio, E.; Felts, A. K., On the Nonpolar Hydration Free Energy of Proteins: Surface Area and Continuum Solvent Models for the Solute−Solvent Interaction Energy. *J. Am. Chem. Soc.* **2003,** *125* (31), 9523-9530.

[98] Wagoner, J. A.; Baker, N. A., Assessing implicit models for nonpolar mean solvation forces: The importance of dispersion and volume terms. *Proc. Natl. Acad. Sci. U. S. A.* **2006,** *103* (22), 8331-8336.

[99] Gallicchio, E.; Kubo, M. M.; Levy, R. M., Enthalpy−Entropy and Cavity Decomposition of Alkane Hydration Free Energies: Numerical Results and Implications for Theories of Hydrophobic Solvation. *J. Phys. Chem. B* **2000,** *104* (26), 6271-6285.

[100] Huang, D. M.; Chandler, D., The Hydrophobic Effect and the Influence of Solute−Solvent Attractions. *J. Phys. Chem. B* **2002,** *106* (8), 2047-2053.

[101] Collins, M. D.; Hummer, G.; Quillin, M. L.; Matthews, B. W.; Gruner, S. M., Cooperative water filling of a nonpolar protein cavity observed by high-pressure crystallography and simulation. *Proc. Natl. Acad. Sci. U. S. A.* **2005,** *102* (46), 16668-16671.

CHAPTER 4

Recent Advances in the Discovery and Development of Protein-Protein Interaction Modulators by Virtual Screening

Dik-Lung Ma[1,*], Li-Juan Liu[2], Sheng Lin[1], Modi Wang[1], Daniel Shiu-Hin Chan[1] and Chung-Hang Leung[2,*]

[1]Department of Chemistry, Hong Kong Baptist University, Kowloon Tong, Hong Kong, China and [2]State Key Laboratory of Quality Research in Chinese Medicine, Institute of Chinese Medical Sciences, University of Macau, Macao, China

Abstract: The expanding knowledge of the critical roles played by protein-protein interactions in cell proliferation, differentiation and apoptosis has highlighted protein-protein interfaces as promising therapeutic targets for the treatment of various human diseases. However, targeting protein-protein interfaces is considered a particularly challenging task as protein interfaces are usually large and featureless, and lack well-defined cavities or binding contacts for small molecule recognition. Furthermore, the flexibility of protein-protein interfaces may lead to the formation of transient binding pockets that may be absent in the static structure of the free protein target or the protein-protein complex. Despite these inherent challenges, virtual screening has recently emerged as a powerful technique complementing traditional high-throughput screening technologies in identifying new protein-protein interaction modulators. The rapid virtual screening of chemical libraries could weed out non-binding candidates *in silico*, thereby greatly reducing the operational costs associated with chemical synthesis and *in vitro* screening. This review aims to provide an introductory framework for the use of virtual screening in drug discovery and serves to highlight successful examples of the identification of novel protein-protein interaction modulators by virtual screening techniques.

Keywords: Computer-aided drug discovery, drug development, molecular docking, protein-protein interactions, virtual screening.

INTRODUCTION

Historically, medicinal chemists have primarily focused on the discovery and

***Corresponding authors Dik-Lung Ma:** Department of Chemistry, Hong Kong Baptist University, Kowloon Tong, Hong Kong, China, Tel: +852 3411-7075, E-mail: edmondma@hkbu.edu.hk
Chung-Hang Leung: State Key Laboratory of Quality Research in Chinese Medicine, Institute of Chinese Medical Sciences, University of Macau, Macao, China, Tel: +853 8397-8518, Fax: +853 2884-1358, E-mail: duncanleung@umac.mo

Zaheer Ul-Haq and Jeffry D. Madura (Eds). **10.1016/B978-1-60805-865-5.50004-6**

development of small molecule inhibitors targeting the active sites of enzymes or protein receptors, which are usually small, well-defined and solvent-shielded [1, 2]. Validated therapeutic proteins such as cell surface receptors and protein kinases have received the lion's share of attention over the past few decades [3]. These therapeutic targets have been estimated to represent over 80% of all current drug targets, with the remaining fraction largely divided between protein ion channels and protein transporters [4].

Recently, however, targeting protein-protein interfaces (PPI) to inhibit cellular signalling and functions has attracted increasing attention due to the roles of protein-protein interactions in controlling cellular proliferation, differentiation and apoptosis. Remarkably, it has been estimated that the human genome may be able to produce up to 100,000 proteins that are involved with up to 650,000 protein-protein interactions, and that only about 10% of all protein-protein interactions have been resolved to the present time [5]. Consequently, the discovery and development of protein-protein interaction modulators (PPIMs) has tremendous potential for the treatment of human diseases [6-8].

The three main classes of PPIMs are peptides, therapeutic antibodies and small molecules. These chemical entities exert their biological functions by acting on the protein-protein interface either through stabilizing the protein-protein complex or by disrupting the interaction between the two proteins. However, while peptides and therapeutic antibodies generally enjoy high affinities and specificities for their cognate targets, they possess intrinsic drawbacks that may limit their further development as potential protein-protein interaction modulating drugs. For example, antibodies and peptides are generally expensive to produce and may exhibit limited oral bioavailability, cell permeability and metabolic stability [9-11]. This has stimulated the discovery and development of small molecule PPIMs as potential candidates for the treatment of human diseases. Besides therapeutic applications, small molecules able to target particular protein-protein interfaces selectively also represent valuable tools for the study of protein-protein interaction networks [12, 13].

Protein-protein interactions are targets of a number of bioactive natural products [14]. For instance, the anti-hypertensive natural product forskolin was found to stabilize the subdomains of adenylyl cyclase (AC) dimer, thus facilitating the formation of an active AC complex that can catalyze the production of cyclic AMP (cAMP), which is an important secondary signaling messenger [15]. Other PPIMs bind to the protein in an allosteric fashion either to increase or decrease the binding affinity of the protein surface with other protein partners. This is best illustrated by the taxane agents paclitaxel (Taxol), a diterpenoid isolated from the bark of the Pacific yew tree (*Taxus brevifolia*) [16], and its semisynthetic derivative docetaxel (Taxotere), that have been approved as anti-mitotic agents for the treatment of a number of cancers [17, 18]. Taxol and Taxotere bind to the β-subunit of the tubulin heterodimer and stabilize the interaction between the heterodimers, thereby facilitating the polymerization of tubulin into a long microtubule [19]. As microtubules normally undergo depolymerization during cell growth, stabilization of microtubules by these small molecules acts to impose cell-cycle arrest and apoptosis.

Although several well-known natural products have been determined to exert their therapeutic effects *via* targeting protein-protein interfaces, the identification of new small molecules as PPIMs is still an immature science. Traditional methods to identify small molecule PPIMs include biophysical and/or biochemical assays such as enzyme-linked immunosorbent assay (ELISA), nuclear magnetic resonance (NMR) spectroscopy, surface plasmon resonance (SPR) spectroscopy and X-ray crystallography [19-23]. However, some of these techniques are too costly and/or time-consuming to adapt to a high-throughput screening (HTS) format. Virtual screening has recently emerged as a powerful technique in drug discovery complementing traditional HTS technologies [24-27]. Virtual screening can be broadly defined as the use of computational analysis of a database of chemical structures to identify potential "hits" against a specific pharmacological target. The rapid virtual screening of chemical libraries could eliminate inactive chemical structures *in silico*, thereby dramatically reducing the costs associated with chemical synthesis and/or biological testing [28]. Consequently, the hit rates of computer-identified molecules in *in vitro* assays are often much higher compared to conventional HTS without preliminary virtual screening.

This chapter is dedicated to introducing the use of virtual screening techniques to identify novel PPIMs as potential agents for the treatment of human diseases. We will first highlight the challenges and difficulties involved with discovering small molecules as PPIMs. We will then briefly describe different *in silico* approaches used in virtual screening, as well as discuss special strategies that are specific for PPIM discovery. Due to the breadth of this field, we will focus especially on structure-based virtual screening by molecular docking. Finally, we will highlight interesting examples of the discovery of PPIMs since 2008 using the structure-based approach, which is particularly useful for discovery of PPIMs. Interested readers are referred to several excellent review articles that comprehensively summarize the progress of small molecule modulators of protein-protein interfaces up to 2008 [6, 7].

CHALLENGES IN TARGETING PROTEIN-PROTEIN INTERFACES

The characteristics of protein-protein interfaces have been studied extensively over the years. Typically, proteins can interact with themselves to form a homodimer, or they can interact with a structurally distinct protein to form a heterodimer [29, 30]. Targeting protein-protein interfaces is considered a particularly challenging task as they usually comprise large areas (~1,500-3,000Å^2) [31] compared to protein-small-molecule binding sites (~300-1,000Å^2) [32], as well as their amorphous nature that lacks well-defined cavities for recognition by small molecules [33]. Many protein-protein interfaces are composed of several small binding pockets that are dispersed throughout the protein structure, which can make the rational design of small molecules targeting protein-protein interfaces significantly more difficult than for enzyme or protein receptor active sites. Furthermore, existing collections of low-molecular weight compounds specifically designed for traditional "druggable" targets such as G-protein coupled receptors and protein kinases may not be well suited for targeting the relatively large binding pockets of protein-protein interactions.

Another problem associated with some protein-protein targets is their inherent flexibility [34]. The movement of side chains and perturbation of loops under the dynamic equilibrium or influence of small molecule modulators may affect the conformation of the protein surface, leading to the formation of "transient" binding pockets that may be absent in the static structure of the free protein target or the protein-protein complex. Taken together, these factors make the discovery of small molecule PPIMs generally more difficult than for traditional molecules targeting enzyme or protein receptors.

In spite of the large surface area of the protein-protein interaction surface, however, it has been shown that a small subset of amino acid residues of the protein-protein interface contributes significantly to the binding affinity of small molecules [35]. These amino acid residues or "hotspots" tend to be clustered together at the centre of protein-protein interfaces and are surrounded by other amino acids that contribute significantly less to binding and probably serve as "gatekeepers" to prevent the entrance of the bulk solvent [36]. In recent years, a number of *in silico* methods, web-servers and databases have been developed to analyze the geometry, energetics and chemical nature of protein-protein interfaces [37-39]. For example, *in silico* models or tools such as iPred [40], PIER [41], KFC2 [42], HotPoint [43, 44], HSPred [45] and APIS [46] are able to predict the binding pocket or hotspot residues within the protein-protein interface. Some databases, such as MINT (http://mint.bio.uniroma2.it/mint) and DOMINE (http://domine.utdallas.edu/cgi-bin/Domine), collate information on experimentally verified protein-protein interaction interfaces and are available for public access. A survey of the available tools and web servers for analysis of protein-protein interactions and interfaces has been compiled by Nussinov and co-workers [38].

WHAT IS VIRTUAL SCREENING?

Virtual screening can be defined as the use of computational techniques in the early phase of drug discovery research [28]. These techniques have gained increasing attention as powerful and valuable tools complementing traditional HTS techniques for the discovery of novel bioactive compounds [47-62]. Regardless of which computational algorithms and scoring tools are used, the ultimate goal of a virtual screening campaign is to identify bioactive chemical entities against a particular biomolecular target, while simultaneously eliminating the majority of non-binding molecules from a chemical database of compounds. The resulting, smaller set of hit compounds can then be synthesized or purchased for biological testing. This integrated, multi-disciplinary approach allows the researcher to explore the interactions between the biomolecular target and a large number of chemical compounds in a systematic and time-effective manner, and can dramatically enhance the hit rate while lowering the experimental cost for the biological testing in a drug discovery project. Virtual screening strategies used in drug discovery can be broadly classified into ligand-based or structure-based approaches, as described below.

Ligand-Based Virtual Screening

In ligand-based virtual screening, prior knowledge of the three-dimensional (3D) structure of the biomolecular target is not required. Within the realm of ligand-based screening techniques, pharmacophore modeling has been a popular strategy used in drug discovery, as described in comprehensive reviews articles [63-66]. In ligand-based pharmacophore screening, the most important common structural features relevant for a given biological activity are extracted from a "training set" of molecules possessing a similar mechanism of action and experimentally determined affinities [67]. Thus, a pharmacophore is a virtual chemical entity containing an ensemble of steric and electronic features that are believed to be necessary (although not sufficient) for activity against the biomolecular target. The pharmacophores can be generated by the following steps: (i) ligand conformational flexibility is sampled and the most favorable conformations of each compound in the training set are retained and (ii) the compounds are aligned to derive the common structural and/or electronic features of the training set in order to produce the pharmacophore model. Computational software used for pharmacophore generation include HypoGen (Accelrys Inc. [68]), HipHop [69], PHASE [70] and DISCO [71].

The quality of pharmacophore models depends greatly on the size of the training set and their chemical diversity. Multiple pharmacophore models can be subjected to cost analysis and scoring, which aim to rank and validate the statistical significance of the hypothesized models. The best pharmacophore model can then be used to screen chemical libraries *in silico* in order to identify ligands that possess the necessary chemical and/or electronic features in the appropriate spatial arrangement for biological activity.

Besides pharmacophore modeling, other ligand-based screening approaches include data mining or machine learning methods (such as support vector machine, Bayesian, and decision tree strategies), in which classification rules are developed based on a training set of active and inactive compounds [72-75], and similarity searching methods, where compound similarity is analysed using molecular property descriptors or "fingerprints" [76-78]. Ligand-based screening techniques are considered to be relatively less computationally demanding as the affinity calculations are based upon the geometric matching of the ligand atoms and groups to the chemical or structural features of the virtual template [79]. However, one drawback of ligand-based virtual screening is that it cannot used for novel or "orphan" pharmacological targets for which the required agonists or antagonists have yet not been identified. For the *in silico* discovery of PPIMs, the structure-based virtual screening approach may be more useful because very few, if any, known small molecule ligands are available for most protein–protein complexes.

Structure-Based Virtual Screening

If an experimentally-determined 3D structure of the target has been obtained from either X-ray crystallography or NMR spectroscopy, structure-based techniques can be utilized in order to study the interactions between the candidate compounds and the biomolecule [80, 81]. Two popular approaches employed in structure-based virtual screening are structure-based pharmacophore modeling and molecular docking.

In structure-based pharmacophore modeling, the structure of the target is analysed to pick out features of the binding site that are important for ligand binding affinity and selectivity. A structure of the biomolecular target complexed with a ligand is preferable since specific features of the ligand-biomolecule interaction can be readily identified. These features can be classified into interactions such as hydrogen bonding, charge transfer, and lipophilic interactions. Some programs, such as LIGANDSCOUT [82], can perform calculations on the relevant interactions between the ligand and the biomolecule and automatically generate a pharmacophore model.

In molecular docking, chemical compounds from the virtual library are docked into the binding pocket of the biomolecular target, and their affinity for the target is evaluated computationally. This process typically involves the examination of the binding interactions of the compounds with the target, followed by a score assignment reflecting the predicted binding energy of the ligand-target complex. Although structure-based strategies are often considered to be more computer-intensive and time-consuming compared to ligand-based screening methods, they do offer several distinct advantages. Firstly, the binding mode of the compound with the target can be predicted, allowing the important features of the ligand-target interaction to be identified. Secondly, structure-based methods can uncover bioactive compounds with entirely different chemical scaffolds from reported ligands (although it should be emphasized that ligand-based pharmacophore searching or 2D/3D similarity methods can also accomplish "scaffold hopping" [83-87]). Finally, these methods can discover ligands for novel biomolecular targets for which no existing inhibitors are available. The following sections discuss chemical library construction and pre-treatment procedures with an emphasis on structure-based molecular docking.

CHEMICAL LIBRARIES FOR VIRTUAL SCREENING

The choice of the chemical library to be used is of crucial importance in every structure-based virtual screening campaign. The result of any high-throughput (virtual) screening exercise is ultimately predicated upon the quality of the compound collection itself. Poorly designed libraries lacking sufficient diversity may result in few hits. Virtual compound libraries are readily available from both commercial and non-commercial sources on the internet, or they may be generated using computational software. Popular compound libraries include those containing marketed drugs [88, 89], or databases of natural products and natural product-like compounds. Furthermore, certain filters can be applied to remove compounds that are unlikely to progress beyond preliminary drug development. For example, some compounds in the virtual library may not have satisfactory ADME (absorption, distribution, metabolism and elimination) properties and toxicological profiles to be developed as potential drugs. These compounds could be filtered out from the virtual library prior to screening on the basis of selective criteria (such as molecular weight, logP, logD, and number of hydrogen bond donors or acceptors) [90-92]. These *in silico* ADME and toxicological prediction methods have proven to be useful in certain screening campaigns [93, 94]. Hetényi *et al.* have recently suggested that filters of molecular weight and lipophilicity may be limited as predictors of general drug-likeness [95]. Interestingly, the filters showed increased performance in specific cases (*e.g.* central nervous system diseases), indicating that specific disease-focused libraries may be more effective for virtual screening campaigns in the future.

PRE-TREATMENT OF BIOMOLECULE AND LIGAND

In structure-based virtual screening by molecular docking, both the biomolecular target and the library of ligands are generally subjected to a pre-treatment procedure in order to improve reliability of the screening results. The inadequate pre-treatment of the ligand library can result in the inaccurate prediction of binding poses and biased score assignments. The atomic coordinates and bond types should be assigned to the ligands appropriately. Furthermore, other factors such as the protonation state and the presence of tautomers and stereoisomers may dramatically affect the interaction between the biomolecular target and ligand [96]. Ligand minimization is performed to identify a stable 3D conformation of the small molecule before docking is carried out. These procedures can be performed by commercially available software such as CORINA (Molecular Networks GmbH) [97] and the ligand preparation tools implemented in most docking suites, such as ICM-pro (Molsoft) [98], GLIDE (Schrodinger Inc.) [99], AutoDock [100, 101], FlexX [102] or ChemAxon (http://www.dockingserver.com).

For the biomolecular target, a 3D molecular model of target is first constructed from available structural data obtained from high-resolution NMR spectroscopy or X-ray crystallography [103], or by homology modelling from related validated structures [104]. Such structural information can be accessed freely from the Protein Data Bank (PDB) [105], if available. Data obtained from X-ray diffraction is generally considered to be most reliable for use in a virtual screening campaign. An X-ray co-crystal structure of a biomolecular target complexed with an inhibitor or ligand (holostructure) is considered advantageous compared to the target without a ligand (apostructure) as the optimal interactions between the biomolecular target and the small molecules can be more easily identified.

In molecular docking, a prior knowledge of critical ligand-target interactions can also help identify false positives that arise during screening. Furthermore, the search area for docking can be restricted to the region around the bound ligand, which both avoids the unnecessary wastage of computational resources and lowers the chance of identifying non-specific molecules that bind outside the binding site. Finally, more accurate docking calculations can be performed since the target is in its active or induced conformation, thus improving the quality of the docking results.

After a suitable molecular structure is chosen for docking, the hydrogen atoms are added to the structure in order to predict the hydrogen bonding interactions between the ligand and target. The tautomeric state of the amino acid residues such as histidine should also be taken into account. Finally, a standard molecular simulation algorithm should be applied to minimize the energy of the whole biomolecule. If only the apostructure is available, computational algorithms such as ICM PocketFinder (Molsoft) and Pocket Finder (http://www.modelling.leeds.ac.uk/pocketfinder/) can be used to identify likely binding pockets for docking. However, it should be noted that the larger the size of the chosen docking site, the longer the time required to perform a single docking experiment, which could reduce the overall efficiency of a large scale virtual screening campaign.

VIRTUAL SCREENING BY MOLECULAR DOCKING

The overall workflow of structure-based virtual screening by molecular docking is depicted in Fig. **1**. In the first stage of the screening process, the docking algorithm generates a number of conformations of each of the ligands that are sequentially inserted into the defined binding pocket of the biomolecular target. Most algorithms incorporate ligand flexibility so that the binding pose of the ligand can be correctly predicted. Three main methods are commonly employed to tackle the issue of ligand flexibility, namely (i) ligand incremental construction, (ii) generation of multiple conformers (rotamer library) before docking and (iii) stochastic methods.

In ligand incremental construction, a fragment of the ligand (the "anchor") is first docked into the binding site. The best conformation of the anchor is then incrementally "grown" and docked again. This process is repeated until all the remaining fragments have been appended in sequence to rebuild the original compound within the binding pocket [102]. Another strategy is to generate a number of low-energy conformations of the ligands that are each directly docked to the biomolecular target [106].

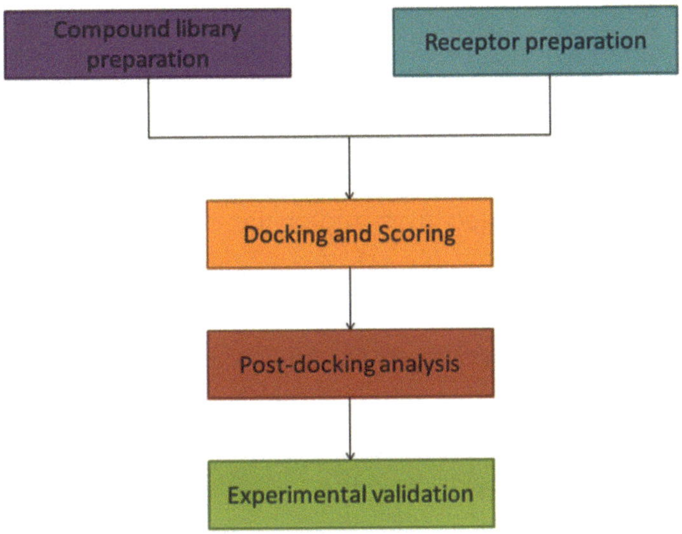

Figure 1: Schematic flowchart showing the major stages of molecular docking.

Other commonly-used strategies to account for ligand flexibility are stochastic methods such as Monte Carlo simulated annealing (MCSA) [107] or genetic algorithms (GA) [108]. An MCSA algorithm operates by stochastically varying one parameter at a time in order to generate new conformations which are accepted or rejected based on Boltzmann considerations. The process is initiated at a high temperature so that there is a significant chance of accepting the next conformation sampled. The temperature is then progressively decreased during docking in order to trap the ligand-target complex in a low energy state as a consequence of the reduced conformational freedom during cooling. On the other hand, genetic algorithms adopt a totally different approach that draws inspiration from Darwin's theory of evolution. The conformations of the ligands begin as a random population of states modeled as a set of chromosomes. The ligand conformations are allowed to perform random crossovers and mutations in order to produce another set of different conformations. The "fittest" conformations, which possess the lowest binding energies with the target, are accepted and used to produce a new generation of conformations. This cycle is iteratively repeated a number of times until the local energy minimum of the target-ligand complex has been reached.

Algorithms are also available that aim to model receptor flexibility in order to increase the success rate of a docking campaign. The first is the use of multiple receptor conformations (MRC) of the biomolecule obtained from different X-ray or NMR structures, or generated from molecular dynamics simulations *in silico* [109]. The compounds in the virtual library are screened against different conformations of the receptors, and the highest-scoring ligand poses from each receptor-ligand complex are combined. Another method to tackle this problem is to use a "soft docking" approach, which tolerates some degree of steric clashes between the ligand and the biomolecules [110]. Finally, some modern docking algorithms are able to explicitly model receptor flexibility, but this is usually constrained to the ligand binding domain as the explicit inclusion of receptor flexibility for the whole protein in the docking calculations would be too computationally demanding [111].

After generating the binding poses of each compound in the database, scoring must be performed to rank or score each binding pose to determine the relative binding affinity of the ligand against the target, and to discriminate the active compounds from the decoys (inactive compounds). Scoring functions are generally classified into force field-based, empirical-based and knowledge-based scoring functions [47, 112]. Scoring functions are subjected to continual improvement and there exist no general rules that specify which scoring functions should be used under certain circumstances. It should be noted that scoring functions currently constitute a weak link in structure-based virtual screening, as their inability to predict binding energy values accurately places a major limitation on the quality of the docking results. One strategy used to tackle this issue is to utilise multiple scoring algorithms (consensus scoring) to provide a more reliable estimation of ligand binding affinity [113]. Furthermore, post-docking analysis can be performed *via* visual inspection or topological filters in order to remove binding poses that contain significant steric and/or electrostatic clashes.

STRATEGIES FOR PPIM DISCOVERY

To tackle the unique challenges presented by protein-protein interaction interfaces, researchers have explored adjustments to the general virtual screening strategies described above. Several groups have attempted to characterise the nature of the protein-protein interface and the PPIM chemical space, the results of which could potentially yield chemical libraries enriched with fragments or sub-substructures with greater propensity to modulate protein-protein interactions [114]. Here, we highlight some recent contributions that could be useful for the application of virtual screening techniques in PPIM discovery.

Several years ago, Morelli, Roche and co-workers presented the 2P2I database (http://2p2idb.cnrs-mrs.fr), which aimed to analyse protein-protein and protein-inhibitor interfaces in terms of geometrical parameters, atom and residue properties, buried accessible surface area and other biophysical parameters [115]. At the time of the study, the 2P2I database was comprised of 17 protein-protein complexes from 14 families, in addition to 56 small molecule inhibitors bound to their cognate targets. Their analysis generated several interesting conclusions. For example, it was observed that protein partners with known PPI inhibitors did not undergo major conformational changes upon heterodimeric complex formation, implying that these types of complexes are easier to target. Additionally, PPIs with known inhibitors displayed more hydrogen bonds, fewer salt bridges and fewer charged residues at the interface compared to typical heterodimers. In a later work, the group analysed the structural features of PPIM in the 2P2I database [116]. A statistical analysis of 39 PPI inhibitors suggested a "rule of four" framework for small molecule PPI inhibitors, where molecular weight > 400, ALogP > 4, number of rings > 4 and number of hydrogen bond acceptors >

4. Interestingly, these criteria are in direct contradiction with Lipinski's classic rule of five (MW $<$ 500, logP $<$ 5, H-bond donors $<$ 5, H-bond acceptors $>$ 10 [117]) for small molecule drugs, suggesting that traditional drug-likeness screens may fall critically short when utilised for PPIM screening.

The group of Sperandio utilised machine-learning methods to design focused chemical libraries enriched in PPI modulators [118]. The study was performed using a chemically diversified learning data set of 66 validated drug-like PPI inhibitors and 557 non-PPI small molecule inhibitors. Their analysis yielded descriptor-based decision trees that managed to positively discriminate PPI inhibitors by using only two molecular descriptors, RDF070m and Ui, which describe a specific molecular shape and the presence of 15-17 multiple bonds in the compound, respectively. A computer package (PPI-HitProfiler) developed to implement these criteria showed robust performance when applied to two commercial compound collections screened against 11 distinct PPI systems, with 70-81% of true PPI inhibitors identified and 42-52% of putative non-PPI inhibitors discarded.

In a recent work, Fry and co-workers have analysed the important binding determinants of the Nutlins, which are a distinct class of PPI inhibitors that bind to the protein MDM2 and block its interaction with p53 [119]. In their study, RG7112, the first member of the Nutlin family to enter clinical trials [120], was systematically constructed into smaller fragments, and the ability of the fragments to bind to MDM2 was analysed using surface plasmon resonance spectroscopy (SPR), NMR, and X-ray crystallography [121]. Interestingly, the smallest fragment capable of binding to MDM2 had a molecular weight of 305 Da, which is a value that is located at the upper end of the molecular weight range of typical fragments. This study supported the use of fragment-based techniques for PPIM discovery, and suggested that the fragment-based (virtual) screening of protein-protein systems may be benefited by a bias towards fragments with higher molecular weight.

The principles of binding hotspots have been extended into the concept of "hot regions", which are clusters of hotspots at PPI interfaces [36]. For hub proteins, which are proteins that bind to multiple protein partners, such hot regions are a characteristic signature for their protein-protein interfaces, with each hot region on the hub protein potentially binding to a different partner protein [122]. Building on this idea, Keskin and co-workers have established HotRegion as a database of predicted hot spot clusters [123]. The study of hot regions may reveal cooperative effects in the contributions of individual hotspots towards the overall stability of the PPI.

A few PPIMs bind to transient pockets on the protein-protein interface that are absent in either the free protein partners or the heterodimeric complex [124]. Gohlke and co-workers devised the first computational method that was able to simultaneously address the energetics and plasticity of PPIM binding at the protein interface, and to identify the determinants of ligand binding, hot spots and transient pockets in a protein [125]. In their study, conformation ensembles of the IL-2-IL-2Rα protein-protein complex were generated using molecular dynamics (MD) and constrained geometric (FRODA) simulations, and hot spot and transient pockets were identified using energetic or geometric criteria. Compounds were docked to the transient pockets, followed by structure selection based on hotspot prediction, RSMD clustering and intermolecular docking energies. This eventually yielded a library enriched in IL-2 PPI inhibitors over decoy compounds. Significantly, this study demonstrated that *in silico* techniques could be used to discover transient binding pockets at protein-protein interfaces even when the structure of the ligand-protein complex is not available. Recently, this method was also used to identify the first small-molecule protein-protein interaction inhibitors of RUNX1/ETO tetramerization [126].

CASE STUDIES

Over the last decade, a number of small molecules that target protein-protein interfaces have been reported, as described in comprehensive review articles [14, 93, 127]. Using both conventional screening and virtual screening techniques, inhibitors of the Bcl-xL-BH3 protein-protein interaction [128-130], the p53-MDM2 interaction [131, 132], the BIR3 domain of XIAP [133, 134] and the IL-2 alpha receptor-IL-2 interaction [135, 136] have been discovered. In this section, we highlight interesting examples of the application of structure-based virtual screening for the discovery of biologically validated PPIMs since 2008. To emphasize the versatility of this technique, interesting examples will be chosen from a variety of protein-protein interactions, such as the linear protein binding domains of transcription factors, or protein-protein homodimeric or heterodimeric interactions.

MODULATORS OF THE TUMOR NECROSIS FACTOR ALPHA (TNF-A) INTERACTION

The tumor necrosis factor-α (TNF-α) trimer is an important human cytokine that is involved in the inflammatory response through regulation of diverse signaling pathways. Aberration in the cellular levels of TNF-α has been implicated in a variety of inflammatory disorders [137]. The clinically-proven biopharmaceutical infliximab targets TNF-α trimerization and is routinely used to treat inflammatory disorders such as rheumatoid arthritis, psoriatic arthritis, and Crohn's disease. However, the use of TNF-α antibodies such as infliximab can elicit an autoimmune anti-antibody response or weaken the body's immune system to opportunistic infections. In 2010, our group applied structure-based, high-throughput virtual screening (HTVS) methods to identify small-molecule inhibitors of TNF-α from a database containing over 20,000 natural products or natural product-like compounds [138]. An X-ray co-crystal structure of the TNF-α dimer bound by the small-molecule inhibitor SPD304 (PDB: 2AZ5) was chosen for the construction of the molecular model [139]. The X-ray structure was thoroughly examined and was energy minimized using the ICM-pro docking suite, and the search area for docking was restricted to the binding cavity that was occupied by SPD304. The compounds from the natural product and natural product-like database were then docked against a grid representation of the receptor using the ICM method and assigned an ICM score reflecting the quality of their binding to the receptor pocket. The high scoring structures were visually inspected and twelve of these compounds were experimentally tested for TNF-α inhibition using an ELISA. Two chemically distinct compounds, the pyrazole-linked quinuclidine **1** and the indolo-[2,3-α]quinolizidine **2**, were identified as the top candidates against the TNF-α protein-protein interaction (Fig. **2a**).

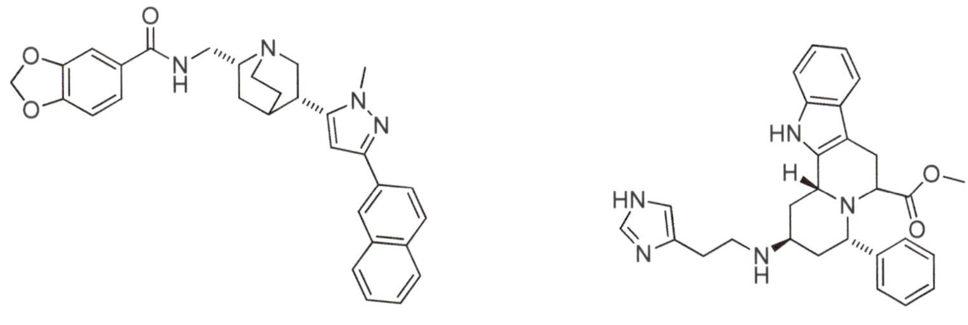

(a) **1** **2**

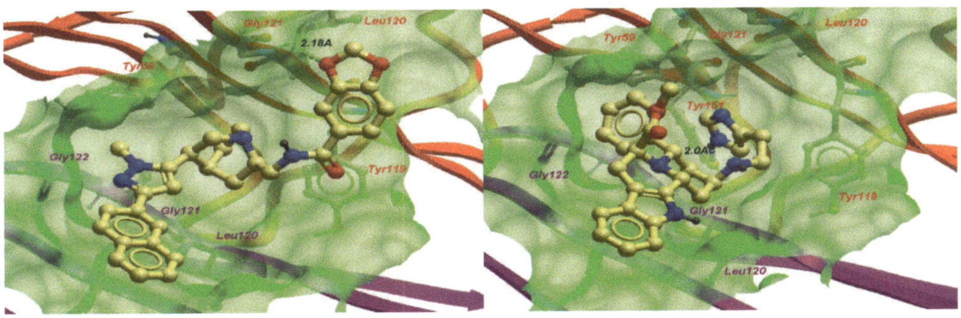

(b)

Figure 2: (a) Chemical structures of the pyrazole-linked quinuclidine **1** and the indolo-[2,3-α]quinolizidine **2**. (b) Low-energy binding conformations of **1** (left) and **2** (right) generated by molecular docking. Hydrogen bonds are depicted as dotted lines. Reproduced from Ref. [139].

In *in vitro* assays, these two compounds were able to disrupt the TNFR1-TNF interaction in an ELISA and down-regulate TNF-α-driven gene expression in human cells. Significantly, compound **2** (IC_{50} = 10 μM) was slightly more potent in ELISA compared to SPD304 (IC_{50} = 22.4 μM), which was the most potent direct TNF-α inhibitor reported to date, and displayed comparable potency to SPD304 in the cell-based luciferase reporter assay. Molecular modeling analysis revealed that compounds **1** and **2** are large and flat enough to interact with the residues from both subunits of the TNF-α dimer, thereby occupying and blocking the binding site for the third TNF-α subunit (Fig. **2b**). Notably, the lack of salt bridges or hydrogen bonding networks in our models of **1** and **2** with TNF-α was consistent with the previous finding that the interaction between SPD304 and TNF-α was primarily hydrophobic and shape-driven [140]. This study highlighted the application of structure-based molecular docking to discover natural product-like inhibitors of the TNF-α protein-protein interaction.

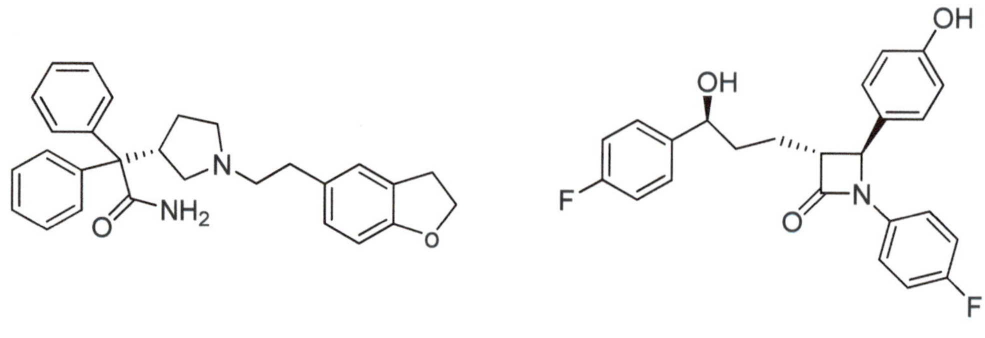

(a) **3** **4**

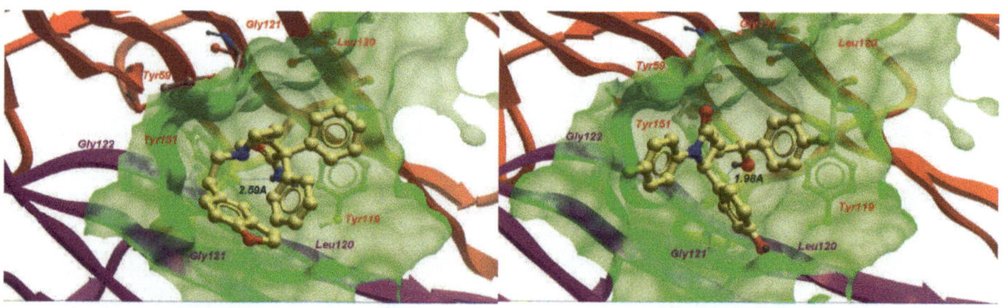

(b)

Figure 3: (a) Chemical structures of the TNF-α PPI inhibitors: darifenacin 3 (overactive bladder syndrome) and ezetimibe 4 (hypercholesterolemia). (b) Low-energy binding conformations of 3 (left) and 4 (right) generated by molecular docking. Hydrogen bonds are depicted as dotted lines. Reproduced from Ref. [141].

Later, we utilized an *in silico* drug repositioning strategy with the aim of discovering existing drugs as TNF-α protein-protein interaction inhibitors. In this approach, over 3,000 compounds from a database of US FDA-approved drugs were docked against the TNF-α molecular model using the ICM method as described above [141]. Two of the hit compounds (darifenacin **3** and ezetimibe **4**, Fig. **3a**) from the FDA database were subsequently demonstrated to disrupt the TNF-α-TNF-α receptor interaction *in vitro* and down-regulate TNF-α-driven gene expression in human cells. Darifenacin **3** (trade name: Enablex) is currently used in the treatment of overactive bladder (OAB) syndrome by targeting the M3 muscarinic acetylcholine receptor, while ezetimibe **4** (trade name: Zedia) is a potent inhibitor of cholesterol absorption in the intestines and is used for the treatment of hypercholesterolemia. Like the natural product-like inhibitors described above, compounds **3** and **4** were predicted to interact with both subunits of the TNF-α dimer largely through hydrophobic interactions, leading to the inhibition of TNF-α trimerization (Fig. **3b**).

MODULATORS OF PROTEIN-PROTEIN INTERACTIONS OF THE TOLL-LIKE RECEPTOR (TLR) SIGNALLING PATHWAY

Toll-like receptors (TLRs) are type I transmembrane proteins that recognize pathogen-derived macromolecules and play a key role in the innate immune system [142-144]. Pathogen-derived macromolecules, which are broadly shared by pathogens but are distinguishable from host molecules, are collectively known as pathogen-associated molecular patterns (PAMPs) [145]. The dimerization of TLR leads to activation of the transcription factor nuclear factor-κB (NF-κB) and interferon regulatory factors (IRFs), which in turn leads to the production of pro-inflammatory cytokines and type I interferons. Thus, dysregulation of TLR activity has been associated with the development of inflammatory diseases [146].

In 2010, Yin and co-workers applied a novel *in silico* screening methodology that included molecular mechanics (MM)/implicit solvent methods to identify inhibitors of the TLR4/MD-2 protein-protein interaction [147]. Toll-like receptor 4 (TLR4) is a membrane-spanning immune receptor that functions in a complex with its accessory protein myeloid differentiation factor 2 (MD-2) [148]. TLR4 detects lipopolysaccharide (LPS), which is a TLR4 agonist and a component of gram-negative bacterial cell walls [127]. TLR4 signaling has been implicated in numerous disease states including acute sepsis and neuropathic pain [149]. Consequently, the TLR4/MD-2 interaction is an attractive therapeutic target as it is essential for TLR4 signaling [150].

Traditional molecular dynamic (MD) stimulations can model the interaction between proteins and small molecules in a fully flexible manner, allowing the relaxation of the binding site residues and the incorporation of explicit water molecules that are generally excluded from most algorithms [151]. However, these techniques are too computationally expensive for application in high-throughput virtual screening. In order to improve the accuracy of the screening process while preserving the computational calculation time within reasonable limits, Yin and colleagues used the molecular docking algorithm GLIDE to generate the binding poses of the screened ligands and MD simulations to score the ligand poses with QUANTUM [152]. The screened compounds were further clustered to ensure that a large variety of chemical compounds were included. The hits were subsequently profiled against around 500 representative human proteins to filter out non-selective binders *in silico*. Compounds **5** and **6** were identified to target TLR4 and MD-2, respectively (Fig. **4a**). These small molecule inhibitors were able to disrupt TLR4 signaling in mouse macrophage cells with complete inhibition at 2 μM of compound **5** or 200 nM of compound **6**, presumably due to the inhibition of the TLR4/MD-2 protein-protein interaction. Compound **6** was further demonstrated to selectively target TLR4 signaling without affecting the signaling of other TLRs. The molecular models of **5**/TLR4 and **6**/MD-2 reveal a high degree of shape complementarity between the small molecules and the protein binding pockets (Fig. **4b**).

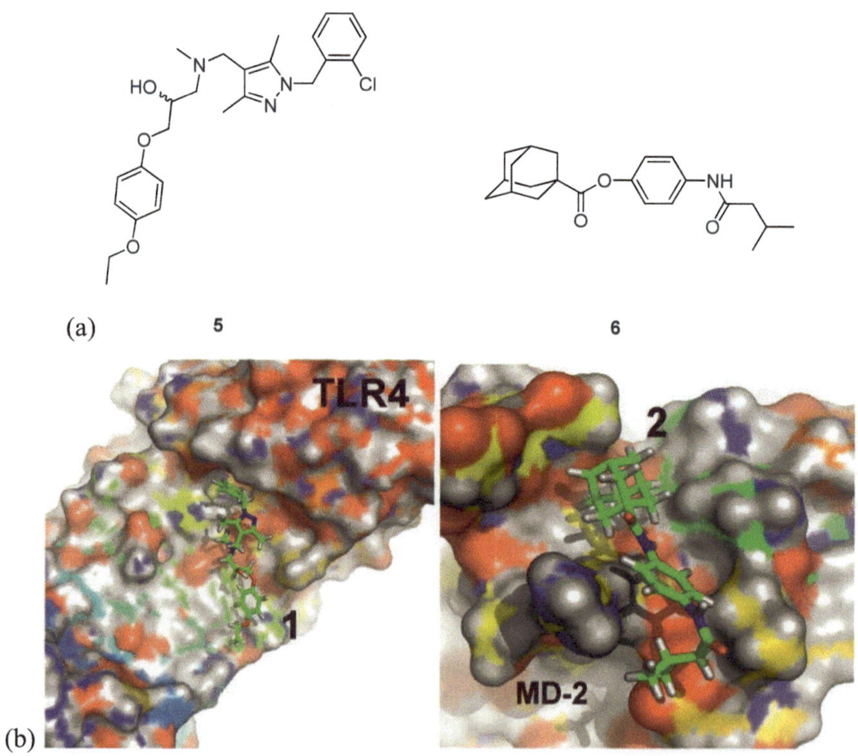

Figure 4: (a) Chemical structures of the TLR4 antagonist 5 and MD2 antagonist and 6. (b) Molecular models of the 5/TLR4 (left) and 6/MD-2 (right) complexes generated by molecular docking. Reproduced from Ref. [152].

MODULATORS OF STAT3 DIMER INTERACTIONS

The STAT3 dimer is a key transcription factor through which receptors of multiple cytokines and growth factors exert their effects. The phosphorylation of STAT3 at the tyrosine 705 residue in the SH2 domain induces the formation of STAT3 homodimers or STAT3-STAT1 heterodimers. These protein complexes subsequently translocate into the nucleus and activate target genes through binding to specific DNA-response elements. STAT3 is constitutively activated in many types of cancer and has been linked to tumor progression through enhancing angiogenesis, metastasis, growth and survival of cancer cells [153]. In 2010, Asai and co-workers identified a new series of STAT3 inhibitors [154] using a docking and consensus scoring approach implemented in CONSENSUS-DOCK [155], a customized version of the DOCK4 program [156]. The X-ray structure of the DNA-bound STAT3 homodimer (PDB: 1BG1) [157] was processed by removal of the DNA and the docking site was restricted to the SH2 domain of the dimer. Approximately 360,000 small molecules were docked to the SH2 domain of STAT3 and 136 compounds were selected for subsequent *in vitro* screening based on the consensus docking scores and from visual inspection. From a preliminary luciferase reporter assay, STX-0119 **7** was identified as a potential STAT3 inhibitor with 99% inhibition of STAT3-driven luciferase activity at 100 μM (Fig. **5a**). A fluorescence resonance energy transfer (FRET) assay indicated that STX-0119 **7** was able to inhibit STAT3 dimerization in cells by 62% at a concentration of 50 μM. The molecular model of STX-0119 **7** bound to the SH2 domain of STAT3 revealed that the 2-phenyl ring of the small molecule was inserted into a hydrophobic cleft where it comes into contact with the phospho-tyrosine binding site (Fig. **5b**). The complex was further stabilized by a hydrogen bonding interaction between the amide group of **7** and the backbone carbonyl group of Ser636, and a hydrophobic interaction between the furan ring of **7** with the indole moiety of Trp623.

Subsequently, Li and co-workers employed a multiple ligand simultaneous docking (MLSD) approach to identify a small molecule inhibitor of STAT3 dimerization [158]. An X-ray structure of the STAT3 SH2 domain (PDB: 1BG1) features three characteristic sub-binding pockets involving the "hotspot" residues

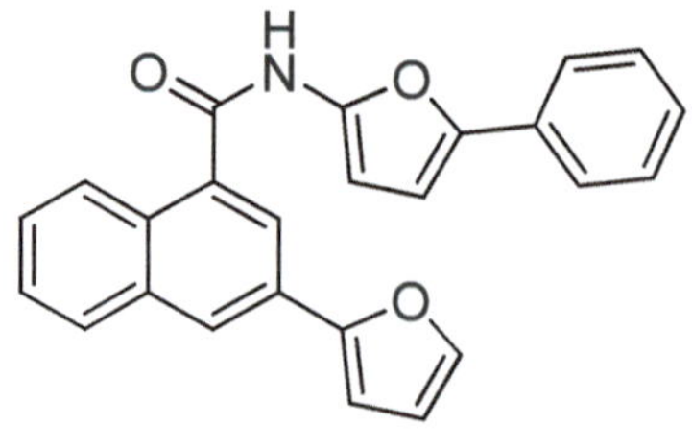

(a) **7**

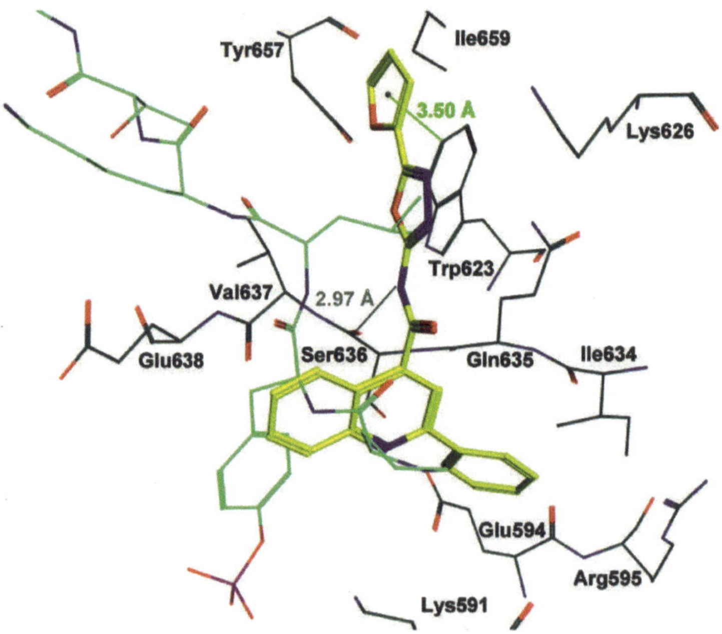

(b)

Figure 5: (a) Chemical structure of the STAT3 dimerization inhibitor STX-0119 7. (b) Molecular model of compound 7 with the STAT3-SH2 domain. Reproduced from Ref. [157].

Tyr705 and Leu706, as well as a side pocket composed of Ile597, Leu607, Thr622 and Ile634. As the binding affinity of small molecules to the STAT3 SH2 domain primarily depends upon their ability to interact with the basic Tyr705 and hydrophobic Leu706 residues within the SH2 domain, the authors first constructed a fragment library based upon reported small molecule inhibitors of the STAT3 SH2 domain. The fragments were then filtered using a similarity search of chemically or structurally similar entities from a drug scaffold database in order to weed out any fragments with undesirable ADMET properties. The fragment library was then further classified into polar and non-polar fragments with potential affinity for the Tyr705 or Leu706 residues within the binding site. In the second stage of the screening, three drug fragments from the library, including one polar fragment and two non-polar fragments, were simultaneously docked against a pre-treated model of the STAT3 SH2 binding pocket and a docking score was assigned to each fragment. Finally, the high-scoring fragments were linked together to generate fifteen virtual templates using various chemical linkers such as amide, amine, ether or alkene groups. The virtual templates were

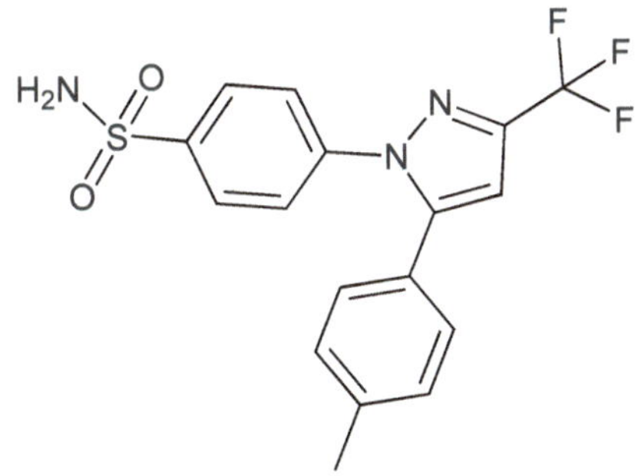

(a) **8**

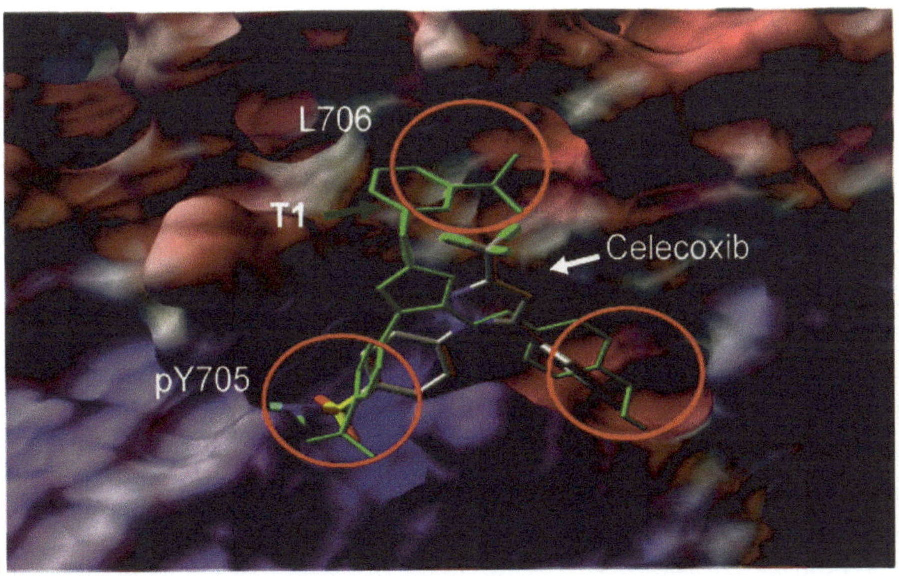

(b)

Figure 6: (a) Chemical structures of the STAT3 dimerization inhibitor celecoxib 8, originally developed as an NSAID. (b) Molecular model of 8 bound to the SH2 domain of STAT3. Reproduced from Ref. [158].

then used as pharmacophores for screening an FDA-approved drug database. Thirteen of the fifteen pharmacophore models identified celecoxib **8** as a top hit for STAT3 inhibition (Fig. **6a**). *In vitro* experiments demonstrated that celecoxib **8** could down-regulate STAT3 phosphorylation in a dose-dependent manner, selectively antagonize interleukin-6 (IL-6)-induced STAT3 phosphorylation, and inhibit cancer cell growth with micromolar potency. Celecoxib **8** (trade name: Celebrex) is a non-steroidal anti-inflammatory drug (NSAID) and a selective COX-2 inhibitor that is mainly used for the treatment of various conditions such as rheumatoid arthritis, acute pain, and colorectal polyps. In the molecular model of celecoxib **8** bound to the STAT3-SH2 domain, the phenylsulfonamide group of **8** was bound to the pTyr705 site, while the non-polar phenylmethyl moiety of **8** occupied the side pocket formed from Ile597, Leu607, Thr622, and Ile634 (Fig. **6b**). This study demonstrated that a virtual screening strategy combining privileged drug fragments, MLSD, and drug repositioning, can be an effective approach to identifying inhibitors of the STAT3 protein-protein interaction.

MODULATORS OF THE IFNA-IFN RECEPTOR INTERACTION

Type I interferons are proinflammatory cytokines that are released in response to viral infection and help coordinate the first line of defense against the pathogens [159]. All type I interferons bind to the IFN receptor (IFNAR) to initiate the positive feedback loop leading to elevated IFN levels [160]. Recent research has suggested that chronically activated plasmacytoid dendritic cells, which are the main producers of type I IFN [161], produce IFN in response to the activation of toll-like receptors and may be implicated in the development of systemic lupus erythematosus.

Recently, Schneider and co-workers utilized an integrated approach involving pharmacophore screening and molecular docking to identify compound **9** as a potential inhibitor of the IFN-IFNAR interaction [162]. The NMR solution structure of human IFN-α was taken from the PDB (PDB: 1ITF, model 16) and the potential small molecule binding pocket was extracted using their in-house tool PocketPicker [163]. A total of 19 candidate pockets were identified and 3 of them were located within the interacting site of the IFN-α-IFN receptor ectodomain. Further analysis using iPred, which is a tool for surface “hot spot” residue identification based on a knowledge-based scoring function adapted from the field of protein folding and small molecule docking [40], revealed six hotspot residues, four (Phe27, Leu30, Lys31 and Arg149) of which surrounded a small pocket with an area of 155 $Å^2$. Subsequently, a structure-based pharmacophore model was generated using the VirtualLigand software [164], and over 500,000 commercially available compounds were screened against the pharmacophore. The 100 top-ranking compounds from the pharmacophore search were visually inspected and six compounds were chosen for further molecular docking analysis using GOLD docking software [165]. These compounds all displayed favorable binding to the target, and compound **9** emerged as the highest-ranking compound (Fig. **7a, b**). *In vitro* assays demonstrated that compound **9** could inhibit IFN-α responses induced by modified *Vaccinia* virus Ankara (MVA) in Flt3-L-differentiated pDC cultures (BM-pDCs) with IC_{50} values of 2-8 μM. The binding of **9** to IFN-α was further investigated using saturation-transfer difference (STD) NMR spectroscopy, which confirmed that compound **9** bound directly to IFN-α as observed by saturation transfer of the NMR signal in the aromatic range of the spectrum. This study highlighted the application of hotspot prediction techniques for the structure-based pharmacophore screening of PPIMs targeting the FN-α-IFNAR interaction.

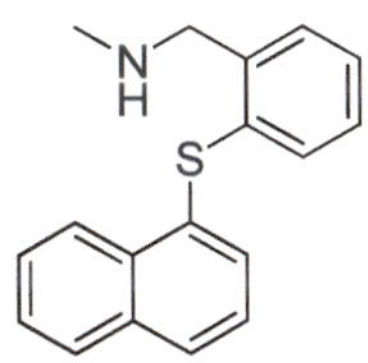

(a) **9**

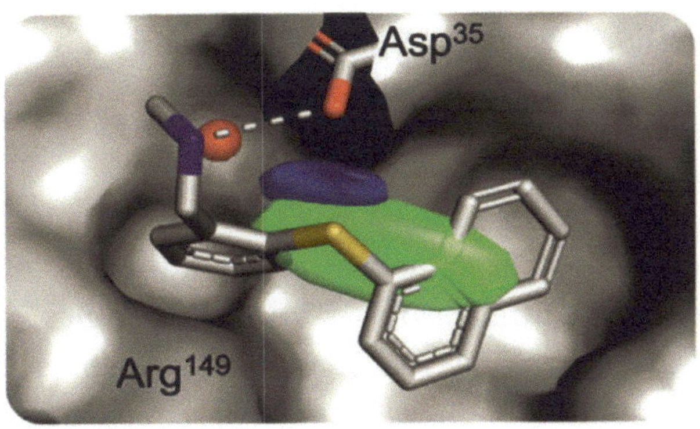

(b)

Figure 7: (a) Chemical structure of IFN-α-IFNAR interaction inhibitor 9. (b) Molecular overlay of the docking pose of compound 9 within the IFN-α binding pocket. Reproduced from Ref. [165].

FUTURE PERSEPECTIVES

Towards the future, we envision that the character and properties of protein-protein interfaces would continue to be clarified, and the mode of action of PPIMs to be further elucidated. PPIMs have been discovered for all of the major protein folding topologies, such as α-helix, β-strand and mixed α/β-type PPI domains [116], and it might be suggested that the different types of topological scaffolds would demand distinct structural requirements in the PPI ligands. Based on analysis of the P2PI database, it has also been suggested that PPI interfaces could be divided into two major classes depending on the degree of structural regularity at the interface [115]. Additionally, while most examples described above have described orthosteric inhibition (where the PPIMs bind at the protein-protein interface), the allosteric inhibition of PPIs has been less explored and deserves further attention [166, 167].

As more PPIMs are discovered, we envisage that ligand-based screening strategies may be able to make a significant contribution. Presently, the use of ligand-based techniques such as similarity searching or ligand-based pharmacophore modeling is restricted by the very few small-molecule ligands that are known for each PPI target. Increasing knowledge on the structural features of such ligands would also aid in the generation of libraries specialised for PPIM discovery. At the same time, improved structural biological understanding and computational algorithms would allow programs to accurately model the binding interface of PPIs and to better evaluate the structural features important for ligand interaction, such as hotspots (or hot regions) and transient binding pockets.

CONCLUDING REMARKS

Virtual screening has established itself as a powerful technique that complements traditional high-throughput experimental screening technologies in early phase drug discovery research. Efficient *in silico* methodologies, in conjunction with experimental validation, can play a significant role in accelerating the drug discovery process by filtering out non-active compounds without excessive economic or temporal investment.

Although protein-protein interactions have been historically considered as challenging targets in pharmaceutical research, studies in recent years have provided rationale for modulating these so-called undruggable targets. For example, the identification of critical hot spots can provide certain hints for the design of low-molecular weight PPIMs that are able to selectively disrupt the protein-protein interaction without having to cover the entire protein-protein interface. Moreover, we anticipate that the further discovery and elucidation of new protein-protein interactions involved with human diseases will provide fertile fields of investigation for the identification of PPIMs from both old and new regions of the chemical space.

In this chapter, we have described the rationale and challenges involved with targeting protein-protein interfaces with small molecules. We have also introduced the basic principles and techniques in the field of computer-assisted drug discovery that can be used for the discovery of novel PPIMs. In particular, the application of structure-based virtual screening techniques for the identification of novel protein-protein interaction modulators deserves further attention as most protein-protein interfaces lack sufficient ligands to allow for ligand-based screening strategies. In addition, we have highlighted several interesting cases of the discovery of PPIMs that target different protein-protein interactions. We envisage that continual refinements in the understanding of protein-protein interfaces and an improved knowledge of the chemical entities that are privileged at such surfaces should eventually lead to a significant breakthrough in this young and exciting field of study.

ACKNOWLEDGEMENTS

This work is supported by Hong Kong Baptist University (FRG2/12-13/021 and FRG2/13-14/008), Centre for Cancer and Inflammation Research, School of Chinese Medicine (CCIR-SCM, HKBU), the Health and Medical Research Fund (HMRF/13121482), the Research Grants Council (HKBU/201811, HKBU/204612, and HKBU/201913), the French National Research Agency/Research Grants Council Joint Research Scheme (A-HKBU201/12), the Science and Technology Development Fund, Macao SAR (103/2012/A3) and the University of Macau (MYRG091(Y3-L2)-ICMS12-LCH, MYRG121(Y3-L2)-ICMS12-LCH and MRG023/LCH/2013/ICMS).

CONFLICT OF INTEREST

The authors declare that they have no competing interests.

REFERENCES

[1] Fletcher, S.; Hamilton, A. D. Protein surface recognition and proteomimetics: mimics of protein surface structure and function. *Curr. Opin. Chem. Biol.,* **2005**, *9*, 632-638.

[2] Fletcher, S.; Hamilton, A. D. Protein surface recognition and proteomimetics: mimics of protein surface structure and function. *Curr. Opin. Chem. Biol.,* **2005**, *9*, 632-638.

[3] Ruffner, H.; Bauer, A.; Bouwmeester, T. Human protein-protein interaction networks and the value for drug discovery. *Drug Discov. Today,* **2007**, *12*, 709-716.

[4] Arcangeli, A.; Becchetti, A. New trends in cancer therapy: targeting ion channels and transporters. *Pharmaceuticals,* **2010**, *3*, 1202-1224.

[5] Stumpf, M. P. H.; Thorne, T.; de, S. E.; Stewart, R.; An, H. J.; Lappe, M.; Wiuf, C. Estimating the size of the human interactome. *Proc. Natl. Acad. Sci. U. S. A.,* **2008**, *105*, 6959-6964.

[6] Yin, H.; Hamilton, A. D. Strategies for Targeting Protein-Protein Interactions With Synthetic Agents. *Angew. Chem. Int. Ed.,* **2005**, *44*, 4130-4163.

[7] Berg, T. Modulation of Protein-Protein Interactions with Small Organic Molecules. *Angew. Chem. Int. Ed.,* **2003**, *42*, 2462-2481.

[8] Wells, J. A.; McClendon, C. L. Reaching for high-hanging fruit in drug discovery at protein-protein interfaces. *Nature,* **2007**, *450*, 1001-1009.

[9] Robinson, J. A.; DeMarco, S.; Gombert, F.; Moehle, K.; Obrecht, D. The design, structures and therapeutic potential of protein epitope mimetics. *Drug Discov. Today,* **2008**, *13*, 944-951.

[10] Baines, I. C.; Colas, P. Peptide aptamers as guides for small-molecule drug discovery. *Drug Discov. Today,* **2006**, *11*, 334-341.

[11] Whitty, A.; Kumaravel, G. Between a rock and a hard place? *Nat. Chem. Biol.,* **2006**, *2*, 112-118.

[12] Lenz, T.; Fischer, J. J.; Dreger, M. Probing small molecule-protein interactions: A new perspective for functional proteomics. *J. Proteomics.,* **2011**, *75*, 100-115.

[13] Spring, D. R. Chemical genetics to chemical genomics: small molecules offer big insights. *Chem. Soc. Rev.,* **2005**, *34*, 472-482.

[14] Thiel, P.; Kaiser, M.; Ottmann, C. Small-Molecule Stabilization of Protein-Protein Interactions: An Underestimated Concept in Drug Discovery? *Angew. Chem. Int. Ed.,* **2012**, *51*, 2012-2018.

[15] Seamon, K. B.; Padgett, W.; Daly, J. W. Forskolin: unique diterpene activator of adenylate cyclase in membranes and in intact cells. *Proc. Natl. Acad. Sci. U. S. A.,* **1981**, *78*, 3363-3367.

[16] Wani, M. C.; Taylor, H. L.; Wall, M. E.; Coggon, P.; McPhail, A. T. Plant antitumor agents. VI. Isolation and structure of taxol, a novel antileukemic and antitumor agent from Taxus brevifolia. *J. Am. Chem. Soc.,* **1971**, *93*, 2325-2327.

[17] Nogales, E.; GrayerWolf, S.; Khan, I. A.; Luduena, R. F.; Downing, K. H. Structure of tubulin at 6.5 A and location of the taxol-binding site. *Nature,* **1995**, *375*, 424-427.

[18] Löwe, J.; Li, H.; Downing, K. H.; Nogales, E. Refined structure of αβ-tubulin at 3.5 Å resolution. *J. Mol. Biol.,* **2001**, *313*, 1045-1057.

[19] Nogales, E.; Whittaker, M.; Milligan, R. A.; Downing, K. H. High-Resolution Model of the Microtubule. *Cell,* **1999**, *96*, 79-88.

[20] Blundell, T. L.; Jhoti, H.; Abell, C. High-throughput crystallography for lead discovery in drug design. *Nat. Rev. Drug Discov.,* **2002**, *1*, 45-54.

[21] Congreve, M.; Murray, C. W.; Blundell, T. L. Keynote review: Structural biology and drug discovery. *Drug Discov. Today,* **2005**, *10*, 895-907.

[22] Congreve, M.; Chessari, G.; Tisi, D.; Woodhead, A. J. Recent Developments in Fragment-Based Drug Discovery. *J. Med. Chem.,* **2008**, *51*, 3661-3680.

[23] Chessari, G.; Woodhead, A. J. From fragment to clinical candidate—a historical perspective. *Drug Discov. Today,* **2009**, *14*, 668-675.

[24] Schneider, G. Virtual screening: an endless staircase? *Nat. Rev. Drug Discov.,* **2010**, *9*, 273-276.

[25] McInnes, C. Virtual screening strategies in drug discovery. *Curr. Opin. Chem. Biol.,* **2007**, *11*, 494-502.

[26] Ghosh, S.; Nie, A.; An, J.; Huang, Z. Structure-based virtual screening of chemical libraries for drug discovery. *Curr. Opin. Chem. Biol.,* **2006**, *10*, 194-202.

[27] Cavasotto, C. N.; Orry, A. J. W. Ligand docking and structure-based virtual screening in drug discovery. *Curr. Top. Med. Chem.,* **2007**, *7*, 1006-1014.

[28] Bajorath, J. Integration of virtual and high-throughput screening. *Nat. Rev. Drug Discov.,* **2002**, *1*, 882-894.

[29] Reichmann, D.; Rahat, O.; Cohen, M.; Neuvirth, H.; Schreiber, G. The molecular architecture of protein-protein binding sites. *Curr. Opin. Struct. Biol.,* **2007**, *17*, 67-76.

[30] Bonvin, A. M. J. J. Flexible protein-protein docking. *Curr. Opin. Struct. Biol.,* **2006**, *16*, 194-200.

[31] Conte, L. L.; Chothia, C.; Janin, J. The atomic structure of protein-protein recognition sites. *J. Mol. Biol.,* **1999**, *285*, 2177-2198.

[32] Cheng, A. C.; Coleman, R. G.; Smyth, K. T.; Cao, Q.; Soulard, P.; Caffrey, D. R.; Salzberg, A. C.; Huang, E. S. Structure-based maximal affinity model predicts small-molecule druggability. *Nat. Biotech.,* **2007**, *25*, 71-75.

[33] Hopkins, A. L.; Groom, C. R. The druggable genome. *Nat. Rev. Drug Discov.,* **2002**, *1*, 727-730.

[34] Mintseris, J.; Wiehe, K.; Pierce, B.; Anderson, R.; Chen, R.; Janin, J.; Weng, Z. Protein-protein docking benchmark 2.0: An update. *Proteins: Struct. Funct. Bioinf.,* **2005**, *60*, 214-216.

[35] Clackson, T.; Wells, J. A. A hot spot of binding energy in a hormone-receptor interface. *Science,* **1995**, *267*, 383-386.

[36] Keskin, O.; Ma, B.; Nussinov, R. Hot Regions in Protein-Protein Interactions: The Organization and Contribution of Structurally Conserved Hot Spot Residues. *J. Mol. Biol.,* **2005**, *345*, 1281-1294.

[37] Grosdidier, S.; Fernández-Recio, J. Docking and scoring: applications to drug discovery in the interactomics era. *Expert Opin. Drug Discov.,* **2009**, *4*, 673-686.

[38] Tuncbag, N.; Kar, G.; Keskin, O.; Gursoy, A.; Nussinov, R. A survey of available tools and web servers for analysis of protein-protein interactions and interfaces. *Brief. Bioinform.,* **2009**, *10*, 217-232.

[39] Cho, K.-i.; Kim, D.; Lee, D. A feature-based approach to modeling protein-protein interaction hot spots. *Nucleic Acids Res.,* **2009**, *37*, 2672-2687.

[40] Geppert, T.; Hoy, B.; Wessler, S.; Schneider, G. Context-Based Identification of Protein-Protein Interfaces and "Hot-Spot" Residues. *Chem. Biol.,* **2011**, *18*, 344-353.

[41] Kufareva, I.; Budagyan, L.; Raush, E.; Totrov, M.; Abagyan, R. PIER: Protein interface recognition for structural proteomics. *Proteins: Struct. Funct. Bioinf.,* **2007**, *67*, 400-417.

[42] Zhu, X.; Mitchell, J. C. KFC2: A knowledge-based hot spot prediction method based on interface solvation, atomic density, and plasticity features. *Proteins: Struct. Funct. Bioinf.,* **2011**, *79*, 2671-2683.

[43] Tuncbag, N.; Keskin, O.; Gursoy, A. HotPoint: hot spot prediction server for protein interfaces. *Nucleic Acids Res.,* **2010**, *38*, W402-W406.

[44] Tuncbag, N.; Gursoy, A.; Keskin, O. Identification of computational hot spots in protein interfaces: combining solvent accessibility and inter-residue potentials improves the accuracy. *Bioinformatics,* **2009**, *25*, 1513-1520.

[45] Lise, S.; Buchan, D.; Pontil, M.; Jones, D. T. Predictions of Hot Spot Residues at Protein-Protein Interfaces Using Support Vector Machines. *PLoS ONE,* **2011**, *6*, e16774.

[46] Xia, J.-F.; Zhao, X.-M.; Song, J.; Huang, D.-S. APIS: accurate prediction of hot spots in protein interfaces by combining protrusion index with solvent accessibility. *BMC Bioinformatics,* **2010**, *11*, 174.

[47] Kitchen, D. B.; Decornez, H.; Furr, J. R.; Bajorath, J. Docking and scoring in virtual screening for drug discovery: methods and applications. *Nat. Rev. Drug Discov.,* **2004**, *3*, 935-949.

[48] Lee, H.-M.; Chan, D. S.-H.; Yang, F.; Lam, H.-Y.; Yan, S.-C.; Che, C.-M.; Ma, D.-L.; Leung, C.-H. Identification of natural product Fonsecin B as a stabilizing ligand of c-myc G-quadruplex DNA by high-throughput virtual screening. *Chem. Commun.,* **2010**, *46*, 4680-4682.

[49] Chan, D. S.-H.; Yang, H.; Kwan, M. H.-T.; Cheng, Z.; Lee, P.; Bai, L.-P.; Jiang, Z.-H.; Wong, C.-Y.; Fong, W.-F.; Leung, C.-H.; Ma, D.-L. Structure-based optimization of FDA-approved drug methylene blue as a c-myc G-quadruplex DNA stabilizer. *Biochimie,* **2011**, *93*, 1055-1064.

[50] Leung, C.-H.; Chan, D. S.-H.; Yang, H.; Abagyan, R.; Lee, S. M.-Y.; Zhu, G.-Y.; Fong, W.-F.; Ma, D.-L. A natural product-like inhibitor of NEDD8-activating enzyme. *Chem. Commun.,* **2011**, *47*, 2511-2513.

[51] Ma, D.-L.; Chan, D. S.-H.; Lee, P.; Kwan, M. H.-T.; Leung, C.-H. Molecular modeling of drug-DNA interactions: Virtual screening to structure-based design. *Biochimie,* **2011**, *93*, 1252-1266.

[52] Ma, D.-L.; Chan, D. S.-H.; Leung, C.-H. Molecular docking for virtual screening of natural product databases. *Chem. Sci.,* **2011**, *2*, 1656-1665.

[53] Ma, D.-L.; Ma, V. P.-Y.; Chan, D. S.-H.; Leung, K.-H.; Zhong, H.-J.; Leung, C.-H. In silico screening of quadruplex-binding ligands. *Methods,* **2012**, *57*, 106-114.

[54] Zhong, H.-J.; Ma, V. P.-Y.; Chan, D. S.-H.; He, H.-Z.; Leung, K.-H.; Ma, D.-L.; Leung, C.-H. Discovery of a natural product inhibitor targeting protein neddylation by structure-based virtual screening. *Biochimie,* **2012**, *94*, 2457-2460.

[55] Leung, C.-H.; Chan, D. S.-H.; Li, Y.-W.; Fong, W.-F.; Ma, D.-L. Hit identification of IKKβ natural product inhibitor. *BMC Pharmacol. Toxicol.,* **2013**, *14*, 3.

[56] Ma, D.-L.; Chan, D. S.-H.; Leung, C.-H. Drug repositioning by structure-based virtual screening. *Chem. Soc. Rev.,* **2013**, *42*, 2130-2141.

[57] Yang, H.; Zhong, H.-J.; Leung, K.-H.; Chan, D. S.-H.; Ma, V. P.-Y.; Fu, W.-C.; Nanjunda, R.; Wilson, W. D.; Ma, D.-L.; Leung, C.-H. Structure-based design of flavone derivatives as c-myc oncogene down-regulators. *Eur. J. Pharm. Sci.,* **2013**, *48*, 130-141.

[58] Liu, L.-J.; Leung, K.-H.; Chan, D. S.-H, Wang, Y.-T.; Ma, D.-L, Leung, C.-H. Identification of a natural product-like STAT3 dimerization inhibitor by structure-based virtual screening. *Cell Death Dis.,* **2014**, *5*, e1293.

[59] Zhong, H.-J.; Liu, L.-J.; Chan, D. S.-H.; Wang, H.-M.; Chan, P. W. H.; Ma, D.-L.; Leung, C.-H. Structure-based repurposing of FDA-approved drugs as inhibitors of NEDD8-activating enzyme. *Biochimie,* **2014**, *102*, 211-215.

[60] Zhong, H.-J.; Liu, L.-J.; Chong, C.-M.; Lu, L.; Wang, M.; Chan, D. S.-H.; Chan, P. W. H.; Lee, S. M.-Y.; Ma, D.-L.; Leung, C.-H. Discovery of a Natural Product-Like iNOS Inhibitor by Molecular Docking with Potential Neuroprotective Effects *In Vivo*. *PLoS ONE,* **2014**, *9*, e92905.

[61] Leung, K.-H.; Liu, L.-J.; Lin, S.; Lu, L.; Zhong, H.-J.; Susanti, D.; Rao, W.; Wang, M.; Che, W. I.; Chan, D. S.-H.; Leung, C.-H.; Chan, P. W. H.; Ma, D.-L. Discovery of a small-molecule inhibitor of STAT3 by ligand-based pharmacophore screening. *Methods*, **2014**, DOI: 10.1016/j.ymeth.2014.07.010.

[62] Zhong, H.-J.; Lin, S.; Tam, I. L.; Lu, L.; Chan, D. S.-H.; Ma, D.-L.; Leung, C.-H. In silico identification of natural product inhibitors of JAK2. *Methods*, **2014**, DOI: 10.1016/j.ymeth.2014.07.003.

[63] Yang, S.-Y. Pharmacophore modeling and applications in drug discovery: challenges and recent advances. *Drug Discov. Today,* **2010**, *15*, 444-450.

[64] Recent advances in pharmacophore modeling and its application to anti-influenza drug discovery. *Expert Opin. Drug Discov.,* **2013**, *8*, 411-426.

[65] Leach, A. R.; Gillet, V. J.; Lewis, R. A.; Taylor, R. Three-Dimensional Pharmacophore Methods in Drug Discovery. *J. Med. Chem.,* **2009**, *53*, 539-558.

[66] Wolber, G.; Seidel, T.; Bendix, F.; Langer, T. Molecule-pharmacophore superpositioning and pattern matching in computational drug design. *Drug Discov. Today,* **2008**, *13*, 23-29.

[67] Van, d. W. H.; Carter, R. E.; Grassy, G.; Kubinyi, H.; Martin, Y. C.; Tute, M. S.; Willett, P. Glossary of terms used in computational drug design. *Pure Appl. Chem.,* **1997**, *69*, 1137-1152.

[68] Li, H.; Sutter, J.; Hoffmann, R. HypoGen: an automated system for generating 3D predictive pharmacophore models. *Pharmacophore Perception, Development, and Use in Drug Design,* **2000**, *2*, 171.

[69] Barnum, D.; Greene, J.; Smellie, A.; Sprague, P. Identification of Common Functional Configurations Among Molecules. *J. Chem. Inf. Comput. Sci.,* **1996**, *36*, 563-71.

[70] Dixon, S.; Smondyrev, A.; Knoll, E.; Rao, S.; Shaw, D.; Friesner, R. PHASE: a new engine for pharmacophore perception, 3D QSAR model development, and 3D database screening: 1. Methodology and preliminary results. *J. Comput. Aided Mol. Des.,* **2006**, *20*, 647-671.

[71] Martin, Y. C. DISCO: what we did right and what we missed. *Pharmacophore Perception, Development, and Use in Drug Design. La Jolla, CA, International University Line,* **2000**, 51-66.

[72] Geppert, H.; Vogt, M.; Bajorath, J. Current Trends in Ligand-Based Virtual Screening: Molecular Representations, Data Mining Methods, New Application Areas, and Performance Evaluation. *J. Chem. Inf. Model.,* **2010**, *50*, 205-216.

[73] Melville, J. L.; Burke, E. K.; Hirst, J. D. Machine Learning in Virtual Screening. *Comb. Chem. High Throughput Screen.,* **2009**, *12*, 332-343.

[74] Ma, X. H.; Jia, J.; Zhu, F.; Xue, Y.; Li, Z. R.; Chen, Y. Z. Comparative Analysis of Machine Learning Methods in Ligand-Based Virtual Screening of Large Compound Libraries. *Comb. Chem. High Throughput Screen.,* **2009**, *12*, 344-357.

[75] Chen, B.; Harrison, R.; Papadatos, G.; Willett, P.; Wood, D.; Lewell, X.; Greenidge, P.; Stiefl, N. Evaluation of machine-learning methods for ligand-based virtual screening. *J. Comput. Aided Mol. Des.,* **2007**, *21*, 53-62.

[76] Willett, P.; Similarity Searching Using 2D Structural Fingerprints. In *Chemoinformatics and Computational Chemical Biology*, Bajorath, J.; Ed. Humana Press: 2011, Vol. 672, pp 133-158.

[77] Willett, P. Similarity-based virtual screening using 2D fingerprints. *Drug Discov. Today,* **2006**, *11*, 1046-1053.

[78] Eckert, H.; Bajorath, J. Molecular similarity analysis in virtual screening: foundations, limitations and novel approaches. *Drug Discov. Today,* **2007**, *12*, 225-233.

[79] Löwer, M.; Proschak, E. Structure-Based Pharmacophores for Virtual Screening. *Mol. Inform.,* **2011**, *30*, 398-404.

[80] Cheng, T.; Li, Q.; Zhou, Z.; Wang, Y.; Bryant, S. Structure-Based Virtual Screening for Drug Discovery: a Problem-Centric Review. *AAPS J.,* **2012**, *14*, 133-141.

[81] Waszkowycz, B.; Clark, D. E.; Gancia, E. Outstanding challenges in protein-ligand docking and structure-based virtual screening. *Wiley Interdiscip. Rev. Comput. Mol. Sci.,* **2011**, *1*, 229-259.

[82] Wolber, G.; Langer, T. LigandScout: 3-D Pharmacophores Derived from Protein-Bound Ligands and Their Use as Virtual Screening Filters. *J. Chem. Inf. Model.,* **2004**, *45*, 160-169.

[83] Lloyd, D. G.; Buenemann, C. L.; Todorov, N. P.; Manallack, D. T.; Dean, P. M. Scaffold Hopping in De Novo Design. Ligand Generation in the Absence of Receptor Information. *J. Med. Chem.,* **2003**, *47*, 493-496.

[84] Feher, M.; Gao, Y.; Baber, J. C.; Shirley, W. A.; Saunders, J. The use of ligand-based *de novo* design for scaffold hopping and sidechain optimization: Two case studies. *Bioorg. Med. Chem.,* **2008**, *16*, 422-427.

[85] Schneider, G.; Schneider, P.; Renner, S. Scaffold-Hopping: How Far Can You Jump? *QSAR Comb. Sci.,* **2006**, *25*, 1162-1171.

[86] Gardiner, E. J.; Holliday, J. D.; O'Dowd, C.; Willett, P. Effectiveness of 2D fingerprints for scaffold hopping. *Future Med. Chem.,* **2011**, *3*, 405-414.

[87] Vogt, M.; Stumpfe, D.; Geppert, H.; Bajorath, J. r. Scaffold Hopping Using Two-Dimensional Fingerprints: True Potential, Black Magic, or a Hopeless Endeavor? Guidelines for Virtual Screening. *J. Med. Chem.,* **2010**, *53*, 5707-5715.

[88] Kanehisa, M.; Goto, S.; Furumichi, M.; Tanabe, M.; Hirakawa, M. KEGG for representation and analysis of molecular networks involving diseases and drugs. *Nucleic Acids Res.,* **2010**, *38*, D355-D360.

[89] Goede, A.; Dunkel, M.; Mester, N.; Frommel, C.; Preissner, R. SuperDrug: a conformational drug database. *Bioinformatics,* **2005**, *21*, 1751-1753.

[90] Lipinski, C. A. Drug-like properties and the causes of poor solubility and poor permeability. *J. Pharmacol. Toxicol. Methods,* **2000**, *44*, 235-249.

[91] Muegge, I. Selection criteria for drug-like compounds. *Med. Res. Rev.,* **2003**, *23*, 302-321.

[92] Rishton, G. M. Nonleadlikeness and leadlikeness in biochemical screening. *Drug Discov. Today,* **2003**, *8*, 86-96.

[93] Roche, O.; Schneider, P.; Zuegge, J.; Guba, W.; Kansy, M.; Alanine, A.; Bleicher, K.; Danel, F.; Gutknecht, E.-M.; Rogers-Evans, M.; Neidhart, W.; Stalder, H.; Dillon, M.; Sjögren, E.; Fotouhi, N.; Gillespie, P.; Goodnow, R.; Harris, W.; Jones, P.; Taniguchi, M.; Tsujii, S.; von der Saal, W.; Zimmermann, G.; Schneider, G. Development of a Virtual Screening Method for Identification of "Frequent Hitters" in Compound Libraries. *J. Med. Chem.,* **2001**, *45*, 137-142.

[94] Good, A. C.; Hermsmeier, M. A. Measuring CAMD Technique Performance. 2. How "Druglike" Are Drugs? Implications of Random Test Set Selection Exemplified Using Druglikeness Classification Models. *J. Chem. Inf. Model.,* **2006**, *47*, 110-114.

[95] Garcia-Sosa, A. T.; Maran, U.; Hetenyi, C. Molecular property filters describing pharmacokinetics and drug binding. *Curr. Med. Chem.,* **2012**, *19*, 1646-1662.

[96] ten Brink, T.; Exner, T. E. Influence of Protonation, Tautomeric, and Stereoisomeric States on Protein−Ligand Docking Results. *J. Chem. Inf. Model.,* **2009**, *49*, 1535-1546.

[97] Gasteiger, J.; Rudolph, C.; Sadowski, J. Automatic generation of 3D-atomic coordinates for organic molecules. *Tetrahedron. Comput. Meth.,* **1990**, *3*, 537-547.

[98] Abagyan, R.; Totrov, M.; Kuznetsov, D. ICM—A new method for protein modeling and design: Applications to docking and structure prediction from the distorted native conformation. *J. Comput. Chem.,* **1994**, *15*, 488-506.

[99] Friesner, R. A.; Banks, J. L.; Murphy, R. B.; Halgren, T. A.; Klicic, J. J.; Mainz, D. T.; Repasky, M. P.; Knoll, E. H.; Shelley, M.; Perry, J. K.; Shaw, D. E.; Francis, P.; Shenkin, P. S. Glide: A New Approach for Rapid, Accurate Docking and Scoring. 1. Method and Assessment of Docking Accuracy. *J. Med. Chem.,* **2004**, *47*, 1739-1749.

[100] Trott, O.; Olson, A. J. AutoDock Vina: Improving the speed and accuracy of docking with a new scoring function, efficient optimization, and multithreading. *J. Comput. Chem.,* **2010**, *31*, 455-461.

[101] Goodsell, D. S.; Morris, G. M.; Olson, A. J. Automated docking of flexible ligands: application of AutoDock. *J. Mol. Recognit.,* **1996**, *9*, 1-5.

[102] Kramer, B.; Rarey, M.; Lengauer, T. Evaluation of the FLEXX incremental construction algorithm for protein-ligand docking. *Proteins: Struct. Funct. Bioinf.,* **1999**, *37*, 228-241.

[103] McGovern, S. L.; Shoichet, B. K. Information Decay in Molecular Docking Screens against Holo, Apo, and Modeled Conformations of Enzymes. *J. Med. Chem.,* **2003**, *46*, 2895-2907.

[104] Ferrara, P.; Jacoby, E. Evaluation of the utility of homology models in high throughput docking. *J. Mol. Model.,* **2007**, *13*, 897-905.

[105] Berman, H. M.; Westbrook, J.; Feng, Z.; Gilliland, G.; Bhat, T. N.; Weissig, H.; Shindyalov, I. N.; Bourne, P. E. The Protein Data Bank. *Nucleic Acids Res.,* **2000**, *28*, 235-242.

[106] Kearsley, S.; Underwood, D.; Sheridan, R.; Miller, M. Flexibases: A way to enhance the use of molecular docking methods. *J. Comput. Aided Mol. Des.,* **1994**, *8*, 565-582.

[107] Hart, T. N.; Read, R. J. A multiple-start Monte Carlo docking method. *Proteins: Struct. Funct. Bioinf.,* **1992**, *13*, 206-222.

[108] Morris, G. M.; Goodsell, D. S.; Halliday, R. S.; Huey, R.; Hart, W. E.; Belew, R. K.; Olson, A. J. Automated docking using a Lamarckian genetic algorithm and an empirical binding free energy function. *J. Comput. Chem.,* **1998**, *19*, 1639-1662.

[109] Totrov, M.; Abagyan, R. Flexible ligand docking to multiple receptor conformations: a practical alternative. *Curr. Opin. Struct. Biol.,* **2008**, *18*, 178-184.

[110] Kokh, D. B.; Wenzel, W. Flexible Side Chain Models Improve Enrichment Rates in *In Silico* Screening. *J. Med. Chem.,* **2008**, *51*, 5919-5931.

[111] Spyrakis, F.; BidonChanal, A.; Barril, X.; Javier Luque, F. Protein Flexibility and Ligand Recognition: Challenges for Molecular Modeling *Curr. Top. Med. Chem.,* **2011**, *11*, 192-210.

[112] Gohlke, H.; Klebe, G. Approaches to the Description and Prediction of the Binding Affinity of Small-Molecule Ligands to Macromolecular Receptors. *Angew. Chem. Int. Ed.,* **2002**, *41*, 2644-2676.

[113] Feher, M. Consensus scoring for protein-ligand interactions. *Drug Discov. Today,* **2006**, *11*, 421-428.

[114] Jubb, H.; Higueruelo, A. P.; Winter, A.; Blundell, T. L. Structural biology and drug discovery for protein protein interactions. *Trends Pharmacol. Sci.,* **2012**, *33*, 241-248.

[115] Bourgeas, R.; Basse, M.-J.; Morelli, X.; Roche, P. Atomic Analysis of Protein-Protein Interfaces with Known Inhibitors: The 2P2I Database. *PLoS ONE,* **2010**, *5*, e9598.

[116] Morelli, X.; Bourgeas, R.; Roche, P. Chemical and structural lessons from recent successes in protein-protein interaction inhibition (2P2I). *Curr. Opin. Chem. Biol.,* **2011**, *15*, 475-481.

[117] Lipinski, C. A.; Lombardo, F.; Dominy, B. W.; Feeney, P. J. Experimental and computational approaches to estimate solubility and permeability in drug discovery and development settings. *Adv. Drug Deliv. Rev.,* **2001**, *46*, 3-26.

[118] Reynès, C.; Host, H.; Camproux, A.-C.; Laconde, G.; Leroux, F.; Mazars, A.; Deprez, B.; Fahraeus, R.; Villoutreix, B. O.; Sperandio, O. Designing Focused Chemical Libraries Enriched in Protein-Protein Interaction Inhibitors using Machine-Learning Methods. *PLoS Comput. Biol.,* **2010**, *6*, e1000695.

[119] Vassilev, L. T.; Vu, B. T.; Graves, B.; Carvajal, D.; Podlaski, F.; Filipovic, Z.; Kong, N.; Kammlott, U.; Lukacs, C.; Klein, C.; Fotouhi, N.; Liu, E. A. *In Vivo* Activation of the p53 Pathway by Small-Molecule Antagonists of MDM2. *Science,* **2004**, *303*, 844-848.

[120] Tovar, C.; Graves, B.; Packman, K.; Filipovic, Z.; Xia, B. H. M.; Tardell, C.; Garrido, R.; Lee, E.; Kolinsky, K.; To, K.-H.; Linn, M.; Podlaski, F.; Wovkulich, P.; Vu, B.; Vassilev, L. T. MDM2 Small-Molecule Antagonist RG7112 Activates p53 Signaling and Regresses Human Tumors in Preclinical Cancer Models. *Cancer Res.,* **2013**, *73*, 2587-2597.

[121] Fry, D. C.; Wartchow, C.; Graves, B.; Janson, C.; Lukacs, C.; Kammlott, U.; Belunis, C.; Palme, S.; Klein, C.; Vu, B. Deconstruction of a Nutlin: Dissecting the Binding Determinants of a Potent Protein-Protein Interaction Inhibitor. *ACS Med. Chem. Lett.,* **2013**, *4*, 660-665.

[122] Cukuroglu, E.; Gursoy, A.; Keskin, O. Analysis of hot region organization in hub proteins. *Ann. Biomed. Eng.,* **2010**, *38*, 2068-78.

[123] Cukuroglu, E.; Gursoy, A.; Keskin, O. HotRegion: a database of predicted hot spot clusters. *Nucleic Acids Res.,* **2012**, *40*, D829-D833.

[124] Eyrisch, S.; Helms, V. What induces pocket openings on protein surface patches involved in protein-protein interactions? *J. Comput. Aided Mol. Des.,* **2009**, *23*, 73-86.

[125] Metz, A.; Pfleger, C.; Kopitz, H.; Pfeiffer-Marek, S.; Baringhaus, K.-H.; Gohlke, H. Hot Spots and Transient Pockets: Predicting the Determinants of Small-Molecule Binding to a Protein-Protein Interface. *J. Chem. Inf. Model.,* **2012**, *52*, 120-133.

[126] Metz, A.; Schanda, J.; Grez, M.; Wichmann, C.; Gohlke, H. From Determinants of RUNX1/ETO Tetramerization to Small-Molecule Protein-Protein Interaction Inhibitors Targeting Acute Myeloid Leukemia. *J. Chem. Inf. Model.,* **2013**, *53*, 2197-2202.

[127] Poltorak, A.; Ricciardi-Castagnoli, P.; Citterio, S.; Beutler, B. Physical contact between lipopolysaccharide and Toll-like receptor 4 revealed by genetic complementation. *Proc. Natl. Acad. Sci. U. S. A.,* **2000**, *97*, 2163-2167.

[128] Oltersdorf, T.; Elmore, S. W.; Shoemaker, A. R.; Armstrong, R. C.; Augeri, D. J.; Belli, B. A.; Bruncko, M.; Deckwerth, T. L.; Dinges, J.; Hajduk, P. J.; Joseph, M. K.; Kitada, S.; Korsmeyer, S. J.; Kunzer, A. R.; Letai, A.; Li, C.; Mitten, M. J.; Nettesheim, D. G.; Ng, S.; Nimmer, P. M.; O'Connor, J. M.; Oleksijew, A.; Petros, A. M.; Reed, J. C.; Shen, W.; Tahir, S. K.; Thompson, C. B.; Tomaselli, K. J.; Wang, B.; Wendt, M. D.; Zhang, H.; Fesik, S. W.; Rosenberg, S. H. An inhibitor of Bcl-2 family proteins induces regression of solid tumours. *Nature,* **2005**, *435*, 677-681.

[129] Pinto, M.; del Mar Orzaez, M.; Delgado-Soler, L.; Perez, J. J.; Rubio-Martinez, J. Rational Design of New Class of BH3-Mimetics As Inhibitors of the Bcl-xL Protein. *J. Chem. Inf. Model.,* **2011**, *51*, 1249-1258.

[130] Mukherjee, P.; Desai, P.; Zhou, Y.-D.; Avery, M. Targeting the BH3 Domain Mediated Protein–Protein Interaction of Bcl-xL through Virtual Screening. *J. Chem. Inf. Model.,* **2010**, *50*, 906-923.

[131] Lawrence, H. R.; Li, Z.; Richard Yip, M. L.; Sung, S.-S.; Lawrence, N. J.; McLaughlin, M. L.; McManus, G. J.; Zaworotko, M. J.; Sebti, S. M.; Chen, J.; Guida, W. C. Identification of a disruptor of the MDM2-p53 protein-protein interaction facilitated by high-throughput *in silico* docking. *Bioorg. Med. Chem. Lett.,* **2009**, *19*, 3756-3759.

[132] Dudkina, A. S.; Lindsley, C. W. Small Molecule Protein-Protein Inhibitors for the p53-MDM2 Interaction *Curr. Top. Med. Chem.,* **2007**, *7*, 952-960.

[133] Schimmer, A. D.; Dalili, S.; Batey, R. A.; Riedl, S. J. Targeting XIAP for the treatment of malignancy. *Cell Death Differ.,* **2005**, *13*, 179-188.

[134] Rajapakse, H. A. Small Molecule Inhibitors of the XIAP Protein-Protein Interaction *Curr. Top. Med. Chem.,* **2007**, *7*, 966-971.

[135] Arkin, M. R.; Wells, J. A. Small-molecule inhibitors of protein-protein interactions: progressing towards the dream. *Nat. Rev. Drug Discov.,* **2004**, *3*, 301-317.

[136] Thanos, C. D.; DeLano, W. L.; Wells, J. A. Hot-spot mimicry of a cytokine receptor by a small molecule. *Proc. Natl. Acad. Sci. U. S. A.,* **2006**, *103*, 15422-15427.

[137] Beutler, B.; Cerami, A. The Biology of Cachectin/TNF -- A Primary Mediator of the Host Response. *Annu. Rev. Immunol.,* **1989**, *7*, 625-655.

[138] Chan, D. S.-H.; Lee, H.-M.; Yang, F.; Che, C.-M.; Wong, C. C. L.; Abagyan, R.; Leung, C.-H.; Ma, D.-L. Structure-Based Discovery of Natural-Product-like TNF-α Inhibitors. *Angew. Chem. Int. Ed.,* **2010**, *49*, 2860-2864.

[139] He, M. M.; Smith, A. S.; Oslob, J. D.; Flanagan, W. M.; Braisted, A. C.; Whitty, A.; Cancilla, M. T.; Wang, J.; Lugovskoy, A. A.; Yoburn, J. C.; Fung, A. D.; Farrington, G.; Eldredge, J. K.; Day, E. S.; Cruz, L. A.; Cachero, T. G.; Miller, S. K.; Friedman, J. E.; Choong, I. C.; Cunningham, B. C. Small-Molecule Inhibition of TNF-α. *Science,* **2005**, *310*, 1022-1025.

[140] He, M. M.; Smith, A. S.; Oslob, J. D.; Flanagan, W. M.; Braisted, A. C.; Whitty, A.; Cancilla, M. T.; Wang, J.; Lugovskoy, A. A.; Yoburn, J. C.; Fung, A. D.; Farrington, G.; Eldredge, J. K.; Day, E. S.; Cruz, L. A.; Cachero, T. G.; Miller, S. K.; Friedman, J. E.; Choong, I. C.; Cunningham, B. C. Small-Molecule Inhibition of TNF-α. *Science (Washington, DC, U. S.),* **2005**, *310*, 1022-1025.

[141] Leung, C.-H.; Chan, D. S.-H.; Kwan, M. H.-T.; Cheng, Z.; Wong, C.-Y.; Zhu, G.-Y.; Fong, W.-F.; Ma, D.-L. Structure-Based Repurposing of FDA-Approved Drugs as TNF-α Inhibitors. *ChemMedChem,* **2011**, *6*, 765-768.

[142] Murray, P. J.; Smale, S. T. Restraint of inflammatory signaling by interdependent strata of negative regulatory pathways. *Nat. Immunol.,* **2012**, *13*, 916-924.

[143] Hennessy, E. J.; Parker, A. E.; O'Neill, L. A. J. Targeting Toll-like receptors: emerging therapeutics? *Nat. Rev. Drug Discov.,* **2010**, *9*, 293-307.

[144] Zak, D. E.; Schmitz, F.; Gold, E. S.; Diercks, A. H.; Peschon, J. J.; Valvo, J. S.; Niemistö, A.; Podolsky, I.; Fallen, S. G.; Suen, R.; Stolyar, T.; Johnson, C. D.; Kennedy, K. A.; Hamilton, M. K.; Siggs, O. M.; Beutler, B.; Aderem, A. Systems analysis identifies an essential role for SHANK-associated RH domain-interacting protein (SHARPIN) in macrophage Toll-like receptor 2 (TLR2) responses. *Proc. Natl. Acad. Sci. U. S. A.,* **2011**, *108*, 11536-11541.

[145] Kawai, T.; Akira, S. The role of pattern-recognition receptors in innate immunity: update on Toll-like receptors. *Nat. Immunol.,* **2010**, *11*, 373-384.

[146] O'Neill, L. A. J.; Bryant, C. E.; Doyle, S. L. Therapeutic Targeting of Toll-Like Receptors for Infectious and Inflammatory Diseases and Cancer. *Pharmacol. Rev.,* **2009**, *61*, 177-197.

[147] Joce, C.; Stahl, J. A.; Shridhar, M.; Hutchinson, M. R.; Watkins, L. R.; Fedichev, P. O.; Yin, H. Application of a novel *in silico* high-throughput screen to identify selective inhibitors for protein-protein interactions. *Bioorg. Med. Chem. Lett.,* **2010**, *20*, 5411-5413.

[148] Park, B. S.; Song, D. H.; Kim, H. M.; Choi, B.-S.; Lee, H.; Lee, J.-O. The structural basis of lipopolysaccharide recognition by the TLR4-MD-2 complex. *Nature,* **2009**, *458*, 1191-1195.

[149] Medzhitov, R.; Janeway, J. C. The Toll receptor family and microbial recognition. *Trends Microbiol.,* **2000**, *8*, 452-456.

[150] Shimazu, R.; Akashi, S.; Ogata, H.; Nagai, Y.; Fukudome, K.; Miyake, K.; Kimoto, M. MD-2, a Molecule that Confers Lipopolysaccharide Responsiveness on Toll-like Receptor 4. *J. Exp. Med.,* **1999**, *189*, 1777-1782.

[151] Åqvist, J.; Luzhkov, V. B.; Brandsdal, B. O. Ligand Binding Affinities from MD Simulations. *Acc. Chem. Res.,* **2002**, *35*, 358-365.

[152] Ferrara, P.; Curioni, A.; Vangrevelinghe, E.; Meyer, T.; Mordasini, T.; Andreoni, W.; Acklin, P.; Jacoby, E. New Scoring Functions for Virtual Screening from Molecular Dynamics Simulations with a Quantum-Refined Force-Field (QRFF-MD). Application to Cyclin-Dependent Kinase 2. *J. Chem. Inf. Model.,* **2005**, *46*, 254-263.

[153] Yuan, Z.-l.; Guan, Y.-j.; Wang, L.; Wei, W.; Kane, A. B.; Chin, Y. E. Central Role of the Threonine Residue within the p+1 Loop of Receptor Tyrosine Kinase in STAT3 Constitutive Phosphorylation in Metastatic Cancer Cells. *Mol. Cell. Biol.,* **2004**, *24*, 9390-9400.

[154] Matsuno, K.; Masuda, Y.; Uehara, Y.; Sato, H.; Muroya, A.; Takahashi, O.; Yokotagawa, T.; Furuya, T.; Okawara, T.; Otsuka, M.; Ogo, N.; Ashizawa, T.; Oshita, C.; Tai, S.; Ishii, H.; Akiyama, Y.; Asai, A. Identification of a New Series of STAT3 Inhibitors by Virtual Screening. *ACS Med. Chem. Lett.,* **2010**, *1*, 371-375.

[155] Okamoto, M.; Takayama, K.; Shimizu, T.; Ishida, K.; Takahashi, O.; Furuya, T. Identification of Death-Associated Protein Kinases Inhibitors Using Structure-Based Virtual Screening. *J. Med. Chem.,* **2009**, *52*, 7323-7327.

[156] Ewing, T. J. A.; Kuntz, I. D. Critical evaluation of search algorithms for automated molecular docking and database screening. *J. Comput. Chem.,* **1997**, *18*, 1175-1189.

[157] Becker, S.; Groner, B.; Muller, C. W. Three-dimensional structure of the Stat3[beta] homodimer bound to DNA. *Nature,* **1998**, *394*, 145-151.

[158] Li, H.; Liu, A.; Zhao, Z.; Xu, Y.; Lin, J.; Jou, D.; Li, C. Fragment-Based Drug Design and Drug Repositioning Using Multiple Ligand Simultaneous Docking (MLSD): Identifying Celecoxib and Template Compounds as Novel Inhibitors of Signal Transducer and Activator of Transcription 3 (STAT3). *J. Med. Chem.,* **2011**, *54*, 5592-5596.

[159] Malmgaard, L. Induction and Regulation of IFNs During Viral Infections *J. Interferon Cytokine Res.,* **2004**, *24*, 439-454.

[160] Mogensen, K. E.; Lewerenz, M.; Reboul, J.; Lutfalla, G.; Uze, G. The Type I Interferon Receptor: Structure, Function, and Evolution of a Family Business *J. Interferon Cytokine Res.,* **2004**, *19*, 1069-1098.

[161] Swiecki, M.; Colonna, M. Unraveling the functions of plasmacytoid dendritic cells during viral infections, autoimmunity, and tolerance. *Immunol. Rev.,* **2010**, *234*, 142-162.

[162] Geppert, T.; Bauer, S.; Hiss, J. A.; Conrad, E.; Reutlinger, M.; Schneider, P.; Weisel, M.; Pfeiffer, B.; Altmann, K.-H.; Waibler, Z.; Schneider, G. Immunosuppressive Small Molecule Discovered by Structure-Based Virtual Screening for Inhibitors of Protein-Protein Interactions. *Angew. Chem. Int. Ed.,* **2012**, *51*, 258-261.

[163] Weisel, M.; Proschak, E.; Schneider, G. PocketPicker: analysis of ligand binding-sites with shape descriptors. *Chem. Cent. J.,* **2007**, *1*, 7.

[164] Klenner, A.; Hartenfeller, M.; Schneider, P.; Schneider, G. 'Fuzziness' in pharmacophore-based virtual screening and *de novo* design. *Drug Discov. Today Technol.,* **2010**, *7*, e237-e244.

[165] Jones, G.; Willett, P.; Glen, R. C. Molecular recognition of receptor sites using a genetic algorithm with a description of desolvation. *J. Mol. Biol.,* **1995**, *245*, 43-53.

[166] Gorczynski, M. J.; Grembecka, J.; Zhou, Y.; Kong, Y.; Roudaia, L.; Douvas, M. G.; Newman, M.; Bielnicka, I.; Baber, G.; Corpora, T.; Shi, J.; Sridharan, M.; Lilien, R.; Donald, B. R.; Speck, N. A.; Brown, M. L.; Bushweller, J. H. Allosteric Inhibition of the Protein-Protein Interaction between the Leukemia-Associated Proteins Runx1 and CBFβ. *Chem. Biol.,* **2007**, *14*, 1186-1197.

[167] Last-Barney, K.; Davidson, W.; Cardozo, M.; Frye, L. L.; Grygon, C. A.; Hopkins, J. L.; Jeanfavre, D. D.; Pav, S.; Qian, C.; Stevenson, J. M.; Tong, L.; Zindell, R.; Kelly, T. A. Binding Site Elucidation of Hydantoin-Based Antagonists of LFA-1 Using Multidisciplinary Technologies: Evidence for the Allosteric Inhibition of a Protein−Protein Interaction. *J. Am. Chem. Soc.,* **2001**, *123*, 5643-5650.

CHAPTER 5

Computational Design of Biological Systems: From Systems to Synthetic Biology

Milsee Mol and Shailza Singh*

National Centre for Cell Science, NCCS Complex, Ganeshkhind, Pune University Campus, Pune 411007, India

Abstract: Today biology is overwhelmed with 'big data', amassed from genomic projects carried out in various laboratories around the world using efficient high throughput technologies. Biologists are co-opting mathematical and computational techniques developed to address these data and derive meaningful interpretations. These developments have led to new disciplines: systems and synthetic biology. To explore these two evolving branches of biology one needs to be familiar with technologies such as genomics, bioinformatics and proteomics, mathematical and computational modeling techniques that help predict the dynamic behavior of the biological system, ruling out the trial-and-error methods of traditional genetic engineering. Systems and synthetic biology have developed hand-in-hand towards building artificial biological devices using engineered biological units as basic building blocks. Systems biology is an integrated approach for studying the dynamic and complex behaviors of biological components, which may be difficult to interpret and predict from properties of individual constituents making up the biological systems. While, synthetic biology aims to engineer biologically inspired devices, such as cellular regulatory circuits that do not exist in nature but are designed using well characterized genes, proteins and other biological components in appropriate combinations to perform a desired function. This is analogous to an electronic circuit board design that is fabricated using well characterized electrical components such as resistors, capacitors and so on. The *in silico* abstractions and predictions should be tightly linked to experimentation to be proved *in vitro* and *in vivo* systems for their successful applications in biotechnology. This chapter focuses on mathematical approaches and computational tools available to engineer biological regulatory circuits and how they can be implemented as next generation therapeutics in infectious disease.

Keywords: Abstraction, bioengineering, bioinspired, biological parts, computational modelling, computational tools, constructs, dynamic, infectious disease, interdisciplinary, linearization, mathematical framework, nextgen therapeutics, omics, ordinary differential equations, parameters, physical systems, reactions, regulatory circuits, simulation.

***Corresponding author Shailza Singh:** National Centre for Cell Science, NCCS Complex, Ganeshkhind, Pune University Campus, Pune 411007, India; Tel: +91 20 25708296/95; Fax: +91 20 25692259; E-mails: shailza_iitd@yahoo.com; singhs@nccs.res.in

Zaheer Ul-Haq and Jeffry D. Madura (Eds). **10.1016/B978-1-60805-865-5.50005-8**

INTRODUCTION

Time and again human life has been intimidated by nature's bountiful mode of evolution and adaptation in organisms that cause serious health complications to him and his cattle. Disease outbreaks caused by unknown agents have been abundant resulting in loss of precious lives and with known agents; resistance to the already proven drugs is an emerging threat. There is an urgent need to initiate next-gen alternative approaches to deal with the scenario. So, can we learn from nature? Can we use nature's engineering tool kit to safeguard us? Can we develop solutions that may be the next generation therapeutics?

With thoughtful thinking to understand the design principles of nature engineering, the day is not far away where humans can re-engineer a pathogen to combat a diseased condition. To this end the onus goes to the technological advances in the last 10 years in biological research that has imparted biology the potential to contribute to real-world problems confronting the world. This has been largely possible due to the integration and collaboration with allied disciplines like mathematics, computation, physics and engineering within biology which is evolving as a "New Age Biology" (Fig. **1**). Today, this integrated new biology may find its application to one of the major societal needs *i.e.* the improvement in human health care and management. The merger of allied disciplines was a result of simultaneous but independent developments as outlined below.

SEQUENCING TO BIOINFORMATICS TO COMPUTATIONAL BIOLOGY

The elucidation of DNA double helix by Watson and Crick in the year 1953 sowed the seeds of a new revolution in deciphering the hidden myths about biology. Since then, many researchers have worked together to seek answers to questions related to how genetic information is encoded for protein formation (Gamow, 1956), the factors governing structural properties of protein molecules (Cohen, 1957; Anfinsen, 1973), evolution of genes and proteins (Ingram, 1961), molecular homology (Florkin, 1962), structural constraints of polypeptide chains (Ramachandran, 1963), origins of genetic code (Crick, 1968), gene regulation (Britten and Davidson, 1969) and so on. Somewhere around the same time, computation was developed to understand biological macromolecules better, complimented with the experimental data that gave an insight into their functional and interaction aspects. Very soon genetic code was understood to be universal in all living forms and therefore the genetic evolution of life based on phylogenetic tree construction (Fitch and Margoliash, 1967) and molecular evolution (Kimura, 1968) aligned with properties of protein sequence alignment (Cantor, 1968) was investigated [1].

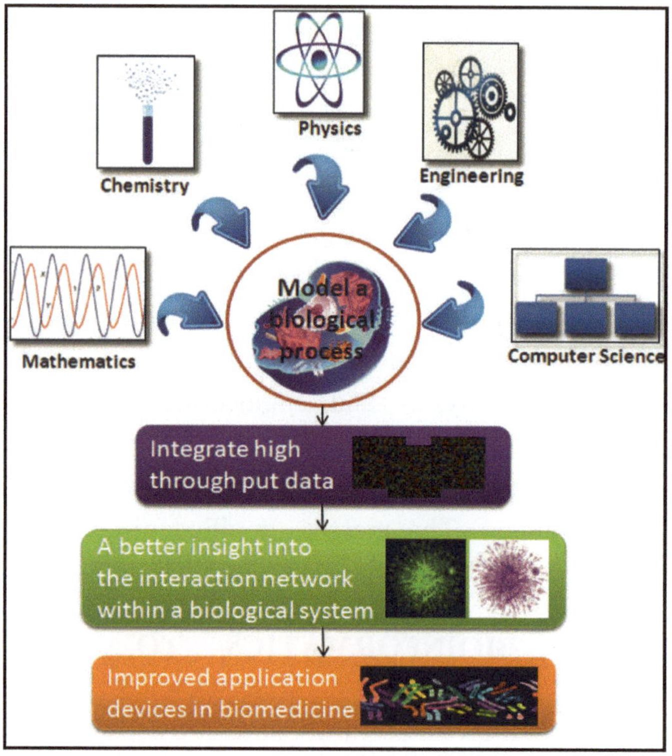

Figure 1: New age biology: An integrated approach.

Sanger (1977) pioneered the task of sequencing short stretches of DNA, which was extended to the whole genome using the Next Generation Sequencing and Microarray technologies. As the number of organisms being studied based on their genetic sequence variation grew, there were technological advances that made collecting gene expression data from the same organism under different conditions and at different time periods possible, giving the way to "the omics

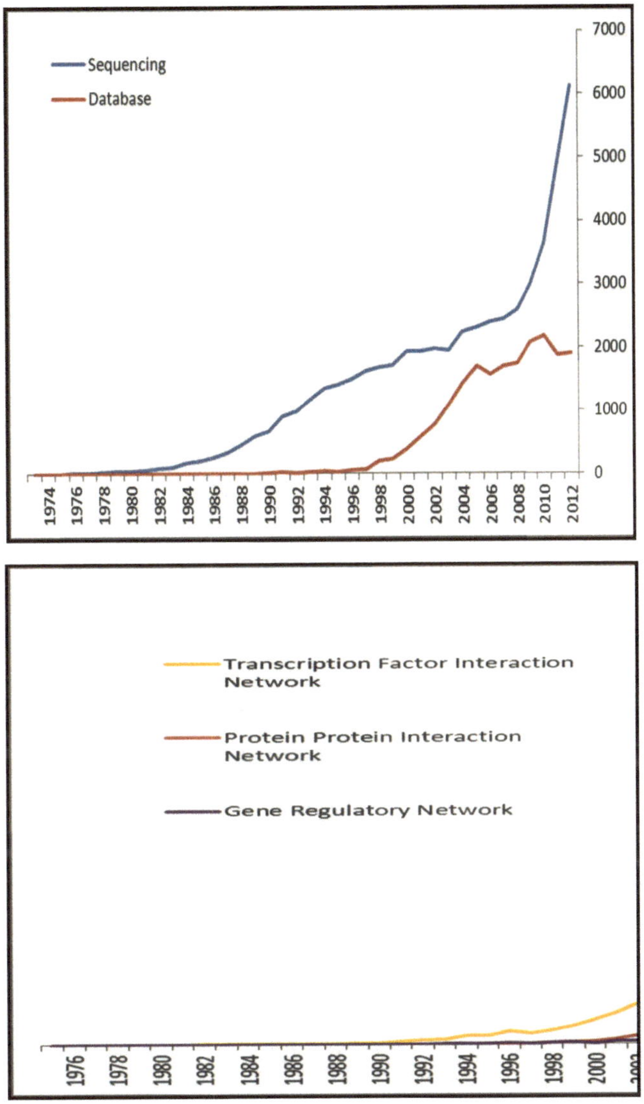

(a)

(b)

Figure 2: (a) Growth of Biological Databases V/s Sequencing; (b) Number of different Interaction Networks (www.ncbi.nlm.nih.gov/pubmed).

data". The massive and parallel data output necessitated management and statistical strategies for biological interpretation. Biological advancements in computation and automation helped data organization into files that could be further stored, processed and retrieved, embarking onto the diverse array of "Bioinformatics". The number of bioinformatics databases (DB) storing information has grown and will continue so exponentially in the coming years as genome sequencing projects expand. As genomes get annotated, novel disease-associated interaction networks, such as protein – protein interaction, transcription factor (TF), and gene regulatory networks grow. (Fig. **2a** and **b**). These databases are expansive repositories of biological data gathered by modern high throughput techniques. The data stored in these databases need statistical culling which needs high computation power. The holistic integration of such data may provide deeper quantitative biological insights that are the current lacunae in biology. Theoretical scientists can fill this gap by describing the characteristic design principles by mathematical formulations that explain the dynamics associated with a biological process, using computers as the most important tool (Fig. **3**). Hence, DBs represent the raw material to formulate theoretical ideas and hypotheses and a source to extract relevant information to the set. Participation of interdisciplinary approaches in biology has changed the purview of biology from a descriptive science to a predictive science, based on which hypotheses are laid and validated through prediction-driven experimental data collection. The next few paragraphs will summarize the basics of mathematical formulations and modeling.

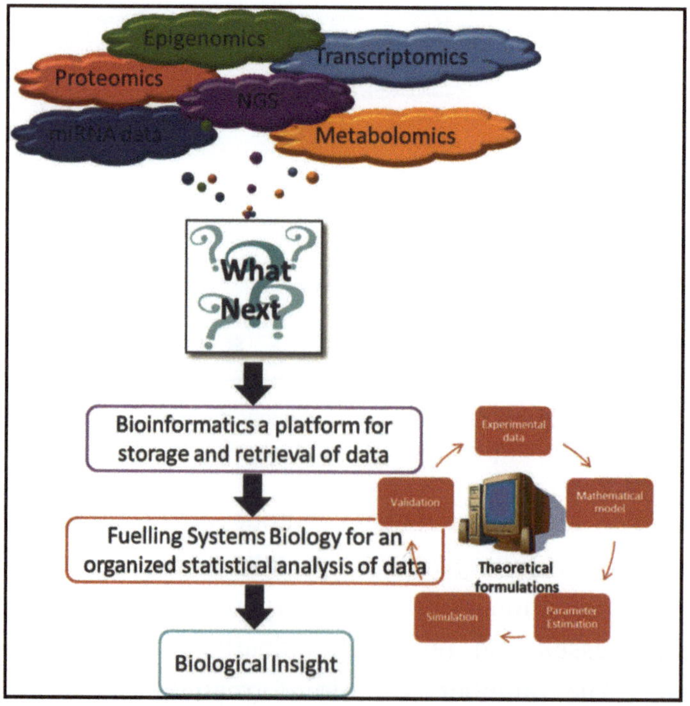

Figure 3: From 'Omics' data to Biological Insight (NGS – Next Generation Sequencing).

A BRIEF INSIGHT TO MATHEMATICAL MODELING AND SIMULATIONS [2, 3]

What is mathematical modeling?: A model is a representation of a process in a system to describe a phenomenon that cannot be observed directly. A biological process can be depicted with drawings or sketches, as well as described with mathematical formulas. Drawings and sketches are static in nature, which does not provide time evolution of the system and therefore are difficult to use for predicting the dynamics of the system. To understand the behavior over time, mathematical formulations are used for modeling, which is common in chemistry and physics. With the availability of enormous amounts of data, mathematical modeling is fitting into the realms of biology. Mathematical models in biology can be built based on observations from the real world that can be measured empirically. These observations are analyzed to describe the behavior of the system and an attempt is made to explain why a behavior occurred and allowed for predictions of the future behavior that are unmeasured or unseen. These predictions are validated by another set of experiments; which can also suggest reasons if the model is inadequate guiding an improvement in the empirical data collection. Thus mathematical modeling is an iterative process that helps to predict and validate real world biological phenomena.

Principles of Mathematical Modeling: The principles of mathematical modeling are philosophical in nature as it asks questions about the intentions and purposes of mathematical modeling. These questions may be the ones that are shown in (Fig. **4**).

The questions posed are not a basis for building a good mathematical model; however they could help in problem formulation. For any model building initiative, it is important to have a clear picture and understanding as to why the model is being built. For an example if an engineer is supposed to estimate the power that could be generated by a dam to be built, then the model formulated would consider its height and flow rate of the river water as an essential parameter, and not its thickness or other physical characteristics (*e.g.* materials, foundations *etc.*), which could be important if the model to be built was to design the actual dam. Thus, defining the task is the first essential step in model building. Next, the engineer should know

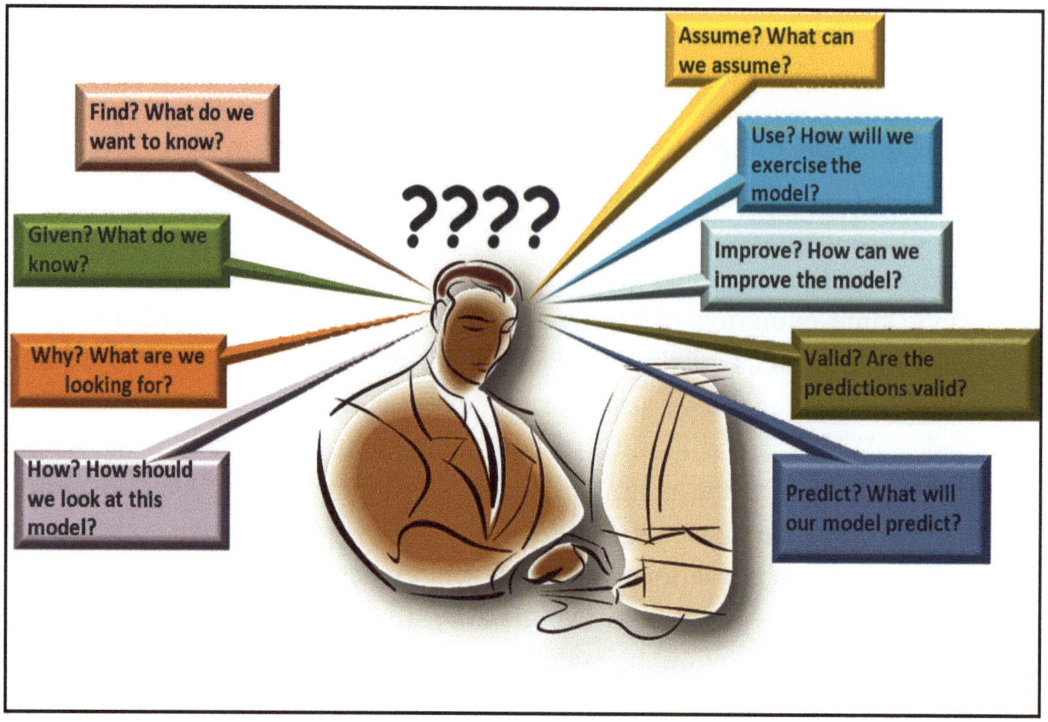

Figure 4: Modeling thought process.

certain quantities, such as the river flow and desired power levels that will be the basis for listing unknown parameters. These quantities should be supported by relevant assumptions; for example, the levels of desired power may be linked to demographic or economic data or consistency of river flows. The physical laws applicable to the model should be listed, including the conservation of mass, momentum of river flows, and energy dissipation and redirection as water flows through turbines in the dam. These parameters can be assigned to an equation that corresponds to the physical laws considered for the model. The model and the data generated can be validated against existing data from already constructed hydroelectric dams. If the model is inadequate or it fails in some way, iterative steps to recheck the assumptions, parameters, principles chosen and the equations can be reworked. Likewise biologists should ask questions, find the physical laws associated with the biological phenomenon and fit the parameters into relevant equations *e.g.*, modeling association or dissociation reaction between protein interactions would require applying the law of mass action, rather than the

Michaelis-Menten equation, which is relevant to an enzyme-catalyzed reaction. The reaction parameters are then validated through wet lab experimentation, which help improve the model. Through this iterative process the models can be improved, corrected, and validated.

Every mathematical model should qualify the following characteristics:

a. **Dimensional Homogeneity and Consistency:** Every equation that is used in the model must be dimensionally homogeneous or consistent *i.e.* physical dimensions that relates a quantity to fundamental physical quantities and units should be numerical expressions of a quantity's dimensions expressed in terms of a given physical standard.

b. **Abstraction and Scaling:** Abstraction is an approach to choose the level of details that needs to be included or excluded from the model. The details incorporated should answer the fundamental questions posed on the model. For example, a linear elastic spring can be used to model more than just the relation between forces, relative extension of a simple coiled spring and also to model the static and dynamic behavior of a tall building, which may be used to analyze the response of the building to an earthquake. The details within the parameters in the model should be such that the behavior of the elastic spring answers the proposed questions. Similarly, in biology a mathematical model of a metabolic network can predict choke point enzymes, effect of an inhibitor on the pathway, its relation to gene expression and so on. Parameter details should be such that the metabolic network can predict the desired behavior. Simultaneously, a model should be fitted to the right scale of abstraction. For example, the spring can be at, a micro scale to model atomic bonds or a macro scale to model a building. Similarly, in biology if one wants to capture protein expression in response to hormone signaling, the time scale would be in minutes. On the other hand a model describing activation of a TF due to the same hormone signaling the time scale would be in pico or micro seconds. Therefore the right scale for a model would be in relation to the "reality" one needs to capture.

c. **Conservation and Balance Principles:** A mathematical model should indicate that some property of an object or system is being conserved. For example, model of a population of an animal colony, individual animals should be balanced across a defined boundary. For such an instance conservation principles are applied to assess the effect of maintaining or conserving levels of important physical properties. The mathematics of balance and conservation laws is straight forward.

For example: The Michaelis-Menten [4] equation is a well-known equation in biochemistry that relates the rate of the enzymatic reaction to the concentration of the substrate available. The dynamics of the enzyme catalyzed reaction can be understood by considering the reaction given below whereby the substrate and enzyme come together to form the enzyme-substrate complex. The enzyme-substrate complex is then able to react to form the product. After the product is formed, it is released to yield the free enzyme again which is able to further react with more substrate. (k1, k2 and k3 are the rate constants in the reaction).

$$\mathrm{E + S \underset{k2}{\overset{k1}{\leftrightarrow}} ES \overset{k3}{\rightarrow} E + P}$$

The change in concentration of each of the components in the reaction can be derived by using differential equations (1, 2, 3, 4) as given below:

$$\frac{d\,[S]}{dt} = -k1\,[S][E] + k2\,[ES] \tag{1}$$

$$\frac{d\,[ES]}{dt} = +\,k1\,[S][E] - \,k2\,[ES] - K3[ES] \tag{2}$$

$$\frac{d\,[P]}{dt} = +\,k3\,[ES] \tag{3}$$

$$\frac{d\,[E]}{dt} = -\,k1\,[S][E] + \,k2\,[ES] + K3[ES] \tag{4}$$

This derivation is based upon the law of mass conservation that says

$$\frac{\mathrm{dx}}{\mathrm{dt}} = [\text{concentration of x produced in the reaction}] -$$

$$[\text{concentration of x consumed in the reaction}]$$

Where 'x' is any species in the reaction system

d. **Constructing Linear Models:** Linearity is one of the most important concepts in mathematical modeling. Models of devices or systems are said to be linear when their basic equations—whether algebraic, differential, or integral—are such that the magnitude of their behavior or response produced is directly proportional to the excitation or input that drives them. This is important in biology as biological systems are inherently nonlinear and for making prediction about biology the approximations should be linearized. Engineers linearize a system to predict the response of a system to a complicated input by decomposing or breaking down that input into a set of simpler inputs that produce known system responses or behaviors.

Lastly, it is most important to remember that mathematical models are representations or descriptions of reality. Thus, if the behavior predicted by the models does not reflect what one sees or measures in the real world then the models needs to be fixed. As rightly said.

Essentially, All Models are Wrong, but some are Useful. – George E. P. Box (1987)

Some methods used to model biological systems are listed in Table **1**.

Table 1: Some methods used in modeling a biological system [5]

Method	Description
ODEs*	Series of reaction-rate equations solved using numerical methods Produces graphs or tables of reagent production and consumption
Stochastic differential equations	Series of reaction rate equations solved using 'master equation' and random number generator Handled using Gillespie algorithm
S-system formalism or power law equations	Uses Taylor approximation to simplify non-linear ODEs Enables steady-state DEs to be transformed to easily solved linear equations
PDE§ or molecular dynamics	Expresses spatial and temporal dependence through partial derivatives Solved using numerical methods\ Produces numeric output of concentrations and x,y,z coordinates

Table 1: contd...

Petri nets	Uses a weighted firing process to activate events from multiple connections that are used as inputs Mimics telephone switchboard or power-grid load handling
Pi calculus	A language for concurrent computational processes Pairs of processes interact by sending and receiving synchronized messages

*Ordinary Differential Equations, § Partial Differential Equations

Representation of biochemical interactions using ODEs are shown in Table **2**.

Table 2: Representation of biochemical interactions using ODEs, where k1 and k2 are the forward and reverse rate constants of the reactions [5]

Reaction	**Reaction Type**	**System of ODEs**
$A \rightarrow B$	Monomolecular conversion	$\frac{d[A]}{dt} = -k1\,[A], \frac{d[B]}{dt} = k1\,[A]$
$A \leftrightarrow B$	Reversible conversion	$\frac{d[A]}{dt} = -k1\,[A] + k2\,[B]$ $\frac{d[B]}{dt} = -k2\,[B] + k1\,[A]$
$A + B \leftrightarrow C + D$	Bimolecular reversible conversion	$\frac{d[A]}{dt} = -k1\,[A][B] + k2\,[C][D]$ $\frac{d[B]}{dt} = -k1\,[A][B] + k2\,[C][D]$ $\frac{d[C]}{dt} = k1\,[A][B] - k2\,[C][D]$ $\frac{d[D]}{dt} = k1\,[A][B] - k2\,[C][D]$
$A + B \rightarrow C$	Production	$\frac{d[A]}{dt} = -k1\,[A][B]$ $\frac{d[B]}{dt} = -k1\,[A][B] + k2\,[C][D]$ $\frac{d[C]}{dt} = k1\,[A][B] - k2\,[C][D]$ $\frac{d[D]}{dt} = k1\,[A][B] - k2\,[C][D]$
$A \rightarrow B + C$	Degradation	$\frac{d[A]}{dt} = -k1\,[A]$ $\frac{d[B]}{dt} = k1\,[A]$ $\frac{d[C]}{dt} = k1\,[A]$

What is Simulation?

Simulation is the imitation of some act or a system. To simulate physical systems one needs to build a mathematical model governed by kinetic laws. Powerful high throughput technologies have given us an extensive parts-list of a cell and have also shed light on a general idea of interactions among genes, proteins, RNA and small molecules. These make up the metabolic and gene regulatory pathways that play crucial roles response to external/internal stimuli that ultimately guide the cellular processes. A high degree of cross-talk between these pathways has resulted in complexity of the system which makes predicting biological behavior almost impossible. Computer simulations of such physical interactions have proved useful for understanding the topology and dynamics of such networks [6]. Fig. **5a** provides a basic overview of different models and simulations that are often used in systems and synthetic biology. Fig. **5b** shows a deterministic simulation output of an enzyme catalysed reaction, predicting the change in concentration of different components in the system for detailed information on modeling and simulations the reader is encouraged to read [4].

(a)

(b)

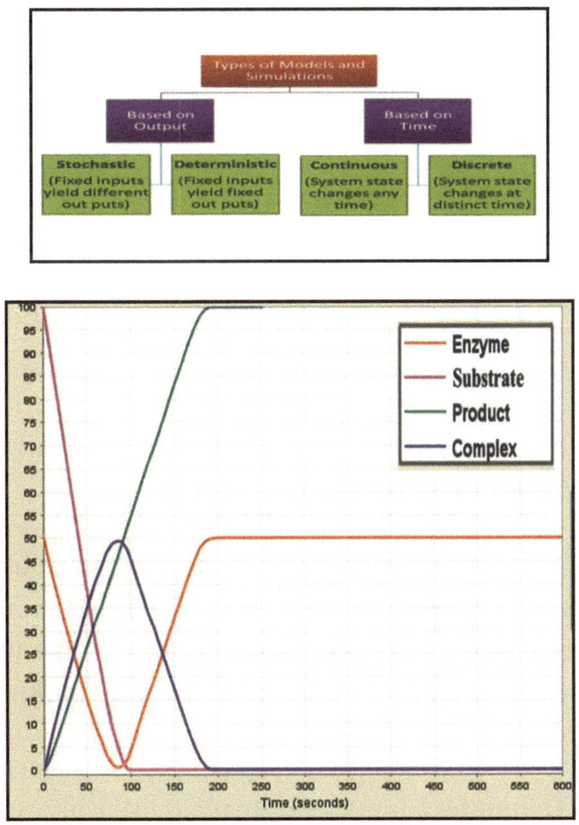

Figure 5: (a) Types of Models and Simulations Commonly Used to Describe Biological Systems; (b) Deterministic Simulation Output of an Enzyme Catalyzed Reaction.

What is Systems and Synthetic Biology?

Systems Biology

As stated earlier, high throughput technologies have given us the parts-list (proteins, genes, transcription factors *etc.*) of what biology is composed of. A system is not just an assembly of single gene or protein, but thousands of them and therefore studying them in isolation using the reductionist approach does not describe the dynamics in biology. A dynamic description and to decode the inherent complexity (due to large numbers of functionally diverse and multifunctional, sets of elements that interact selectively and nonlinearly to produce a complex behaviors) one needs to understand the interaction of these parts at the systems level. For example, the p53 tumor suppressor is activated, inhibited and degraded by modifications, such as phosphorylation, dephosphorylation and proteolytic degradation. The targets of p53 are selected based on its modification state and cannot function in isolation. Such systems level (*i.e.* a top down approach) understanding will give us insights into what are the interaction, how different interaction patterns emerge and how can we control them. We can also seek answers to questions such as: What are the reaction parameters of the interactions in a metabolic or a signaling pathway? How are these pathways regulated? Which genes are involved in the regulation? How can we stabilize parameters against noise and external fluctuations? How do the interactions change when a malfunction/perturbation (disease) occurs in the system? Is there a possibility of a hidden interaction? What are the design principles and possible interaction patterns, and how can we modify them to improve system performance?

A system-level understanding of a biological system can be discerned by considering four key properties [7]:

a. **System Structures:** The mechanism of interaction in the network such as the gene interaction, protein-protein interaction and biochemical pathway and how these interactions modulate the physical properties of intracellular and multicellular structures, constitute the system structure.

b. **System Dynamics:** Metabolic analysis, sensitivity analysis, dynamic analysis methods such as phase portrait and bifurcation analysis help understand the system behavior over time under various conditions. Bifurcation analysis traces time-varying change(s) at a particular state of the system in a multidimensional space where each dimension represents a particular concentration of the biochemical factor involved.

c. **System Control:** A mechanism of modulation that help minimize malfunction of a potential therapeutic target for disease treatment can be understood by systematically introducing control elements in the system.

d. **System Design:** Biological systems can be constructed and modified to a desired property based on definite design principles.

A systems level understanding of the interaction network with its emergent properties in a diseased condition can help drug target prediction. These targets can be tested to predict different target positions, treatment strengths, target combinations or temporal combination scenarios. There are some examples among signaling pathway models like the study of the ErbB network using sensitivity analysis which identified ErbB3 as a key node in response to ligands [8]. Sahin *et al.,* [9] have combined computer simulations and experimental testing to reverse engineer a protein interaction network to define a potential therapeutic strategy for *de novo* trastuzumab resistant breast cancer. They combined ERBB2 and EGFR signal in conjunction with G1/s transition (cell cycle) and identified that c-MYC as a novel target in treatment of trastuzumab-resistant breast cancer.

Similarly, drug toxicity could be assessed by modeling organ specific metabolic stress responses. Drug distribution and metabolism could be modeled through multi-organ, multi-tissue pharmacokinetic (PK) methods, and drug dosing regimens could be determined by modeling responses to different drug concentrations or different dose frequencies [5].

For systems and computational biology to be applied to therapeutics and diagnosis, some standards should be developed in areas of cell and molecular biology by defining how data should be gathered, how they should be modeled, and how results should be further described.

New tools and data formats have been developed during the last few years, which have tremendously streamlined the way data are handled and represented for better application to biological understanding. In the following section, examples of tools currently assisting the research development in systems biology are discussed.

Computational Tools to Assist Systems Biology

System biology models are described with standard formats like the Systems Biology Markup Language (SBML), the Cell Markup Language (CellML), the Biological Pathway Exchange (BioPAX) format, and The Proteomics Standards Initiative—Molecular Interaction exchange format (PSI-MI). Currently the Systems Biology community widely uses the SBML format. It has a large user base that has created over 180 software systems that can create, modify, simulate and analyze information using SBML as a base for exchanging information [10].

Many curated DBs can serve as data sources for model building, which have summarized experimental results performed by the scientific community. Some of the common DBs have been listed in Table **3**.

Table 3: Examples of databases with experimental data sets [5]

Class	**Database**	
Protein-Protein Interactions	BIND	Biomolecular Interaction Network DB
	DIP	Database of Interacting Proteins
	MINT	Molecular Interaction DB
Metabolic Pathways	EcoCyc	Encyclopedia of *E. coli* Genes and Metabolism
	KEGG	Kyoto Encyclopedia of Genes and Genomes
	BRENDA	Braunschweig Enzyme Database
	Reactome	Reactome Knowledge Base
Signaling Pathways	SigPath	Signalling Pathway Information System
	STKE	Signal Transduction Knowledge Environment
Genetic Interaction Networks	BIND	Biomolecular Interaction Network Database
	GeneNet	Genetic Networks
Protein information	UNI-PROT	Universal Protein Knowledge Base
	SGD	Saccharomyces cerveiace Genome Database

Once a systems biologist has used the data from the above mentioned databases to model a biological phenomenon, the results of simulation and [11] model analysis can also be supplied to databases such as the BioModels database, a repository for curated SBML models. It is an international effort to:

- define standards for model curation
- define vocabularies for annotating models with connections to biological data resources
- provide a free, centralized publicly-accessible database of annotated computational models in SBML and other structured formats

Before a model is accepted to the database, it is checked manually for proper numerical simulation results and curation. The components of the models are annotated with terms from controlled vocabularies and link to other relevant data resources. This allows the users to search accurately for the models they need.

Specialized and general-purpose modelling tools (Table **4**) are available to enable one to perform tasks, such as:

- Model definition and building in the form of set of equations or mathematical expressions, or graphical representation
- Model analysis, including calculation of steady states and sensitivities, stability, parameter scans, *etc*.
- Parameter estimation from available data
- Model simulation
- Output of results (graphical and textual form)

Table 4: Tools for mathematical modeling in systems biology

Tools	Description
XPP-AUT [12]	Analysis and simulation for mathematical model of various types; bifurcation analysis
Cell Designer [13]	Graphical representation of biochemical networks; implementation and simulation of ODE models
Cell Net Analyzer [14]	Analysis of regulatory networks based on network topology; network visualization
COPASI [15]	Implementation, simulation and analysis of biochemical models in ODE format. Enables also stochastic simulations

Table 4: contd...

SB Toolbox [16]	MATLAB toolbox for simulation of ODE models as well as parameter estimation, steady state and bifurcation analysis, model reduction.
SBML [17]	Toolbox Collection of functionalities for SBML models based on MATLAB, including reading, writing, graphical presentation, simulation, and many more
Virtual Cell [18]	Deterministic and stochastic simulation of cell processes including diffusion and membrane transport; uses client-server system

Synthetic Biology

The discovery of mathematical logic in gene regulation (*e.g.* the lac operon; Monod and Jacob, 1961) and early achievements in genetic engineering (*e.g.* recombinant DNA technology) paved the way for synthetic biology [19]. Synthetic biology brings together engineering concepts and biology to design and build novel molecular components, interacting networks and pathways. The general theme in synthetic biology is to apply electrical circuit analogies to biological pathways by constructing biological circuits by a bottom up approach [20]. They can be constructed from a list of basic parts (Fig. **6**); *e.g.*, to construct a simple transcription unit, a synthetic

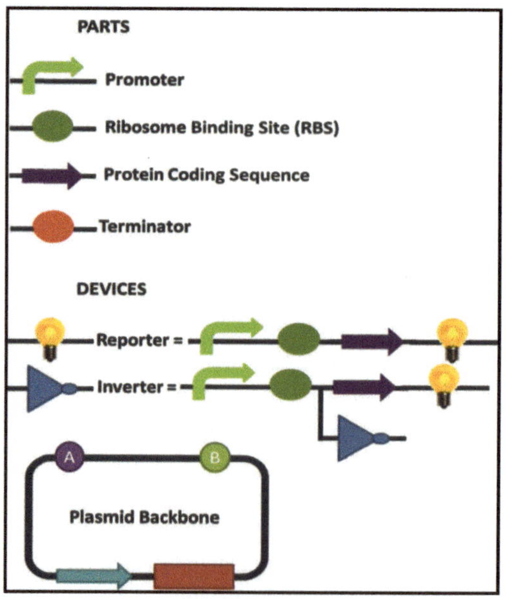

Figure 6: Standard biological parts made of DNA, performing tasks like transcription and translation of the desired protein sequence can be put together to form devices which can be assembled together in a plasmid.

biologist would require promoters, ribosome binding sites, protein coding regions and terminators that are well characterized DNA-based synthetic or natural building blocks. These parts are available in the repository of synthetic parts 'MIT Registry of Standard Biological Parts' (http://partsregistry.org/)'. This repository also contains information relevant about their structures and functions. Other parts, such as spacers or stem-loop RNAs that can be used to fine-tune gene regulation are also part of the repository [21]. During the last decade, researchers have taken an advantage of the comprehensive catalogue of biomolecular parts and the latest molecular biology techniques to move from devices, such as simple transgene switches in single cells to designer networks that program cell-cell communications that respond to specific input signals [22]. These devices are being used in *in vivo* disease models to gauge their application in therapeutics and diagnostics, and are rapidly evolving in the direction of clinical trials.

Computational Tools to Assist Synthetic Biology

As for any modern engineering design, computational tools are indispensable. The process of synthetic circuit design employs mathematical modeling. *In silico* predictions of the circuit can help understand the undesired behavior of the circuit, and also aid in redesign by providing alternative design approaches. Like for systems biology, mathematical models can be built in synthetic biology considering the kinetic laws that govern transcription, translation, promoter affinity and binding to generate ODEs, which can be solved to predict the circuit behavior. Examples of tools that are used for circuit design in synthetic biology are listed in Table **5**.

Table 5: Computational tools to aid synthetic biology

Tools	Description
Bio JADE [23]	Specify a system abstractly, tune it, simulate its behavior using a variety of simulators
Gene Designer [24]	Graphically rich molecular view to display, annotate and edit synthetic constructs. Customizable database to quickly store, manage, and track genetic element, genes and constructs.
Ro Ver Ge Ne [25]	Tool for the analysis of genetic regulatory networks under parameter uncertainty
UNA Fold [26]	Collection of programs that simulate folding, hybridization, and melting pathways for one or two single-stranded nucleic acid sequences
Tinker Cell [27]	Combines visual interface with programming API (Python, Octave, C, Ruby) and simulates the system

Table 5: contd….

Pro Mot [28]	Provides capabilities for the development of dynamic models based on differential-algebraic equations, and their simulation and further analysis.
Geno CAD [29]	Open-source computer-assisted-design, it also includes a large database of annotated genetic parts

Many of these tools share a standardized format for input/output files such as the System Biology Markup Language (SBML), XML-based format for the exchange of mathematical models in biology. This provides a concise representation of the chemical reactions in biological system that can be translated into systems of ordinary differential equations [30].

BIOINSPIRED SYNTHETIC DEVICES AND THEIR MODELING INSIGHTS

The payoff for systems biology research is not merely abstract mathematical understanding that gives a quantitative description, but the empowerment to design new and improved biological functions *via* 'synthetic biology' to better unveil the complexities of the system [31]. Hence, systems and synthetic biology go hand-in-hand towards developing novel circuitry for better understanding and applications in biology. The basic principles of arsenal of circuits available that are genetically encoded with highly complex functionality are further discussed.

Toggle Switch [32]

A toggle switch is composed of two repressors and two constitutive promoters which are in juxtaposition to each other such that each promoter is inhibited by the repressor that is transcribed by the opposing promoter (Fig. 7) *i.e.* repressor 1 binds to promoter 1 that lies upstream of repressor 2 and repressor 2 binds to promoter 2 that lies upstream of repressor 1. The repression on the promoters by the repressor protein is removed by addition of an inducer such as IPTG (isopropylthio-β-galactoside). Green fluorescent protein gene was used as the reporter of gene expression (read out), placed downstream of repressor 1. The toggle switch designed by Gardner requires fewest genes and cis-regulatory elements to achieve robust bistable behavior.

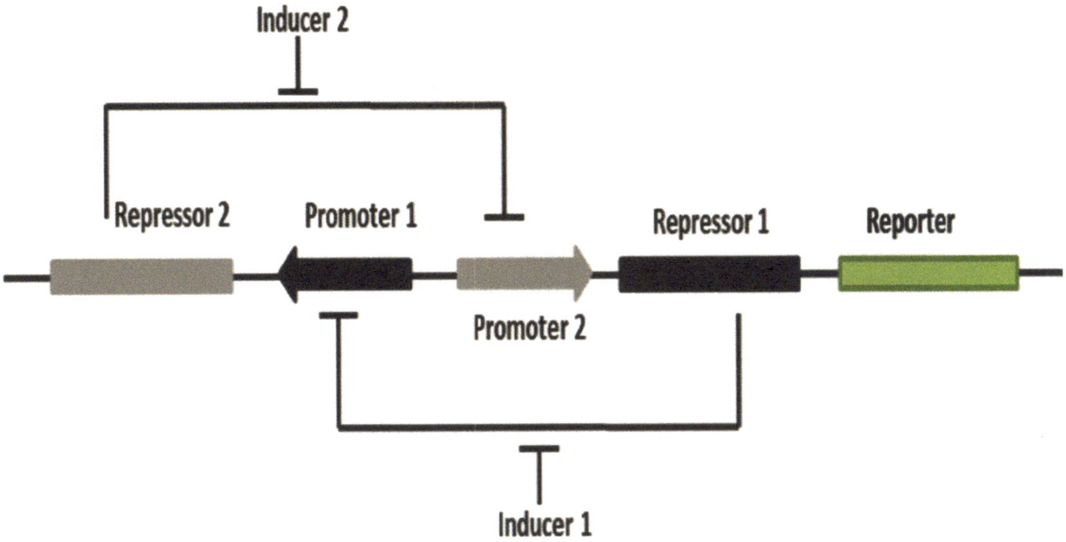

Figure 7: Toggle switch constructed by Gardner *et al.,* 2000.

Robustness of a toggle switch or any circuit corresponds to the stability exhibited by the circuit over a wide range of parameter values. For the toggle switch bistability was seen over a wide range of parameter values and the two states were tolerant to the fluctuations that are inherent in gene expression. This suggests that the toggle switch does not flip randomly between states. The bistable toggle design does not require any specialized promoters, but can be achieved using any set of promoters and repressors as long as they fulfill the conditions given below (equation 5, 6):

$$\frac{du}{dt} = \frac{\alpha_1}{1+v^{\beta}} - u \tag{5}$$

$$\frac{dv}{dt} = \frac{\alpha_1}{1+u^{\gamma}} - v \tag{6}$$

where:

u is the concentration of repressor 1,

v is the concentration of repressor 2,

$\alpha 1$ is the effective rate of synthesis of repressor 1,

$\alpha 2$ is the effective rate of synthesis of repressor 2, (α describes the net process of gene expression *i.e.* effect of RNA polymerase binding, open-complex formation, transcript elongation, transcript termination, repressor binding, ribosome binding and polypeptide elongation).

β is the cooperativity of repression of promoter 2, and

γ is the cooperativity of repression of promoter 1.

(Described cooperativity can arise from the multimerization of the repressor proteins and the cooperative binding of repressor multimers to multiple operator sites in the promoter).

The first term in the equation describes the cooperative repression of constitutively transcribed promoters, whereas the second term describes the degradation of the repressor. Equations 1a and 1b are modified to explain the effect of inducer on the repressors, where K is the dissociation constant and η is the cooperative binding constant for the inducer.

Further geometric structure analysis of equation 1a and 1b (Fig. **8a**) reveals the origin of the bistability: The nullclines ($du/dt = 0$ and $dv/dt = 0$) intersect at three points, producing one unstable and two stable steady states. Three key features of the system become apparent from Fig. **5** that: a. the nullclines intersect three times because of their sigmoidal shape, which arises for $\beta, \gamma > 1$. Thus, the bistability of the system depends on the cooperative repression of transcription. b. The rates of synthesis of the two repressors if not balanced the nullclines will intersect only once, producing a single stable steady state. c. The structure of the toggle network creates two basins of attraction; state 1 and state 2. It was also shown that as the rates of repressor synthesis are increased, the size of the bistable region increases. Furthermore, the slopes of the bifurcation lines, for large $\alpha 1$ and $\alpha 2$, are determined by β and γ (Fig. **8b** and **c**). Thus, to obtain bistability at least one of the inhibitors must repress expression with cooperativity greater than one. Moreover, higher-order cooperativity will increase the robustness of the system, allowing weaker promoters to achieve bistability and producing a broader bistable region.

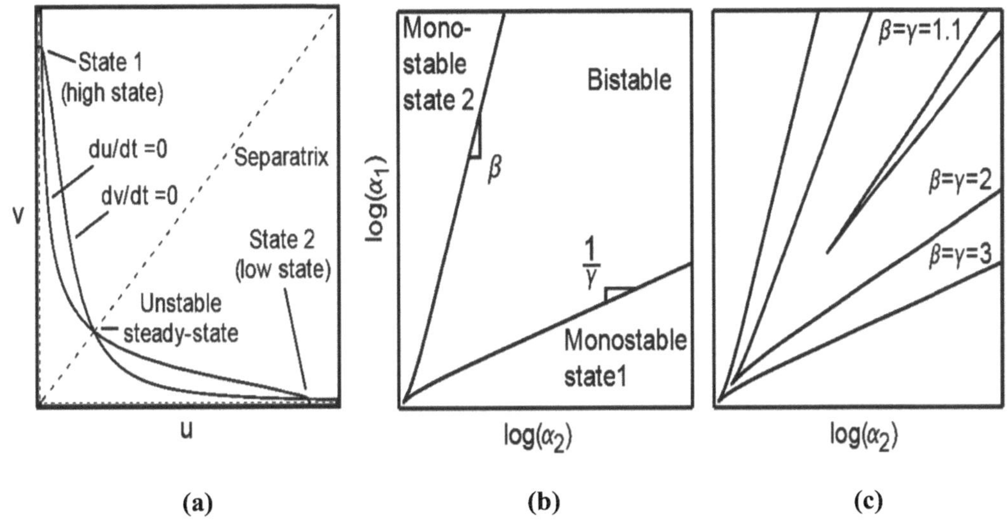

Figure 8: a. Nullclines showing the bistable nature of the toggle switch. b and c. Range of β and γ that decides the region of bistability (Reprinted after due permission from Gardner *et al.*, 2000).

Based on the toggle switch constructed by Gardner, Atkinson *et al.*, (2003) [33] developed a toggle switch composed of the Lac and Ntr systems to design a genetic clock in *E. coli*. They constructed two modules: the activator module and the repressor module (Fig. **9**). The activator module consisted of a modified glnA promoter with lac operators that drive the expression of the activator, NRI that, in turn, activates the glnA promoter, creating an autoactivated circuit repressible by LacI expressed by the repressor module. The "repressor module" acts as an oscillator, which consists of NRI-activated glnK promoter driving LacI expression. This circuit design produced synchronous damped oscillations in turbidostat *E. coli* cultures lasting much longer than the cell cycle. The level of active repressor was controlled by using a lacY mutant and varying the concentration of IPTG giving a discontinuous expression of the activator.

Repressilator [34]

Elowitz and Leibler in 2000 built a three transcriptional repressor system, which shows an oscillatory behavior and named it as repressilator (Fig. **10**). The system read out (*i.e.* the oscillation in the repressilator) was read in the form of induction of green fluorescent protein (GFP). The first repressor protein, LacI, inhibits the

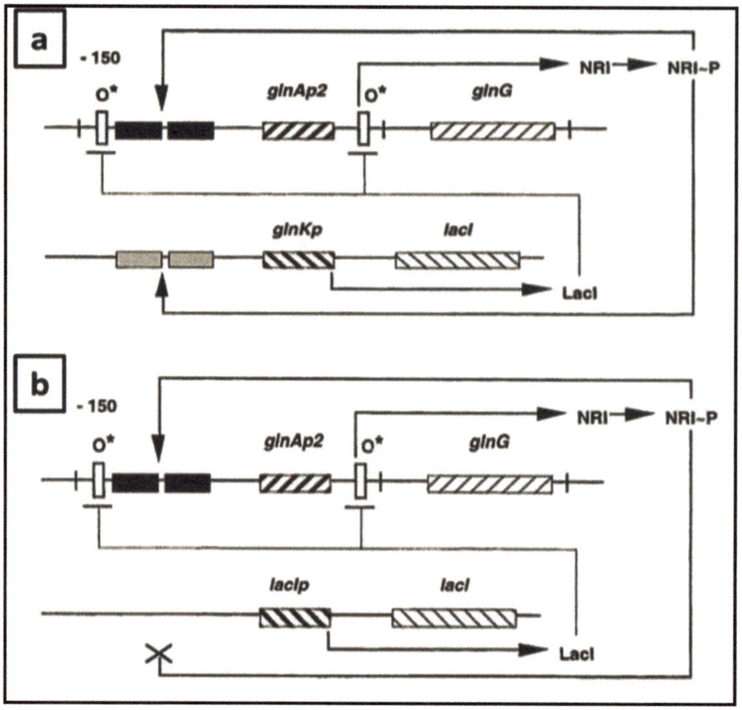

Figure 9: Activator and repressor module of the genetic clock constructed by Atkinson *et al.*, (2003) (Reprinted after due permission from Atkinson *et al.*, 2003).

transcription of the second repressor gene tetR, whose protein product in turn inhibits the expression of a third gene cI. Finally, CI inhibits lacI expression, completing the cycle. These negative feedback loops lead to temporal oscillations in the concentrations of each of its components that can be seen from a simple model of transcriptional regulation, which was used to design the repressilator and study its possible behaviors.

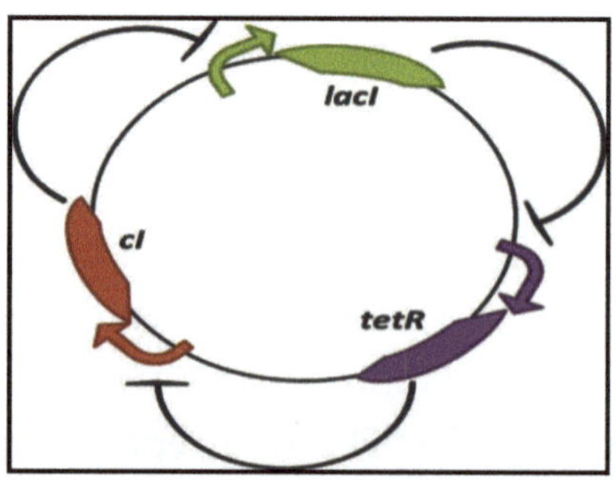

Figure 10: A Repressilator.

The mathematical model used a deterministic, continuous approximation, where three repressor-protein concentrations, pi, and their corresponding mRNA concentrations, mi were treated as continuous dynamical variables; where i is lacI, tetR or cI. Each of these six molecular entities participate in transcription, translation and degradation reactions which results in the formulation of six coupled first-order differential equations (7 and 8) that determine the kinetics of the system.

$$\frac{dm_i}{dt} = -m_i + \frac{\alpha}{1+p_j^n} + \alpha_0 \tag{7}$$

$$\frac{dp_i}{dt} = -\beta\,(p_i - m_i) \tag{8}$$

where:

α0 is the number of protein copies per cell produced from a given promoter type during continuous growth in the presence of saturating amounts of repressor (owing to the "leakiness" of the promoter)

α + α0 is the number of protein copies per cell produced from a given promoter type during continuous growth in the absence of saturating amounts of repressor

b denotes the ratio of the protein decay rate to the mRNA decay rate

n is a Hill coefficient

time t is rescaled in units of the mRNA lifetime

protein concentrations are written in units of KM *i.e.* the number ofrepressors necessary to half-maximally repress a promoter

mRNA concentrations are rescaled by their translation efficiency to the average number of proteins produced per mRNA molecule.

Brusselator

A brusselator is a theoretical (Fig. **11**), minimal mathematical model that explains the oscillating behavior in an autocatalytic reaction system (where species in the reaction system can also act as a catalyst of the reaction), explained by Prigogine and Lefever in 1968 [35]. The word is a morpheme of Brussel and oscillator and was named by their colleague Tyson from the Free University of Brussel.

It is explained by the following set of chemical reactions (considering it as an open system):

$$A \rightarrow X$$

$$B + X \rightarrow Y + C$$

$$2X + Y \rightarrow 3X$$

$$X \rightarrow D$$

where:

X and Y are the autocatalytic species

Reacting species are A and B, which are present in large excess

C and D are produced in the reaction and simultaneously being removed from the system

The presence of an autocatalytic species in the system results in periodic oscillations, which is analysed using the Hoff bifurcation.

The basic mathematical expression in terms of X and Y would be as follows (equations 9 and 10)

$$\frac{dX}{dt} = A - (B+1)x + x^2y \tag{9}$$

$$\frac{dY}{dt} = Bx - x^2y \tag{10}$$

The brusellator model was modified by Tyson [37] to study mitosis and this modified model was used by Toner *et al.*, (2013) [38] to show the effect of "bursty" protein production on downstream pathways. They investigated the effect of burstiness in protein expression and import on downstream pathways.

They considered two identical pathways with equal mean input rates, with an exception that in one pathway proteins are input one at a time and in the other proteins are input in bursts. They showed that deterministically, the dynamics of these two pathways are indistinguishable, but stochastic behavior shows that both pathways may or may not display noise-induced oscillations; the non-bursty input pathway displays noise-induced oscillations whereas the bursty one does not and *vice versa*. Their results suggest that single cell rhythms can be controlled by regulating burstiness in protein production.

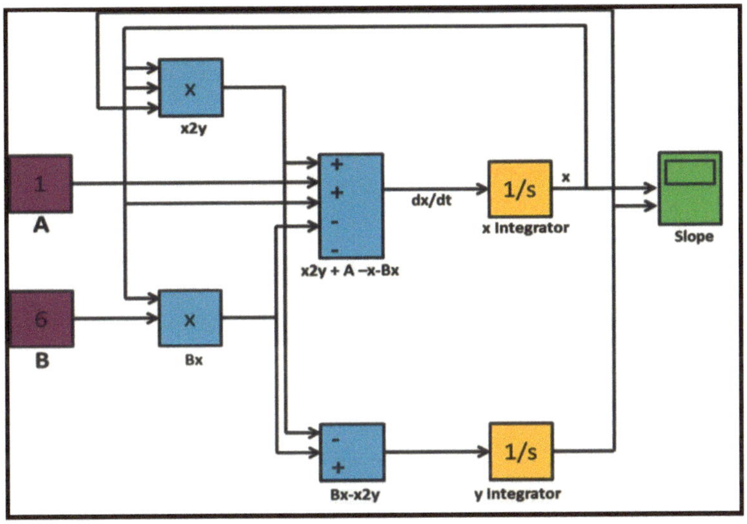

Figure 11: A Simulink model of the brusselator (Adapted from [36]).

Cell–Cell Communicators

Quorum sensing is a way of communication between bacteria based on cellular density. In multicellular organisms cell–cell communication leads to synchronized response to a stimulus. Also there is division of labor (Fig. **12**), making the system more efficient. Synthetic circuits can be modeled and designed for effective cell–cell communication as exemplified by Hoffmann-Sommer *et al.,* (2012) [40], and implemented in Saccharomyces cerevisiae by Regot *et al.,* (2011) [41]. A theoretical quantitative analysis of a synthetic cellular logic-gate system was developed to exploit the endogenous MAP kinase signaling pathways. The developed system is novel as the concept of compartmentalization was used in such a way that each designed circuit based on basic logic gates was implemented in independent single cells that can be then co-cultured to perform complex logic functions. Kinetic models of the multicellular IDENTITY, NOT, OR, and IMPLIES logic gates, using both deterministic and stochastic frameworks were constructed [40].

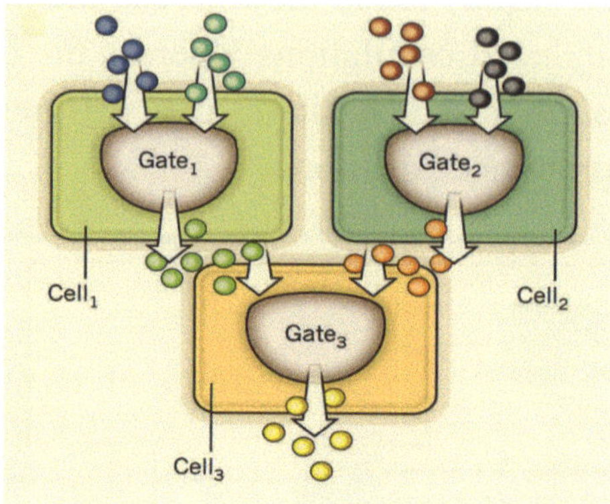

Figure 12: Compartmentalization of logic gates within different cells (Reprinted after due permission from Li *et al.,* 2011 [39]).

Memory Circuits [42, 43]

A sustained cellular response to transient stimulus is referred to as biological memory. Cells maintain these sustained responses to different stimuli by regulating the molecules involved in gene expression (*i.e.* through different transcriptional states). For constructing a synthetic memory circuit, transcriptional responses can be considered as the information processing in cellular responses is through transcription which is a well characterized biological behavior. The output for a stimulus will depend on the interaction and binding capabilities between the transcription factors with their promoters, also whether the transcription factor is an activator or repressor of the gene product. A gene product will increase in concentration if transcription factor is an activator or decrease if it is a repressor; therefore the concentration of the gene product finally is a function of the transcription factor. The production of gene product is further balanced by its degradation and dilution rates. Because output depends on the binding of the transcription factor to the promoter; Hill function has to be considered, which describes the equilibrium binding of a transcription factor to its target promoter.

So,

$$\boldsymbol{Y = s + f(X) - \alpha . f(x)} \tag{11}$$

$$\boldsymbol{Y = s + \frac{\beta X^n}{K^n + X^n} - \alpha . f(X)} \tag{12}$$

$$\boldsymbol{Y = s + \frac{\beta}{1 + \left(\frac{X}{K}\right)^n} - \alpha . f(X)} \tag{13}$$

where;

Y is the gene expression output *i.e.* a protein

X is the transcription factor

α is the degradation or dilution rate

K is the activation/repression coefficient that defines the threshold concentration of X needed for activation or repression of Y (it depends on the affinity of binding of X on the promoter for Y)

β is the maximal expression level of Y obtained when an activator is bound or a repressor is unbound

n is the Hill coefficient, that governs how a network responds to transcriptional stimuli: a larger n produces a bistable, switch-like response (Fig. **13**), which is essential to biological memory as this response allows a shift to an alternative steady state that might persist over time.

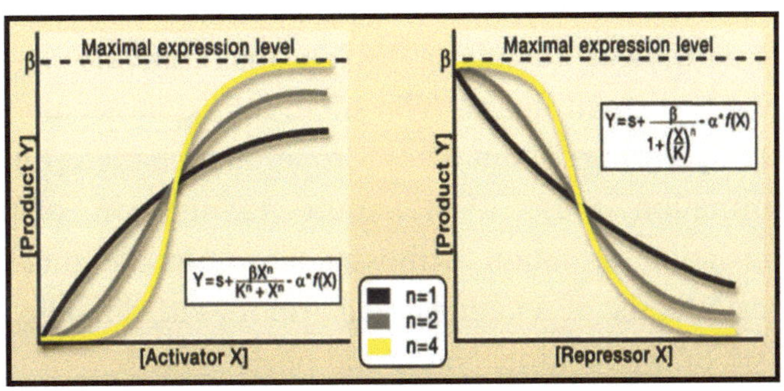

Figure 13: Gene Circuit for Cellular Memory defined by Hill function (Burrile *et al.*, 2010).

Burrill *et al.*, (2012) [44] have applied this memory circuit to track cell fate during cell differentiation in a mammalian sub population on global stimulation. Endogenous stimuli, including hypoxia and DNA damage were used for the activation of the device as these stimuli produces heterogeneous responses at the single-cell level. As hypoxia and DNA damage are the benchmarks of tumor development, this circuit could help delineate the fate and responses in tumor cells in their microenvironment. During hypoxia the HIF-1 transcription factor can activate or silence target genes as well as increase genomic instability by bypassing the DNA repair checkpoints. The synthetic, HIF-1-activated memory device detects and tracks the subpopulations within the heterogeneous tumor microenvironment which helps in determining their specific contributions toward tumor development and metastasis. Similarly, it is known that DNA damage produces an array of physiologic responses at the single-cell level. When the synthetic memory device was linked to native DNA damage pathways it helped in identifying how DNA damage responses are transmitted to subsequent generations and what could be the impact on long-term cellular behavior. Tumor suppressor p53 is activated in response to DNA damage and a memory device triggered by differential levels of p53-induced repair factor, such as ribonucleotide reductase (p53R2) has helped the isolation and tracking of progeny whose ancestors underwent a repair response thus revealing cell's history of DNA damage. Such endogenous stimuli activated memory circuits can be used to analyze epigenetic responsive elements that decides the future of a subpopulation of cells which may help in garnering deeper insight into cell development in different environmental conditions.

Riboswitches and Aptamers

Riboswitches are structures found in mRNA that regulate gene expression in bacteria on binding to a small ligand. The small ligand molecule binds to a region called the aptamer that brings about a conformational change (Fig. **14**) such that a repressing conformation cause a premature termination of transcription or inhibition of translation initiation. Riboswitches regulate metabolic pathways, including the biosynthesis of vitamins (*e.g.* riboflavin, thiamin and cobalamin) and the metabolism of methionine, lysine and purines [45].

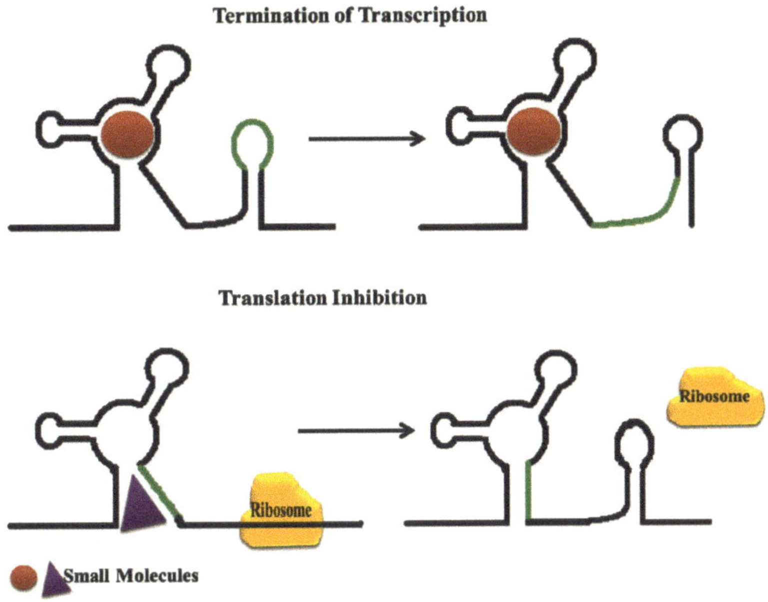

Figure 14: A riboswitch involved in transcription termination and translation inhibition.

Synthetic riboswitches or aptamers can be designed to antagonize the natural one, targeting a key metabolic process for a desired system output and tunable gene expression.

Beisel *et al.,* (2009) [46] constructed a kinetic model of a riboswitch function based on its detailed molecular mechanism that accounts for folding and ligand binding during transcription. They considered translational repression, transcriptional termination, and mRNA destabilization. Rate constant to each mechanistic step in the models were assigned to yield a relationship between ligand concentration (L) and protein levels (P). The equation (14):

$$KM = KMA + KTA = KMB + KTA \tag{14}$$

describes the transcriptional termination mechanism in terms of A and B are the two conformations of the riboswitches that reflect transcriptional folding

KMA and KMB are the rate constants for extension

KTA + KTA are the rate constants for termination

KM is the single parameter that accounts for summed up rate constants of termination and extension

And likewise, rate constants have been assigned for each mechanism considered for model building, exploring how each rate constant contributes to the riboswitch performance.

MEDICAL APPLICATIONS AND NEXT GENERATION THERAPEUTICS

Medical applications of systems and synthetic biology driven devices are particularly difficult because incorporating a new technology at the clinical level requires that the device successfully pass all the critical phases of clinical trials, which in itself can be a long and laborious process. Thereafter, the device should meet the stringent government policies for certifications. Nevertheless, these devices have shown promises as possible next generation therapeutics (therapeutics that are far more superior than the current available therapeutics). In particular, these developments are relevant to the challenges associated with the detection, surveillance, and responses to emerging infectious diseases. It is important to divert attention to infectious disease because of the rising evidence of antibiotic resistance that is making the treatment of these diseases increasingly difficult. With every report of development of antibiotic resistance in pathogens, new antibiotics are being discovered and moved to the clinics. With each new antibiotic the chances of undesired and significant perturbations in the human microbiome have become relevant to the destruction of the normal flora. Therefore selective targeting of pathogens becomes important. Discussed here are the most impressive steps that systems and synthetic biology have taken towards improving human health by rationally re-engineering biological systems *via* the introduction of bio-devices with selectivity.

A synthetic genetic system was developed by Saeidi *et al.,* (2011) [47]. This system comprises the quorum sensing machinery (Fig. **15**), which when activated, activates the kill and lysing devices enabling the chassis organism Escherichia coli to sense and kill pathogenic *Pseudomonas aeruginosa* by producing and releasing pyocin. They also demonstrated this *in vitro* by showing a 99%

reduction in the viable cells. They also showed that the engineered *E. coli* inhibited the formation of *P. aeruginosa* biofilm by close to 90%, leading to much sparser and thinner biofilm matrices. These results suggest that a novel synthetic biology-driven antimicrobial strategy may be applied to fighting *P. aeruginosa* and other infectious pathogens.

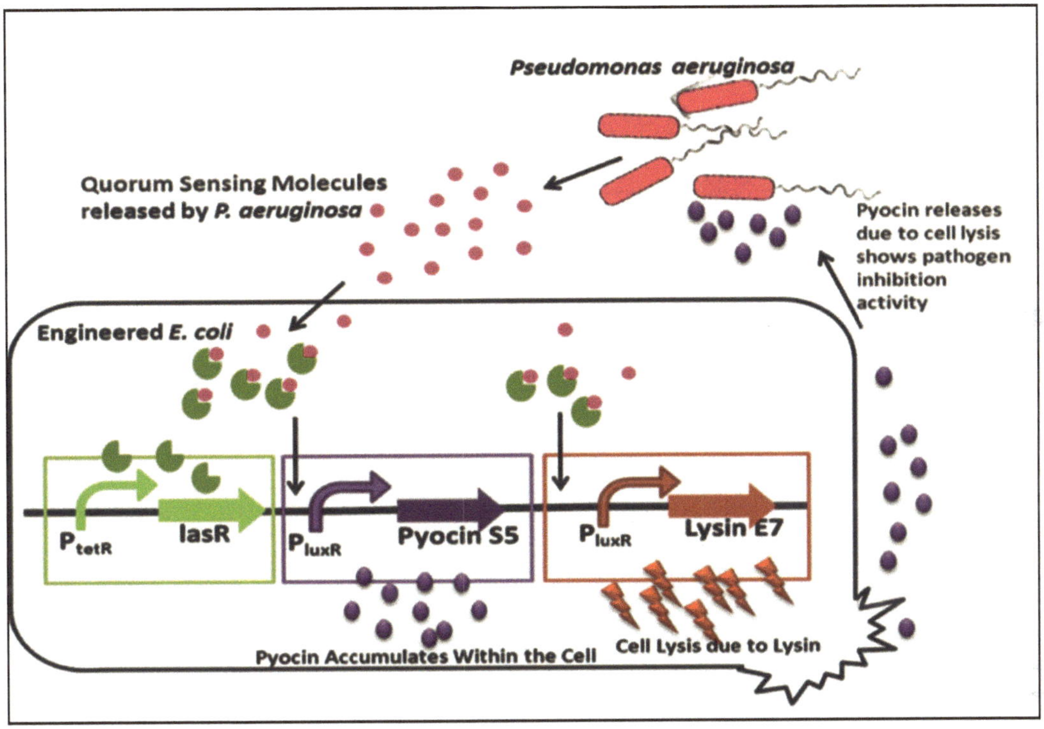

Figure 15: Upon activation of luxR promoter by LasR-3OC12HSL complex, initiates the production of E7 lysis protein and S5 pyocin within *E. coli* chassis. At the threshold concentration E7 lyses the chassis, releasing the accumulated S5 killing *P. aeruginosa*. (Adapted from Saeidi *et al.,* (2011)).

As another example, Lu *et al.,* (2007) [48] had engineered a bacteriophage that expresses a biofilm-degrading enzyme, which can degrade the bacterial extracellular polymeric biofilm matrix. This is relevant to clinically important bacterial infection during which formation of biofilm is crucial for pathogenesis and is difficult to eradicate due to resistance to antimicrobial treatments and removal by the host's immune system. They showed that the efficacy of biofilm removal by this two-pronged enzymatic bacteriophage strategy (Fig. **16**) is significantly greater than that of non-enzymatic bacteriophage treatment. This engineered enzymatic phage substantially reduces bacterial biofilm cell counts by 4.5 orders of magnitude (99.997% removal), which was about two orders of magnitude better than that of non-enzymatic phage.

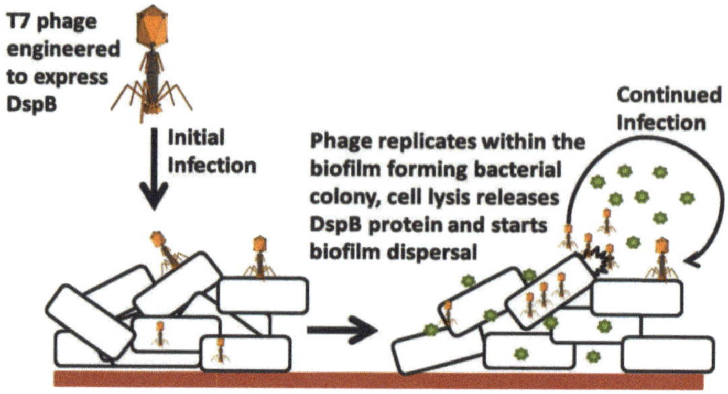

Figure 16: Engineered T7DspB phage expressing biofilm removal with enzymatically active DspB-expressing. It is a two pronged strategy as initial infection of *E. coli* biofilm results in rapid multiplication of phage and expression of DspB, which are released upon lysis, leading to subsequent infection as well as degradation of the biofilm (Adapted from Lu *et al.,* 2007).

The influenza virus envelope has two major immunogenic surface glycoproteins: hemagglutinin (HA) 1 and neuraminidase. The HA plays a key role in initiating viral infection by binding to sialic acid-containing receptors on host cells and thus mediating the subsequent viral entry and membrane fusion. Jeon *et al.,* (2004) [49] constructed a novel aptamer to complement the receptor-binding region of the influenza hemagglutinin molecule. It was constructed by screening a DNA library and processing by the selective evolution of ligands by exponential enrichment (SELEX) procedure [50, 51] (Fig. **17**). This DNA aptamer is capable of inhibiting the hemagglutinin capacity of the virus as proved by infectivity *in vitro*, in tissue culture. Furthermore, influenza animal models showed a 90–99% reduction of virus burden in the lungs of treated mice. The aptamer acts by blocking the binding of influenza virus to host cell receptors, hence preventing the invasion of the viral particle into the host cells.

Another infectious disease with increasing resistance to chemotherapeutics is leishmaniasis caused by a digenetic intracellular protozoan parasite belonging to the species Leishmania. Mandlik *et al.,* (2013) [52] have constructed a bistable

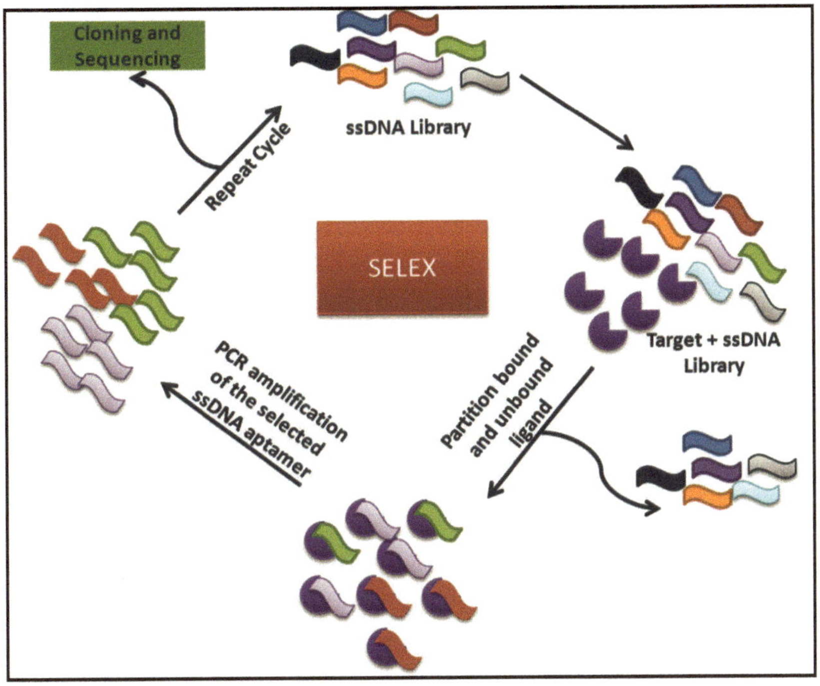

Figure 17: Schematics of the SELEX procedure which is an *in vitro* selection or *in vitro* evolution, for producing oligonucleotides (single-stranded DNA or RNA) that specifically bind to a target ligand or ligands.

genetic circuit targeting an enzyme belonging to sphingolipid metabolism - Inositol Phosphorylceramide synthase (IPCS). This target protein was identified by a systems biology approach by reconstructing the sphingolipid metabolism and developing a mathematical model followed by numerical simulation of the model (Mandlik *et al.,* 2012) [53]. The circuit was validated by Bayesian and Boolean approaches. In this model the bistable circuit was coupled to the repressilator by transcription repression by the first gene product in the repressilator to that of the gene in the toggle switch. The coupling leads to sustained oscillations, bistable behavior, together with intrinsic robustness. Similarly, Mol *et al.,* 2014 [54] have built a hypothetical model, were the modulated immune signal namely CD14 and TNF in leishmaniasis can be linked to EGFR pathway that is shown to be involved in wound healing. This integration of pathways is through MAPK crosstalk points in the isolated pathways. They propose to link the reconstructed signaling network to a gene circuit with a positive feedback loop, that may intervene an anti-inflammatory response in leishmaniasis to a proinflammatory response and also to initiate cell–cell communication, resulting in synchronized response in the immune cell population for disease resolving effect in leishmaniasis.

CONCLUDING REMARKS

Systems and synthetic biology together have certainly reduced the time span for identifying a drug target and designing strategies to deal with such targets. But what remains to be seen is how quickly these two approaches can be translated into the clinics. This requires the development of reliable strategies that help realize higher order networks with predictable behavior. These may be achieved by improving the design cycle. Most circuits are implemented in bacterial or yeast systems. However, the next step is to begin applying these approaches to mammalian systems in to address human diseases. Therefore, a synchronized effort should be strengthened to extend synthetic biology circuit construction strategies, which will prove crucial for the development of next generation therapeutics.

ACKNOWLEDGEMENTS

The authors thank the Department of Biotechnology, Government of India for funding the work and supporting the Bioinformatics and High Performance Computing Facility at National Centre for Cell Science, for, Pune, India.

CONFLICT OF INTEREST

The authors confirm that this chapter contents have no conflict of interest.

ABBREVIATIONS

API = Application Programming Interface

CAD = Computer Assisted Design

CD = Cluster Determinant

CellML = Cell Markup Language

COPASI	=	COmplex PAthway Simulator
CVs	=	Controlled Vocabularies
DBs	=	Databases
DE	=	Differential Equation
EGFR	=	Epidermal Growth Factor Receptor
GFP	=	Green Fluorescent Protein
HA	=	Heme Agglutinin
IPCS	=	Inositol PhosphorylCeramide Synthase
IPTG	=	Isopropylthio-β-Galactoside
MAP	=	Mitogen Activated Protein
NGS	=	Next Generation Sequencing
ODEs	=	Ordinary Differential Equations
PDEs	=	Partial Differential Equations
PK	=	Pharmacokinetics
ProMot	=	Process Modeling Tool
PSIMI	=	Proteomics Standard Initiatives Molecular Interaction
RBS	=	Ribosome Binding Site
RoVerGeNe	=	Robust Verification of Gene Networks
SB	=	Simbiology
SBML	=	Systems Biology Markup Language

SELEX = Systematic Evolution of Ligands by Exponential Enrichment

TFs = Transcription Factors

TNF = Tumor Necrosis Factor

UNAFold = Unified Nucleic Acid Folding

REFERENCES

[1] Ouzounis, C.; Valencia, A. Early bioinformatics: the birth of a discipline--a personal view. Bioinformatics., **2003,** 19(17), 2176–90.

[2] Cha, P.D.; Rosenberg, J.J.; Dym, C.L . Fundamentals of Modeling and Analyzing Engineering Systems; Cambridge University Press, NY, **2000**.

[3] Carson, E.; Cobelli, C. Modelling Methodology for Physiology and Medicine; Academic Press: San Diego, CA, **2001**.

[4] Murray, J.D. Mathematical Biology II; Springer, **2002**.

[5] Wayne, M.; David, S.W. Computational systems biology in drug discovery and development: Methods and applications. Drug Discovery Today., **2007,** 12(7/8), 295 – 303.

[6] Meng, T.C.; Sandeep, S.; Pawan, D. Modeling and simulation of biological systems with stochasticity. *In silico* Biology., **2004,** 4(3),293-309.

[7] Kitano, H. Systems biology: a brief overview. Science. **2002,** 295(5560), 1662–1664.

[8] Chen, W.W.; Schoeberl, B.; Jasper, P.J.; Niepel, M.; Nielsen, U.B.; Lauffenburger, D.A.; Sorger; P.K. Input-output behavior of ErbB signaling pathways as revealed by a mass action model trained against dynamic data. Molecular Systems Biology., **2009,** 5(1), 239.

[9] Sahin, O.; Frohlich, H.; Lobke, C.; Korf, U.; Burmester, S.; Majety, M.; Jens, M.; Ingo, S.; Claudine, C.; Denis, T.; Annemarie, P.; Stefan, W.; Tim, B.; Dorit, A. Modeling ERBB receptor regulated G1/S transition to find novel targets for *de novo* trastuzumab resistance. BMC Systems Biology., **2009,** 3(1).

[10] Klipp, E.; Krause, F. Computational Tools for Systems Biology. In Cancer Systems Biology, Bioinformatics and Medicine. Springer, Netherlands, **2011,** pp. 213-243.

[11] Vijayalakshmi, C.; Camille, L.; Nicolas, Le. N. BioModels Database: A Repository of Mathematical Models of Biological Processes. Methods in Molecular Biology., **2013,** 1021, 189-199.

[12] Ermentrout, B. Simulating, analyzing, and animating dynamical systems: a guide to XPPAUT for researchers and students. Society for Industrial Mathematics, Philadelphia **2002**.

[13] Funahashi, A.; Morohashi, M.; Kitano, H. CellDesigner: A Process Diagram Editor for Gene Regulatory and Biochemical Networks. BioSilico., **2003,** 1, 159-162.

[14] Klamt, S.; Saez-Rodriguez, J.; Lindquist, J. A.; Simeoni, L.; Gilles, E. D. A methodology for the structural and functional analysis of signaling and regulatory networks. BMC Bioinformatics., **2006,** 7(56).

[15] Hoops, S.; Sahle, S.; Gauges, R.; Lee, C.; Pahle, J.; Simus, N.; Singhal, M.; Xu, L.; Mendes, P.; Kummer, U. COPASI—a COmplex PAthway Simulator. Bioinformatics., **2006,** 55(9), 2561-2567.

[16] Schmidt, H.; Jirstrand, M. Systems biology toolbox for MATLAB: a computational platform for research in systems biology. Bioinformatics., **2006,** 22, 514–515.

[17] Keating, S. M.; Bornstein, B. J.; Finney, A.; Hucka, M. SBMLToolbox: an SBML toolbox for MATLAB users. Bioinformatics., **2006,** 22, 1275–1277.

[18] Moraru, I. I.; Schaff, J. C.; Slepchenko, B. M.; Blinov, M. L.; Morgan, F.; Lakshminarayana, A.; Gao, F.; Li, Y.; Leslie M. L. Virtual cell modelling and simulation software environment. IET Systems Biology., **2008,** 2, 352–362.

[19] Ernesto, A.; Subhayu, B.; David, K.K.; Ron, W. Synthetic biology: new engineering rules for an emerging discipline. Molecular Systems Biology., **2006,** 2(1).

[20] Ahmad, S.K.; James, J.C. Synthetic biology: applications come of age. Nature Genetics., **2010,** 11, 367-379.

[21] Marchisio, M.A.; Stelling, J. Computational design of synthetic gene circuits with composable parts. Bioinformatics., **2008,** 24(17), 1903–1910.

[22] Marc, F.; Martin, F. Synthetic biology advancing clinical applications. Current Opinion in Chemical Biology., **2012,** 16, 345–354.

[23] Goler, J. A. BioJADE: A Design and Simulation Tool for Synthetic Biological Systems, DSpace@MIT: Massachusetts Institute of Technology. Open Educational Resources (OER) 2004; <http://www.temoa.info/node/59969>

[24] Villalobos, A.; Ness, J.E.; Gustafsson, C.; Minshull, J.; Govindarajan, S. Gene Designer: a synthetic biology tool for constructing artificial DNA segments. BMC Bioinformatics **2006,** 7(1).

[25] Batt, G.; Yordanov, B.; Weiss, R.; Belta, C. Robustness analysis and tuning of synthetic gene networks. Bioinformatics., **2007,** 23(18),2415-2422.

[26] Markham, N.R.; Zuker, M. UNAFold: software for nucleic acid folding and hybridization. Bioinformatics, Volume II. Structure, Function and Applications, Methods in Molecular Biology, Humana Press, Totowa, **2008,** pp 3–31.

[27] Deepak, C.; Frank, T.B.; Herbert, M.S. TinkerCell: modular CAD tool for synthetic biology. Journal of Biological Engineering., **2009,** 3(19).

[28] Mirschel, S.; Steinmetz, K.; Rempel, M.; Ginkel, M.; Gilles, E.D. ProMoT: Modular Modeling for Systems Biology. Bioinformatics., **2009,** 25(5), 687-689.

[29] Yizhi, C.; Mandy, L.W.; Jean, P. GenoCAD for iGEM: a grammatical approach to the design of standard-compliant constructs. Nucleic Acids Research., **2010,** 38(8), 2637–2644.

[30] Marchisio, M. A.; Stelling, J. Computational design tools for synthetic biology. Current Opinion in Biotechnology., **2009,** 20, 479–485.

[31] Silver, P.; Way, J. Cells by design. Scientist., **2004,** 18, 30–31.

[32] Gardner, T.S.; Charles, R.C.; James, J.C. Construction of a genetic toggle switch in Escherichia coli. Nature., **2000,** 403(6767), 339-342.

[33] Atkinson, M. R.; Savageau, M.A.; Myers, J.T.; Ninfa, A.J. Development of genetic circuitry exhibiting toggle switch or oscillatory behavior in Escherichia coli. Cell., **2003,** 113(5), 597-607.

[34] Elowitz, M.B.; Leibler, S. A synthetic oscillatory network of transcriptional regulators. Nature **2000,** 403(6767), 335-338.

[35] Lefever, R.; Gregoire, N.; Pierre, B. The Brusselator: it does oscillate all the same. Journal of the Chemical Society, Faraday Transactions 1: Physical Chemistry in Condensed Phases., **1988,** 84(4), 1013-1023.

[36] Stefan S. Oscillations in Chemical Systems 2006 <http://www.biosym.uzh.ch/modules/models/Oscillation/oscillation.xhtml#Brusselator>

[37] Tyson, J. J. Modeling the cell division cycle: cdc2 and cyclin interactions. Proceedings of the National Academy of Science, USA., **1991,** 88, 7328-7332.

[38] Toner, D.L.K.; Grima, R. Effects of bursty protein production on the noisy oscillatory properties of downstream pathways. Scientific Reports **2013**; 3.

[39] Li, B.; You, L. Synthetic biology: Division of logic labour. Nature., **2011,** 469(7329), 171-172.

[40] Hoffman-Sommer, M.; Supady, A.; Klipp, E. Cell-to-cell communication circuits: quantitative analysis of synthetic logic gates. Frontiers in Physiology., **2012,** 3.

[41] Regot, S.; Macia, J.; Conde, N.; Furukawa, K.; Kjellen, J.; Peeters, T. Distributed biological computation with multicellular engineered networks. Nature., **2010,** 469(7329), 207-211.

[42] Alon, U. An Introduction to Systems Biology: Design Principles of Biological Circuits. Mathematical and Computational Biology Series, Chapman and Hall, CRC Press, 2007.

[43] Burrill, D. R.; Silver, P. A. Making cellular memories. Cell., **2010,** 140(1), 13-18.

[44] Burrill, D. R.; Inniss, M. C.; Boyle, P. M.; Silver, P. A. Synthetic memory circuits for tracking human cell fate. Genes and Development., **2012,** 26(13), 1486-1497.

[45] Vitreschak, A. G.; Rodionov, D. A.; Mironov, A. A.; Gelfand, M. S. Riboswitches: the oldest mechanism for the regulation of gene expression?. Trends in Genetics., **2004,** 20(1), 44-50.

[46] Beisel, C. L.; Smolke, C. D. Design principles for riboswitch function. PLoS Computational Biology., **2009,** 5(4), e1000363.

[47] Saeidi, N.; Wong, C.K.; Lo, T. M.; Nguyen, H. X.; Ling, H.; Leong, S. S.; Poh, C. L.; Chang, M. W. Engineering microbes to sense and eradicate *Pseudomonas aeruginosa*, a human pathogen. Molecular Systems Biology., **2011,** 7(1).

[48] Lu, T. K.; James, J. C. Dispersing biofilms with engineered enzymatic bacteriophage. Proceedings of the National Academy of Sciences, USA., **2007,** 104(27), 11197-11202.

[49] Jeon, S. H.; Kayhan, B.; Ben-Yedidia, T.; Arnon, R. A. DNA aptamer prevents influenza infection by blocking the receptor binding region of the viral hemagglutinin. Journal of Biological Chemistry., **2004,** 279(46), 48410-48419.

[50] Ellington, A. D.; Szostak, J. W. *In vitro* selection of RNA molecules that bind specific ligands. Nature., **1990,** 346(6287), 818-822.

[51] Stoltenburg, R.; Reinemann, C.; Strehlitz, B. SELEX—a (r) evolutionary method to generate high-affinity nucleic acid ligands. Biomolecular Engineering., **2007,** 24(4), 381-403.

[52] Mandlik, V.; Limbachiya, D.; Shinde, S.; Mol, M.; Singh, S. Synthetic circuit of inositol phosphorylceramide synthase in Leishmania: a chemical biology approach. Journal of Chemical Biology., **2013,** 6(2), 51-62.

[53] Mandlik, V.; Shinde, S.; Chaudhary, A.; Singh, S. Biological network modeling identifies IPCS in Leishmania as a therapeutic target. RSC Integrated Biology., **2012,** 4(9), 1130-1142.

[54] Mol, M.; Patole, M. S.; Singh, S. Immune signal transduction in leishmaniasis from natural to artificial systems: Role of feedback loop insertion. Biochimica et Biophysica Acta (BBA)-General Subjects., **2014,** 1840(1), 71-79.

CHAPTER 6

Considering the Medium when Studying Biologically Active Molecules: Motivation, Options and Challenges

Liliana Mammino[1,*] and Mwadham M. Kabanda[2]

[1]Department of Chemistry, University of Venda, South Africa and [2]Department of Chemistry, North-West University (Mafikeng Campus), South Africa

Abstract: The computational study of biologically active molecules plays important roles in drug development, as it provides information on molecular properties which, in turn, determine the biological activities of compounds. Within a living organism, molecules are within a medium and, therefore, their activity is exerted in a medium. Because of this, knowing how the presence of a medium influences the properties of a given molecule is important for drug development. This chapter aims at providing a comprehensive overview of the aspects relevant to the computational study of biologically active compounds in a medium. It outlines the main models currently utilised to take into account solute-solvent interactions and the solvent effects on the molecular properties of the solute, considering also the information abilities and limitations of each model and the challenges for further research. It discusses relevant criteria for the selection of the preferable solvents to consider in the study of a given molecule. Information, analyses and discussions are extensively supported by the consideration of examples from literature and from the authors' direct experience.

Keywords: Acylphloroglucinols, Biologically active molecules, Discrete models, Drug design, Polarisable continuum model, Solute-solvent interactions, Solvent effects.

INTRODUCTION

Biologically active substances are substances which can stimulate a response from a living organism, when introduced into it. When this response results in gradual decrease of the effects of a disease until the disease is treated, the biologically active compound is called *drug*. The main objectives of drug research concern the development of:

***Corresponding author Liliana Mammino:** Department of Chemistry, University of Venda, South Africa; E-mail: sasdestria@yahoo.com

Zaheer Ul-Haq and Jeffry D. Madura (Eds). **10.1016/B978-1-60805-865-5.50006-X**

- New drugs for the treatment of diseases for which no effective drugs have yet been identified;

- Improved drugs to replace drugs in current clinical use. The "improvement" may involve more potent activity enabling dosage-reduction, or the decrease of unwanted side effects [1, 2];

- New drugs to replace drugs in current clinical use, for which the pathogen has developed resistance. This problem is presently particularly serious for malaria, tuberculosis and cancer [1-3].

The development of effective drugs for the treatment of diseases has been one of the major objectives of chemistry since its very birth, when it was emerging from alchemy. Early investigators involved in the transition – like Paracelsus – considered drug development the major mission of chemistry, and *iatrochemistry* (what would now be called *medicinal chemistry*) was one of the first forms of chemistry entering European universities in the XVII and XVIII centuries. The progress of chemical knowledge through centuries has enabled parallel continuous progress in the discovery and design of new drugs. In recent decades, computational chemistry has brought new perspectives into drug design, thanks to its ability to provide information about molecular properties and about the relationships between the properties of molecules and their biological activities.

The properties of a substance depend on the properties of its molecules. The more we know about the properties of its molecules, the more we can understand about the properties of a given substance. Furthermore, the knowledge of the properties of a sufficiently representative number of different molecules of a given class of compounds enables reliable predictions of several properties of other not-yet-synthesised molecules of the same class. All these aspects play important roles in drug design. The extent of attained information is the key factor both for understanding the action of already known drugs and for predicting possible actions of new drugs.

The biological activity may be related to the finest details of a molecule's properties [4] and, therefore, the computational study of biologically active molecules aims at obtaining information on as many details as possible. Part of this information (including electronic and geometric descriptors) is utilised in the search for *Quantitative Structure Activity Relationships* (QSAR). QSAR aims at relating molecular properties to biological activity. The molecular properties are expressed through descriptors such as geometry parameters, frontier orbitals energy gaps, dipole moments, and other measurable quantities. The values of the descriptors that have proved more relevant for the biological activity of a given molecule or class of molecules are introduced into a regression equation, whose other term is the value of the measured activity.

The conformer responsible for the biological activity is not always the lowest-energy conformer of a given molecule; it may be another conformer that is sufficiently populated. Therefore, it is important to find information about all the conformers with sufficiently low relative energy. Selecting a cautious threshold (such as "relative energy ≤ 3.5 kcal/mol") ensures that no conformer that may be involved in the biological activity is overlooked.

Biological activities may involve a variety of mechanisms: inhibition of the active site of an enzyme vital for the pathogen; intercalation with the DNA of the pathogen, or of sick cells like cancer cells, preventing their reproduction; inhibition of pre-existing ion channels or formation of new membrane pores that disrupt cellular ion-balance in the membrane plasma of pathogens; and others. The knowledge of the mechanism through which existing drugs act is useful in the design of new drugs.

A molecule's biological activity is exerted within a living organism and, therefore, it is exerted in a medium. When a molecule dissolves in a medium, interactions between the molecule (solute) and the medium (solvent) are established. These interactions may influence the properties of the molecule with respect to when it is isolated (gas phase or *in vacuo*) and, therefore, they may also influence its biological activity. Because of this, it is important to study a biologically active molecule in media that may provide sufficiently good approximations for the medium in which it exerts its activity within a living organism.

The medium is the major component of living organisms (for instance, water constitutes about 70% of the human body). It is crucial for the "chemistry" within the organism both directly, by actively participating in biological processes, and indirectly, by stabilizing biologically active conformations of macromolecules (*e.g.*, proteins and nucleic acids). Studying the interactions between biomolecules and the medium in a living system is thus relevant for a better understanding of biological processes [5-10], through the elucidation of the role of the solvent in these processes. Such studies involve both extensive experimental investigation and the utilisation of theoretical and computational methods [11]. Given the size of biomolecules such as proteins and nucleic acids, the computational modelling of their interactions with the medium may initially investigate the interactions with that medium of the molecules that are "building blocks" of the bigger structure (*e.g.*, pyrimidine as the parent compound of the pyrimidinic bases in nucleic acids [12]), or it may select suitable portions of the bigger structures, such as representative segments of DNA, or the region of the active site of an enzyme.

In the study of drugs and in drug design, it is important to investigate how solute-solvent interactions influence the properties of the biologically active molecule considered [11, 13], to elucidate the properties that the molecule will have in the conditions in which it exerts its action. This chapter aims at providing a comprehensive overview of the aspects pertinent to the study of biologically active molecules in a medium. After a presentation of the models currently in use for the study of solute-solvent interactions and a brief outline of their description abilities, the chapter focuses on the features more closely related to the study of biologically active molecules. Since biologically active molecule act in a fundamentally liquid medium, the solutions considered are liquid solutions.

SOLUTE-SOLVENT INTERACTIONS AND THEIR EFFECTS

Main Features of the Dissolution Process

The dissolution process is the process by which the molecules of a substance (called *solute*) disperse within another substance (called *solvent*) to give a homogeneous mixture (called *solution*), in which each solute molecule is completed surrounded by solvent molecules. A substance usually does not dissolve in all types of solvent. Each substance preferably dissolves in some solvents and not in others (often according to the empirical rule that *like dissolves like*, *i.e.*, polar substances preferably dissolve in polar solvents and non-polar substances in non-polar solvents).

The dissolution process is determined by the nature and strength of the intermolecular forces in the pure solute (solute-solute interactions), in the pure solvent (solvent-solvent interactions) and between the solute and the solvent (solute-solvent interactions, which are established when the solution is formed). The solute-solute interactions and some of the solvent-solvent interactions must be broken for solute-solvent interactions to be established. The breaking of the solute-solute interactions and solvent-solvent interactions requires energy, whereas the establishing of the solute-solvent attractive interactions releases energy. A solute A dissolves in a solvent B if the A-B solute-solvent attractive interactions are strong enough to overcome the A-A solute-solute and B-B solvent-solvent attractive interactions. The overall changes in thermodynamic quantities comprise the contributions of these three components of the dissolution process. For instance, the overall enthalpy change (ΔH_{soln}) accompanying the dissolution process is the sum of the enthalpy change accompanying the separation of the solute molecules from each other ($\Delta H_{A\text{-}A}$), the enthalpy change accompanying the separation of the solvent molecules from each other in the places where the solute molecules insert themselves ($\Delta H_{B\text{-}B}$), and the enthalpy change accompanying the establishing of the solute-solvent interactions ($\Delta H_{A\text{-}B}$):

$$\Delta H_{soln} = \Delta H_{A\text{-}A} + \Delta H_{B\text{-}B} + \Delta H_{A\text{-}B} \quad (1)$$

The outcome of the dissolution process is governed by both the enthalpy changes and the entropy changes involved. Therefore, it is more convenient to consider the Gibbs free energy function (G), which incorporates both functions:

$$G = H - TS \quad (2)$$

where H is the enthalpy, T is the absolute temperature and S is the entropy. The Gibbs free energy change (ΔG) accompanying the process must be negative for net dissolution to take place (consistently with the $\Delta G < 0$ condition for a process to be spontaneous). The surrounding of a solute molecule by solvent molecules is called *solvation*. Therefore, the ΔG accompanying the dissolution process is called *free energy of solvation* (ΔG_{solv}). When water is the solvent, the term *hydration* is often used in place of *solvation*.

Similarly to any mixing process, the dissolution process implies an entropy increase. A major contribution to the entropy change is due to the local changes in the solvent as a result of the insertion of the solute molecules [14].

Main Types of Interactions Between Solute Molecules and Solvent Molecules

A solvated solute molecule is surrounded by many solvent molecules. The solvent molecules closer to it (one could say, "in direct contact" with it) constitute the *first solvation layer* (or *first solvation shell*). The solvent molecules surrounding the first solvation layer constitute a *second solvation layer*, in turn surrounded by many other layers of solvent molecules. Actually, the *solvation layer* concept becomes rapidly fuzzy for solvent molecules further away from the solute than the first or, maximum, the second layer, because of the dynamic situation inherent in a liquid. In a liquid, molecules move continuously, sliding over each other. The molecules constituting a solvation layer are not the same over a significantly long period of time. They interchange continuously and rapidly. This dynamic nature is one of the greatest challenges for the development of satisfactory models of the liquid state in general, or of liquid solutions.

The type and strength of the interactions between a solute molecule and the solvent molecules depend on the nature of the solute molecule and on the nature of the solvent molecules. It is possible to view the effect of the solvent on the solute molecule without considering individual solvent molecules, but considering an overall effect called the *bulk solvent effect*. An important component of the bulk effect of the solvent on the solute molecule is the *solute polarisation* on dissolution. It is simpler to analyse it with reference to the isolated solute molecules (gas phase). When a solute molecule moves from the gas phase (with dielectric constant 1) into a solution (where the dielectric constant of the solvent is higher than 1), the geometry and charge distribution of the solute molecule relax to allow greater charge separation within the molecule itself and better interactions with the solvent molecules [14]. The outcome is a distortion of the geometry of the solute molecule with respect to the optimal gas-phase geometry, implying an increase in the internal energy of the solute. In turn, the solute polarises the surrounding solvent, raising its free energy. These two factors (increase in the internal energy of the solute and in the free energy of the solvent) partially cancel the free energy lowering resulting from more favourable interactions of the polarised solvent and polarised solute. The relaxation of the geometry of the solute molecule proceeds as long as further polarisation favours better solute-solvent interactions; it does not proceed further when the favourable consequences are overcome by the intrasolute and intrasolvent costs of the geometry distortion [14]. The extent of the solute polarisation depends on the solvent dielectric constant and is greater when the dielectric constant is higher. By affecting the geometry of the solute molecule, the solute polarisation may also affect its electronic and magnetic properties.

Other effects – besides the solute polarisation – are closely related to the interactions with the solvent molecules in direct contact with the solute molecules (the first solvation layer). The generation of a fresh solvent surface around the solute molecule when this inserts itself into the solvent requires free energy; it is called *cavitation energy* and is one of the non-electrostatic components of ΔG_{solv}. The insertion of the solute molecule into the solvent brings local structural changes in the portion of solvent surrounding the solute [14]; the extent of these changes depends on the types of solute-solvent interactions and is greater when the interactions are stronger, like in the case of solute-solvent hydrogen bonds (H-bonds). The interactions at the solute-solvent interface include also attractive dispersion forces between the solute molecule and the surrounding solvent molecules [14].

When both the solute molecule and the solvent molecules have H-bond donor or acceptor sites, solute-solvent intermolecular H-bonds constitute the strongest solute-solvent interactions. H-bonds are directional, as they involve specific atoms in each of the interacting molecules. (Fig. **1**) shows an example considering water solution. When the solvent molecules can also H-bond to each other, the formation of solute-solvent H-bonds brings important changes in the solvent-solvent H-bonding pattern [14], above all in the first solvation layer (Fig. **2**). Fig. **3** compares the case of a solvent (acetonitrile) in which solvent-solvent H-bonds are not possible and the case of water, in which they are possible. (Fig. **4**) shows the general structure of acylphloroglucinols – a class of compounds largely utilised for illustrative examples in this chapter because of the presence of several H-bond donors and acceptor sites in their molecules.

Effects of the Solvent on the Properties of a Solute Molecule

Solute-solvent interactions may significantly change the properties of the solute molecule, such as geometry parameters (bond lengths, bond angles and torsion angles) of the equilibrium geometry of the molecule's individual conformers, conformational preferences (relative energies of the conformers [16]), charge distribution [17], dipole moment [18], vibrational frequencies [19], electronic transition energies [20-22], NMR constants [18], chemical reactivity [23-25], and others. By depending on the solute-solvent interactions, the extent of properties-changes with respect to the gas phase depends on the nature of the solvent and on the nature of the solute. Tables **1-4** provide illustrative examples for a selected acylphloroglucinol molecule (Fig. **5**) in three solvents - chloroform (ε = 4.90), acetonitrile (ε = 36.64) and water (ε = 78.39) - considering the following effects: changes in the conformers' relative energy [26]; changes in the geometry parameters of the intramolecular hydrogen bond (IHB) between by the sp^2 O atom of the acyl chain and a neighbouring OH [27]; changes in the (harmonic) vibrational frequency for the stretching of the O-H bonds [27]; and changes in the Mulliken atomic charges on the O atoms engaged in the IHBs [26].

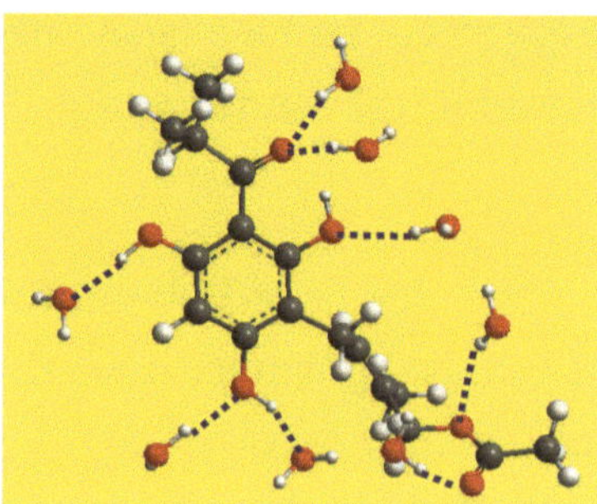

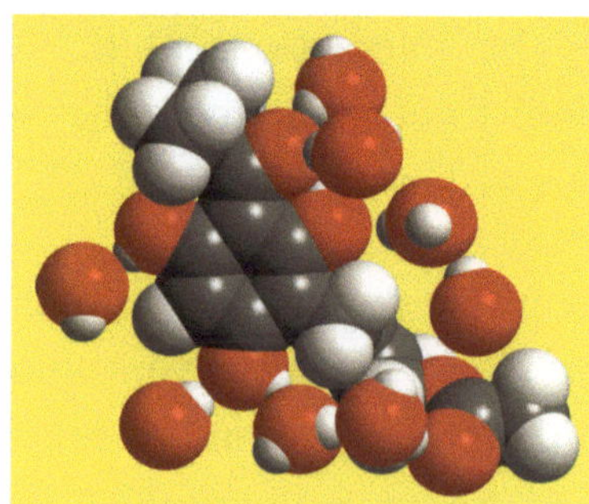

Figure 1: Intermolecular hydrogen bonds between a low-energy conformer of the caespitate molecule and water molecules.
Caespitate is an acylphloroglucinol (Fig. **4**) with an ester function in the side chain in meta to the acyl chain. The solute-solvent H-bonds are denoted by blue dotted segments (left). The image on the right shows a "space filling" model of the same complex. The overall interaction energy between the central molecule and the water molecules is -28.180 kcal/mol (from RHF/6-31G(d,p) calculations).

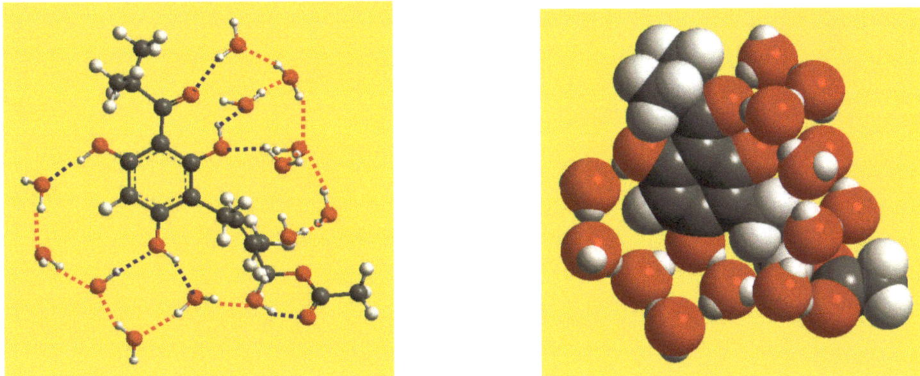

Figure 2: Adduct of a low energy conformer of the caespitate molecule with 13 water molecules [15].

The figure shows the same conformer as in (Fig. **1**), and the solute-water H-bonds are the same as in (Fig. **1**). Differently from (Fig. **1**), this adduct includes water molecules bridging the water molecules directly H-bonded to the solute molecule, thus showing how the presence of solute-solvent H-bonds influences the water-water H-bonding patterns in the vicinity of the solute molecule. In the image on the left, the solute-water intermolecular H-bonds are denoted by blue dotted segments and the water-water intermolecular H-bonds by red dotted segments. The image on the right shows a "space filling" model of the same complex, to better highlight the "contact" between water molecules and the donor or acceptor sites in the solute molecule, and between one water molecule and another. The overall interaction energy between the solute molecule and the water molecules is -38.231 kcal/mol (from RHF/6-31G(d,p) calculations).

(a) (b)

Figure 3: Adducts of phloroglucinol with acetonitrile molecules (a) and with water molecules (b). [16].

In case (a), only solute-solvent H-bonds are possible. In case (b), both solute-solvent H-bonds and solvent-solvent H-bonds are possible, which brings the water molecules bridging those directly H-bonded to the solute molecule into the first solvation layer. Solute-solvent H-bonds are denoted by blue dotted segments and solvent-solvent H-bonds by red dotted segments.

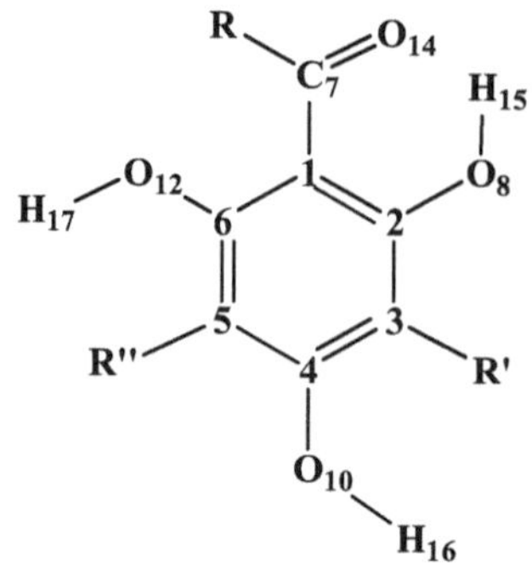

Figure 4: General structure of acylphloroglucinols.
The structure shows the atom numbering utilised in this chapter (*e.g.*, Tables **3**, **4**), which is consistent with the one utilised in works on acylphloroglucinols [26-28]. It also highlights the H-bonding abilities of acylphloroglucinols with solvent molecules. In the interactions with water molecules, O8, O10, O12 and O14 may act as H-bond acceptors and H15, H16 and H17 may each bond to the O atom of a water molecule. In the interaction with acetonitrile molecules, H15, H16 and H17 may each bond to the N atom of an acetonitrile molecule.

The property of the solvent that is more generally taken into account as a source of bulk effects on the solute molecules is its ability to polarise the solute molecule. This is often discussed in terms of *solvent polarity*. However, the solvent effect appears to be related to the solvent dielectric constant more than to its dipole moment. This is clearly evident when the polarities and the dielectric constants of two or more solvents do not have parallel trends. For instance, the acetonitrile molecule has a dipole moment of 3.9 D (4.6 D in the liquid), whereas the dipole moment of water is 1.85 D (2.6 D in the liquid). The dielectric constant of acetonitrile ($\varepsilon = 36.64$) is less than half that of water ($\varepsilon = 78.39$). The effect of acetonitrile on solute molecules is intermediate between that of chloroform (a non-polar solvent with $\varepsilon = 4.90$) and that of water – actually somewhat closer to that of chloroform; this is consistent with the dielectric constant of acetonitrile being intermediate between that of chloroform and that of water.

Intramolecular electrostatic effects may play relevant roles in determining some conformational features of a molecule in the gas phase. If the same molecule is dissolved in a solvent with high dielectric constant, the strength of the intramolecular electrostatic forces may decrease, and this influences conformational preferences [29]. For organic compounds, the effect of a non-polar solvent on conformers' relative energies and conformational preferences may be minimal, whereas a polar solvent may have greater effects by modifying the strength of the electrostatic forces within the solute molecule. However, as long as stronger solute-solvent interactions such as H-bonding or other donor-acceptor interactions are not present, the solvent effect is often reasonably well related to the dielectric constant of the solvent [29].

The gaps between the relative energies of the conformers usually decrease in solution with respect to in vacuo, and the decrease is greater as the solvent dielectric constant increases (as illustrated by the values in Table **4**). This phenomenon influences the range of conformers which may be considered as possible responsibles for the biological activity, as the selected threshold (like the previously-mentioned 3.5 kcal/mol) applies also to the situation in solution; thus, *e.g.*, a conformer whose relative energy is above the threshold in vacuo, but below the threshold in some solvent mimicking one of the media in a living organism, needs to be taken into account.

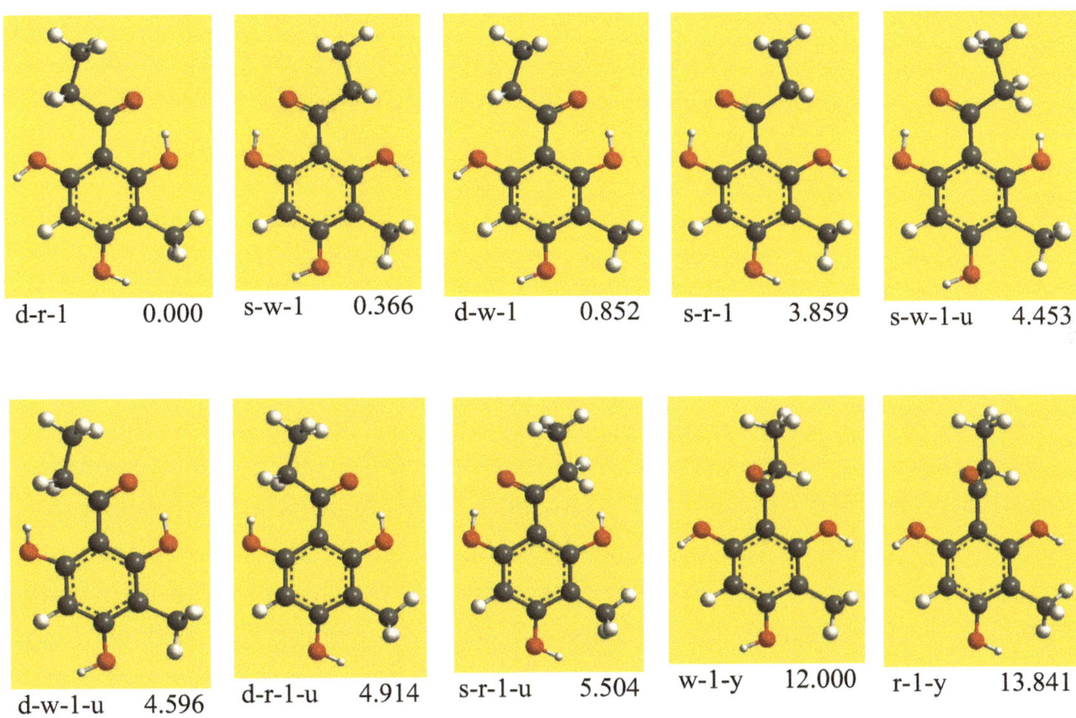

Figure 5: Conformers of the acylphloroglucinol molecule considered in Tables **1**-**4**.
The conformers are denoted with the same acronyms utilised in [15, 26, 27] to keep track of relevant geometry features, and are arranged in order of increasing relative energy, whose values (kcal/mol, from HF/6-31G(d,p) calculations) are reported under each image. The compound is denoted as D in [15, 26, 27].

The possibility of stronger and directional interactions such as solute-solvent H-bonds implies greater effects on the solute. The type of effect depends on the types of solute-solvent H-bonds that can be established, on whether the solvent molecules may H-bond to each other or not (Fig. **3**), and on the presence (Figs. **1**, **2**) or absence (Fig. **3**) of IHBs in the solute molecule in the gas phase. When the solute molecule contains IHBs in the gas phase, there may be a competition between intramolecular and intermolecular (solute-solvent) H-bonding for each of the relevant sites in the solute molecule. The main possible outcomes are the following:

- The IHB is sufficiently strong not to break in solution. If the structure of the molecular system is sufficiently rigid to prevent stable geometries with different orientations of the groups forming the IHBs, the area in the vicinity of the IHB may behave as prevalently hydrophobic (Fig. **6**, [15, 16, 26, 30]).

- The IHB does not break in solution, but the atoms forming it engage also in intermolecular H-bonds with the solvent, resulting in cooperative H-bonds (Fig. **7**, [31-34]), often with some modifications in the geometry of the IHB.

- The IHB breaks in solution, and its donor and acceptor atoms engage in intermolecular H-bonds with the solvent molecules (Fig. **8**, [15, 28, 35, 36]).

What happens for each solute depends on the characteristics of the solute molecule and can be determined both experimentally and through theoretical modelling. The study of acylphloroglucinols (Fig. **4**) in water solution offers illustrative examples of various effects [15, 26, 28, 33, 35, 36]. Considering water solution is particularly interesting because water is the solvent more abundantly present in living organisms, and because it is capable of being both H-bond donor and H-bond acceptor. The results for acylphloroglucinols showed that:

- The IHB between the sp^2 O of the acyl chain and an ortho OH (termed *first IHB*) does not break in water solution, and the region in its vicinity behaves as hydrophobic. The IHB permanence is likely related both to the greater strength of an H-bond involving an sp^2 O and to geometry constrains for which the sp^2 O and the ortho OH cannot move sufficiently far apart from each other for their mutual interaction to vanish.

- The IHB involving a phenol OH and some donor or acceptor in a substituent chain (*second IHB*, [28]) often breaks in water solution to favour the formation of intermolecular H-bonds with water molecules (Fig. **8**).

- The conformers' relative energies change significantly in water solution. For example, a certain conformer A having both the first and the second IHB may be more stable *in vacuo* than a conformer B without the second IHB; however, in water solution, conformer A may have higher relative energy than conformer B, because conformer B is able to form more intermolecular H-bonds with water molecules than conformer A (Fig. **9**).

Table 1: Relative energies of the conformers of the acylphloroglucinol molecule shown in (Fig. **5**), in different media. HF/6-31G(d,p) results, with PCM full reoptimization in solution [26]

Conformer	**Relative energy (kcal/mol)**			
	In vacuo	**In chloroform**	**In acetonitrile**	**In water**
d-r-1	0.000	0.534	1.152	2.225
s-w-1	0.366	0.787	1.321	2.562
d-w-1	0.852	0.000	0.000	0.000
d-r-2	1.386	1.775	2.356	3.432
s-w-2	1.79	2.082	2.568	3.877
d-w-2	2.263	1.257	1.208	1.259
s-r-1	3.859	3.867	4.326	5.584
s-w-1-u	4.453	4.577	4.831	-[a]
d-w-1-u	4.596	4.356	5.384	4.393
d-r-1-u	4.914	6.047	6.946	6.704
s-r-1-u	5.504	5.939	-[a]	6.901
w-1-y	12.000	9.275	8.394	6.334
r-1-y	13.841	11.351	10.760	8.616

[a] Conformer not obtained from PCM optimization in the given solution.

Table 2: Parameters of the intramolecular hydrogen bond between the sp^2 O of the acyl chain and a neighbouring OH in the conformers of the acylphloroglucinol molecule shown in Fig. **5**, in different media. HF/6-31G(d,p) results, with PCM full reoptimization in solution [27]
The dihedral angle of O14 with the plane of the benzene ring is also included, to provide a complete description of the geometry aspects related to the IHB considered.

Conformer	Medium	Intramolecular H-bond Parameters [a]			∠O14 with the plane
		H···O (Å)	O···O (Å)	$O\hat{H}O$	
d-r	in vacuo	1.666	2.517	145.8	0.035
	in chloroform	1.660	2.515	146.5	0.010
	in acetonitrile	1.658	2.515	146.7	0.013
	in water	1.657	2.514	146.7	0.062
d-w	in vacuo	1.682	2.527	145.0	0.013
	in chloroform	1.676	2.526	145.7	0.055
	in acetonitrile	1.673	2.525	146.0	0.033
	in water	1.674	2.525	145.9	0.428
s-r	in vacuo	1.698	2.534	143.9	0.708
	in chloroform	1.689	2.530	144.7	0.900
	in acetonitrile	1.685	2.529	144.9	1.385
	in water	1.692	2.532	144.4	4.517
s-w	in vacuo	1.696	2.533	143.9	0.001
	in chloroform	1.688	2.530	144.7	0.006
	in acetonitrile	1.686	2.530	144.9	0.021
	in water	1.689	2.530	144.5	0.007
d-r-u	in vacuo	1.667	2.514	145.4	0.000
	in chloroform	1.657	2.509	146.0	0.054
	in acetonitrile	1.652	2.506	146.4	0.002
	in water	1.696	2.541	145.1	15.698
d-w-u	in vacuo	1.687	2.555	144.5	0.004
	in chloroform	1.701	2.542	144.7	13.345
	in acetonitrile	1.669	2.517	145.7	0.008
	in water	1.711	2.550	144.4	15.419
s-r-u	in vacuo	1.729	2.554	142.8	15.070
	in chloroform	1.724	2.554	143.4	16.206
	in acetonitrile	1.685	2.529	144.9	1.410
	in water	1.729	2.558	143.1	15.968

Table 2: contd...

s-w-u	in vacuo	1.725	2.526	142.8	13.535
	in chloroform	1.721	2.552	143.4	15.157
	in acetonitrile	1.718	2.551	143.7	15.040

[a] The phenol atoms engaged are O8 and H15 (Fig. 4) for conformers whose name starts with d, and O12 and H16 for conformers whose name starts with s.

Table 3: Calculated (harmonic) vibrational frequencies (ν, cm^{-1}) of the stretching of the O—H bonds in the conformers of the acylphloroglucinol molecule shown in Fig. **5**. and relevant changes when the H atom is engaged in an intramolecular hydrogen bond. HF/6-31G(d,p) results, with PCM full reoptimization in solution [27]
The frequency values have been scaled by 0.903, which is the scaling factor corresponding to HF/6-31G(d,p) calculations [37].
For the cases where the H atom is engaged in an IHB, the following quantities are reported: decrease in the given solvent with respect to *in vacuo* (Δν, cm^{-1}); percent decrease in the given solvent with respect to *in vacuo* (%Δν); red shift (RS, cm^{-1}), calculated in the conventional way (frequency of a free O—H minus frequency of the O—H engaged in the IHB); and percent red shift (%RS, cm^{-1}).
The values of Δν and %Δν are reported to highlight the low effect of the solvent on the frequencies of the O—H engaged in the IHB. The values of RS and %RS are referred to the frequency of O9—H15; they are reported to highlight how their consideration might lead to overestimation of the solvent effect on the frequency of the O—H engaged in the IHB, due to the considerable change in the frequency of the reference (the free O—H).
The media are denoted with the following acronyms: vac (vacuum), chlrf (chloroform), actn (acetonitrile) and aq (water). The atom numbering is shown in (Fig. **4**).

Conformer and medium	**O8—H15**					**O12—H17**					**O10—H16**
	ν	Δν	%Δν	RS	%RS	ν	Δν	%Δν	RS	%RS	ν
d-r											
vac	3442			352	9.3	3780					3794
chlrf	3435	7	0.20	334	8.9	3721					3769
actn	3431	11	0.32	325	8.7	3692					3756
aq	3418	24	0.70	230	6.3	3498					3648
d-w											
vac	3476			315	8.3	3784					3791
chlrf	3463	13	0.37	268	7.2	3723					3731
actn	3456	20	0.58	244	6.6	3693					3700
aq	3447	29	0.83	57	1.6	3497					3504
s-r											
vac	3813					3504			313	8.2	3817
chlrf	3782					3485	19	0.54	299	7.9	3784
actn	3768					3477	27	0.77	289	7.7	3766
aq	3635					3475	29	0.83	152	4.2	3627
s-w											

Table 3: contd…

vac	3798					3490			293	7.7	3783
chlrf	3775					3476	14	0.40	248	6.7	3724
actn	3763					3470	20	0.57	221	6.0	3691
aq	3674					3461	29	0.83	38	1.1	3499

Table 4: Calculated Mulliken charges (atomic units) on the O atoms forming the intramolecular hydrogen bond between the sp^2 O of the acyl chain and a neighbouring OH, in the conformers of the acylphloroglucinol molecule shown in (Fig. **5**). HF/6-31G(d,p) results, with PCM full reoptimization in solution [27]

Conformer	**Charge on O14**				**Charge on the phenol O** [a]			
	vac	**chlrf**	**actn**	**aq**	**vac**	**chlrf**	**actn**	**aq**
d-r-1	-0.652	-0.668	-0.674	-0.689	-0.675	-0.682	-0.685	-0.693
d-w-1	-0.648	-0.665	-0.672	-0.688	-0.665	-0.676	-0.680	-0.690
s-r-1	-0.645	-0.662	-0.669	-0.683	-0.663	-0.677	-0.682	-0.692
s-w-1	-0.648	-0.665	-0.671	-0.686	-0.665	-0.677	-0.681	-0.691
d-r-1-u	-0.648	-0.660	-0.665	-0.673	-0.675	-0.681	-0.683	-0.690
d-w-1-u	-0.643	-0.651	-0.663	-0.672	-0.665	-0.674	-0.679	-0.687
s-r-1-u	-0.633	-0.648	-0.669	-0.670	-0.661	-0.674	-0.682	-0.690
w-1-y [b]	-0.512	-0.542	-0.553	-0.576	-0.667	-0.675	-0.679	-0.698
r-1-y [b]	-0.509	-0.540	-0.552	-0.576	-0.665	-0.675	-0.677	-0.704

[a] The phenol O is O8 (Fig. **4**) for conformers whose name starts with d, and O12 for conformers whose name starts with s.
[b] These conformers do not have any IHB. They are included as reference, to compare with the charges in the cases when the O atoms are engaged in the IHB.

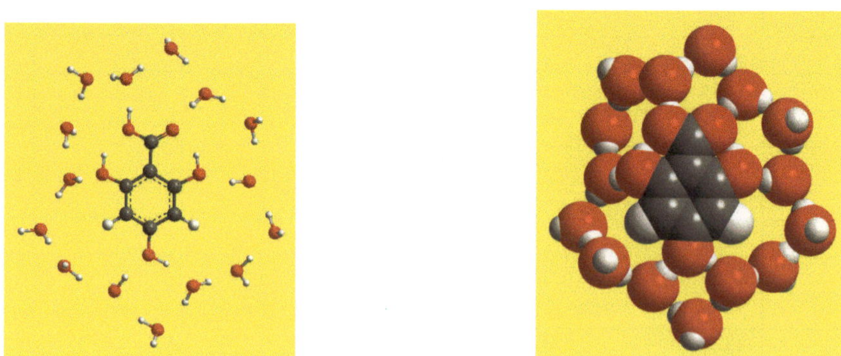

Figure 6: Adduct with explicit water molecules of the lowest energy conformer of the carboxylic acid of phloroglucinol. HF/6-31G(d,p) results [30].

In this case, the rigidity of the structure and the strength of the IHB limit the possibility of the groups forming the IHBs to move away from each other enough to make a situation without the IHBs, or with weakened IHBs, stable. Thus, the IHBs in the solute molecule remain in water solution and the areas in their vicinity have a prevalently hydrophobic character.

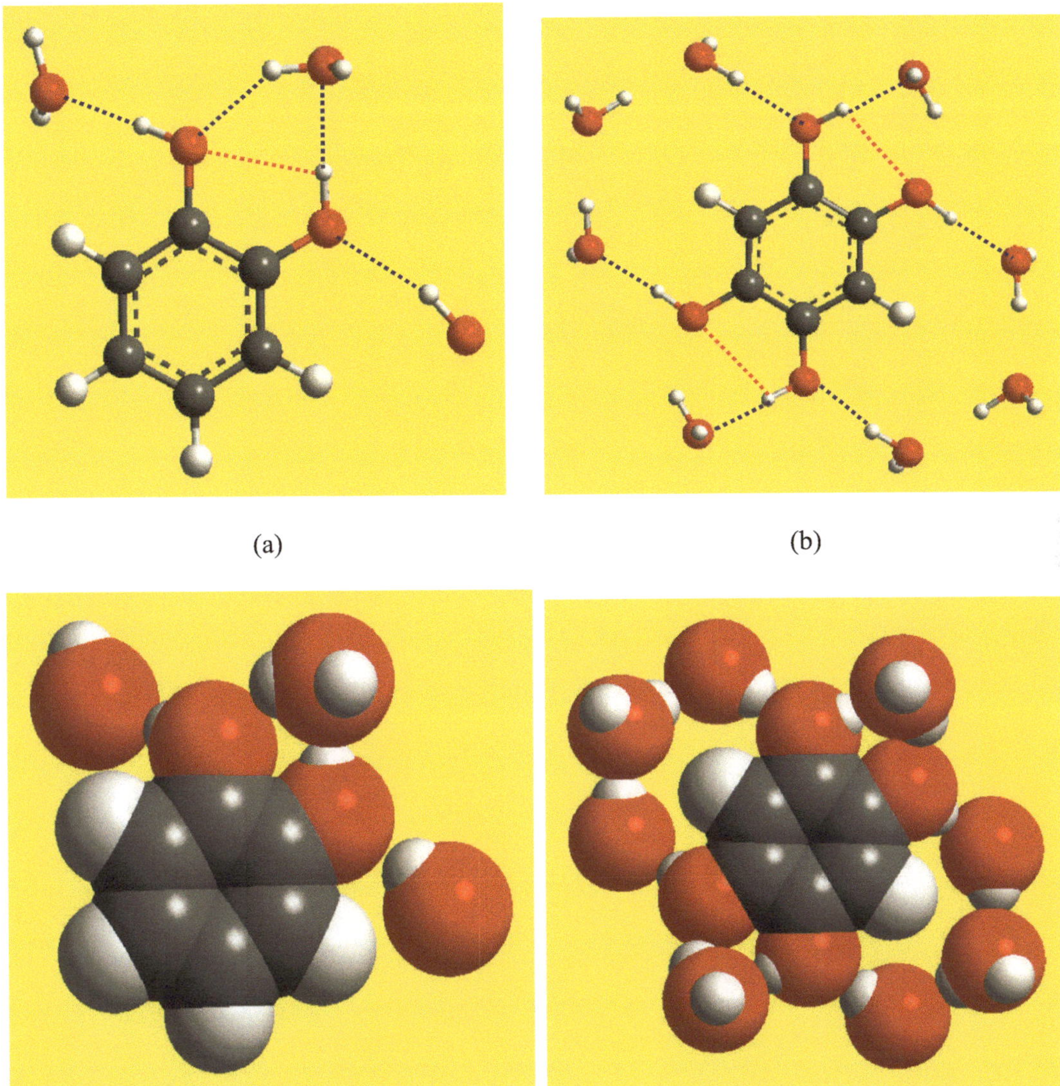

Figure 7: Simultaneous presence of intramolecular hydrogen bonds and solute-solvent intermolecular hydrogen bonds in the case of 1,2-dihydroxybenzene (a) and 1,2,4,5-tetrahydroxybenzenes (b) in water solution. HF/6-31G(d,p) results [33].
In this case, the IHBs are not very strong. The rigidity of the structure does not favour their breaking. The atoms forming them engage also in intermolecular H-bonds with the solvent molecules. In the ball-and-stick models, intramolecular H-bonds are represented by red dotted segments and intermolecular H-bonds by blue dotted segments. The space-filling models highlight the "contacts" between the solute molecule and the solvent molecules and, therefore, also the cooperativity of intramolecular and intermolecular H-bonds.

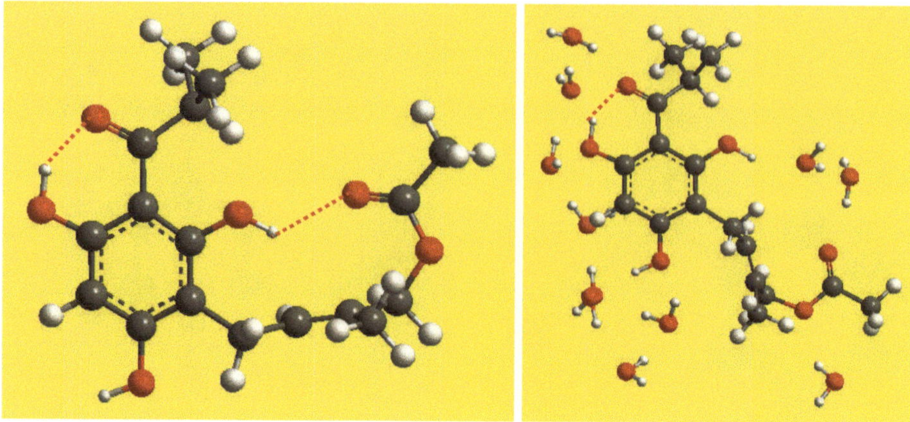

(a) (b)

Figure 8: Breaking, in water solution, of the second intramolecular hydrogen bond in the second lowest-energy conformer of caespitate. HF/6-31G(d,p) results [35].
The IHB engaging H15 and the carbonyl O of the ester function in the chain attached at C3 (second IHB) is weaker than the IHB engaging O14 (first IHB). It is present in the gas phase (a), but calculations of adducts with explicit water molecules (b) show that it breaks in water solution, and its donor and acceptor atoms form intermolecular H-bonds with water molecules.

(a) (b)

Figure 9: Different conformational preferences in non-polar and polar media for the Z isomer of the caespitate molecule [36].
Conformer (a), with two IHBs, is the lowest-energy conformer in vacuo, chloroform and acetonitrile, whereas conformer (b), in which H16 and the sp^2 O in the ester function of the side chain are available for intermolecular H-bonds with water molecules, is the lowest-energy conformer in water. The relative energy values (kcal/mol, HF/6-31G(d,p) result with full reoptimization for PCM calculations in solution) for conformer (a) are 0.000 in vacuo, chloroform and acetonitrile and 2.237 in water; for conformer (b), they are 5.523/vacuum, 3.257/chloroform, 2.15/acetonitrile and 0.000/water.

MODELS FOR THE STUDY OF SOLUTE-SOLVENT INTERACTIONS

Various models have been developed for the description of bulk and specific solvent effects on the properties of a solute molecule. They differ by the modelling of the physical interactions during the solvation process, the representation of the solute molecule, and the modelling of the interactions between the solute and the solvent molecules [38-47]. They can be divided into two broad categories:

- Implicit solvation models, such as continuum solvation models, in which the solvent properties are described in terms of average values (bulk solvent effects);
- Discrete/explicit solvation models, in which a limited number of solvent molecules are included explicitly (as individual molecules) in the study.

The selection of a model for the study of the solvation process of a given molecule depends on a reasonable compromise between computational costs and accuracy in the estimation of the properties of interest. Implicit solvation models are computationally faster and can utilise quantum mechanical calculations for the dissolved solute, with a perturbation formalism to take into account the effects of the solvent. Explicit solvation models can provide better information on several aspects, including the outcomes of the competition between intermolecular and intramolecular H-bonding. However, the explicit presence of solvent molecules increases the total number of atoms in the overall system considered, thus increasing computational costs, which, in turn, limits the number of solvent molecules that can be included in a quantum mechanical calculation. This implies a discrepancy between the model, with a limited number of explicit solvent molecules, and the reality in solution, where the solute molecule is surrounded by a high number of solvent molecules. (The solvent is usually present in much larger amount than the solute; for instance, in a 0.1 M solution of glucose in water (18 g glucose dissolved in enough water to give 1 litre solution), there are roughly 555 water molecules for each glucose molecule).

Continuum Solvent Models

Continuum solvent models are the most common implicit solvation models. The solvent is viewed as a polarizable dielectric continuum characterised by its dielectric constant ε. The solute molecule is considered to be embedded in a cavity in this continuum solvent and is represented by the charge distribution $\rho(r)$ on the surface of this cavity [40-42]. The charge distribution of the solute polarises the dielectric continuum (the solvent) around the cavity, generating a dipole in the medium. This produces a reaction-field potential in the continuum solvent which, in turn, polarises the solute charge distribution leading to some changes with respect to the $\rho^0(r)$ distribution of the solute in the gaseous state, with consequent modifications of the energy and properties of the solute molecule [40-42].

The reaction field may be incorporated into *ab initio* methods which utilise quantum mechanics, leading to methods that are commonly referred to as *self-consistent-reaction field* (SCRF) methods. In these methods, the effects of solute-solvent interactions are considered as a perturbation with respect to the situation of the solute *in vacuo*. A perturbation operator, $V_{int}(\rho_M)$, is added to the Hamiltonian, $\hat{H}^0_M$, of the non-perturbed solute M *in vacuo*, and the Schrödinger equation for the molecule in solution has the form

$$[\hat{H}^0_M + V_{int}(\rho_M)]\,\psi = E'\,\psi \tag{3}$$

where ψ is the wavefunction of the solute molecule in solution and E' is its energy. The wavefunction may be used to interpret and predict solvation effects on the solute observables, while the energy E' is used to evaluate the changes in Gibbs's free energy (ΔG), enthalpy (ΔH) and entropy (ΔS) accompanying the dissolution process. The free energy of solvation (ΔG_{solv}) is defined as the change in the free energy of a solute upon going from the gas phase to the solution phase [38]. Within the theoretical framework of a continuum model, ΔG_{solv} is estimated as a sum of different contributions, each arising from specific types of solute-solvent interactions: an electrostatic contribution G_{el} and a non-electrostatic contribution $G_{non\text{-}el}$ comprising all the non-electrostatic types of interactions:

$$\Delta G_{solv} = G_{el} + G_{non\text{-}el} \tag{4}$$

The electrostatic contribution G_{el} is generally the leading component of ΔG_{solv} in a polar medium and in situations where ionic species are involved. It is related to the dependence of the electrostatic potential on the charge density and the dielectric constant, described by the classic electrostatic Poisson equation:

$$\nabla^2 \varphi(\mathbf{r}) = -\frac{4\pi\rho(\mathbf{r})}{\varepsilon} \tag{5}$$

It is obtained by taking into account the Hartree-Fock solution of the Schrödinger equation in solution (eqn. 3) and the corresponding equation *in vacuo*.

The non-electrostatic contribution is the sum of three contributions [42, 48, 49]:

- a cavitation contribution, G_{cav}, that is the reversible work needed to modify the distribution of the pure solvent in order to create a cavity within which the solute molecule M accommodates itself;
- a repulsion contribution, G_{rep}, that describes the Pauli repulsion between M and the solvent molecules within the framework of a continuous distribution which takes into account the existence of the cavity;
- a dispersion contribution, G_{dis}, that is the contribution due to dispersion interactions between the solute molecule and the solvent.

Therefore, $G_{non\text{-}el}$ can be written as:

$$G_{non\text{-}el} = G_{cav} + G_{rep} + G_{dis} \tag{6}$$

The values of the contributions to ΔG_{solv} enable an analysis of the relative importance of the different types of solute-solvent interactions in determining the solvation free energy [42].

Popular approaches based on the continuum model include [42]:

- The *apparent surface charge* (ASC) methods, in which the polarisation of the medium outside the cavity – generated by the charge distribution inside the cavity – is modelled by a system of apparent surface charges spread on the surface of the cavity.

- The *multipole expansion methods*, in which the electrostatic component of ΔG_{solv} is determined from the individual Born solvation of each atom, corrected for the perturbing effect of the other atoms in the solute molecule [50]. These methods are also referred to as *Generalized Born model* methods. Examples are the series of SMx methods developed by the Cramer and Truhlar group [11, 51-53].

- The *direct field methods*, including the finite elements and finite difference methods, in which the reaction field operator dependent on the solute charge distribution is replaced by an operator based only on individual solute particles [54, 55].

The polarizable continuum model (PCM) is an important example of ASC approach. The cavity has a physical meaning, as it excludes the solvent while including the largest possible part of the solute charge distribution. The cavity surface is divided into a large number of small surface elements (called *tesserae*), and an apparent charge (point charge) is associated with each *tessera*. Since the solvent is treated as a homogeneous isotropic dielectric, the value of the dielectric constant is 1 inside the cavity and ε outside it.

The cavity is most commonly defined by means of a set of intersecting spheres with radii equal to the van der Waals radii of the atoms in the solute molecule [49, 56-61]; the exposed surface of the spheres constitutes the *van der Waals surface* (outlined in red in Fig. **10-a**). This surface presents regions (close to the spheres' intersections) that are not accessible by solvent molecules. These regions are smoothed by considering a probe sphere with size roughly approximating the size of a solvent molecule, rolling in contact with the solute molecule. The surface traced by the inward-facing surface of the probe sphere is called *molecular surface* and is the cavity surface (outlined in red in Fig. **10-b**).; it is formed by the contact surface (the part of the van der Waals surface that can come into direct contact with the probe sphere) and the re-entrant surface (the inward-facing part of the probe sphere when the sphere is in contact with more than one atom). The volume from which the probe sphere is excluded when it rolls around the van der Waals surface is called *solvent excluded surface* (SES, [58-63]). The surface traced by the centre of the probe sphere as it rolls over the solute molecule is the *solvent-accessible surface* (SAS, [58-62]), and its size depends markedly on the solvent molecules' size (Fig. **11**).

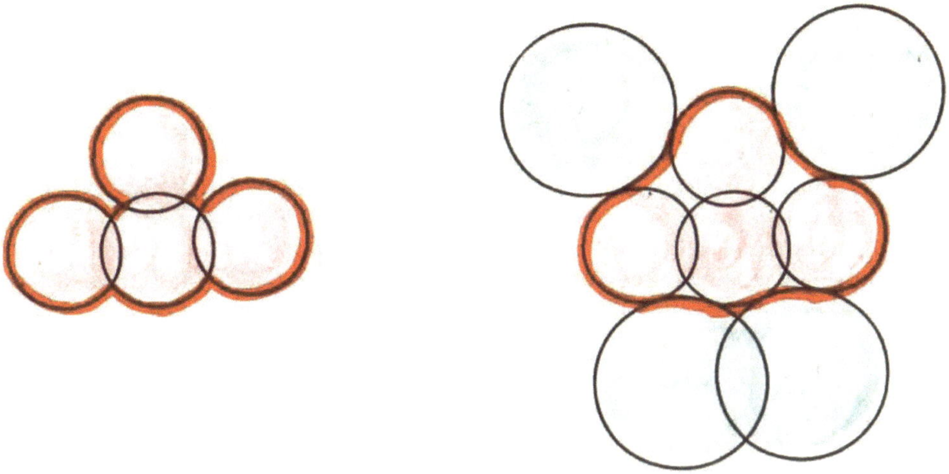

modelling of the solute molecule smoothing of the intersections

(a) (b)

Figure 10: Definition of the cavity surface in the PCM method.
The figure shows the approach to the modelling of the surface of a solute molecule in contact with the solvent, to define the surface of the cavity in which the solvent is embedded. The solute molecule (a) is represented considering its atoms as intersecting spheres, with radii equal to their van der Waals radii. The surface thus obtained – the van der Waals surface, represented by the thick red line in (a) – has sharp intersections between spheres, whose depths cannot be accessed by the solvent molecules. These sharp intersections are smoothed (b) by considering the extent to which solvent molecules (represented by the blue spheres) can come in contact with the atoms of the solute molecule for each intersection. The surface thus obtained – represented by the thick red line in (b) – can be regarded as the surface of the cavity representing the solute molecule. The shape of this surface depends on the size of the solvent molecules.
The blue spheres in (b) are bigger than the spheres representing the atoms of the solute molecule to recall that a solvent molecule is bigger than the individual atoms of the solute molecule. It is also important to note that, while the spheres representing the atoms of the solute molecules are intersecting because this corresponds to how the solute molecule is modelled, the intersection of the two blue spheres in (b) does not represent an intersection of solvent molecules (which would not have a physical meaning), but simply different positions that a solvent molecule can take while "rolling" on the surface of the solute molecule.

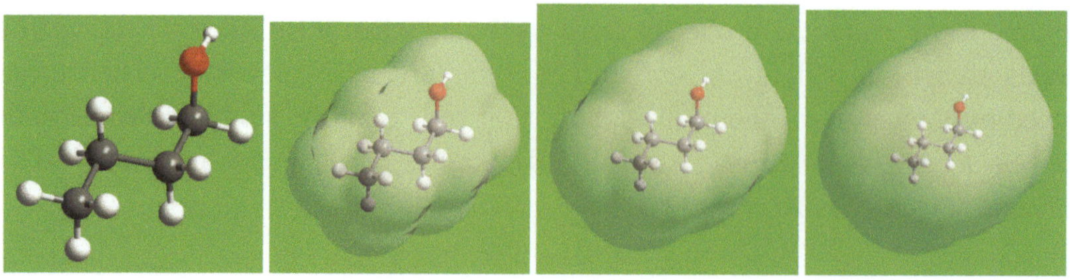

(a) (b) (c) (d)

Figure 11: Solvent accessible surface and solvent molecular size.
The figure shows the 1-butanol molecule in the gas phase (a), and its solvent accessible surface for different solvent-molecule sizes. The solvent accessible surface is obtained by rolling a probe sphere representing the solvent molecule (like the blue sphere in Fig. **10**-b) on the surface of the intersecting spheres representing the atoms of the solute molecule and considering the surface traced by the centre of the probe sphere. In the images, the radius of the probe sphere is respectively 1.4 Å (a), 3.0 Å (b) and 5.0 Å (c). As the size of the solvent molecule increases, its centre is further away from the solute molecule; in particular, the solvent molecule approaches the intersections of the spheres representing the atoms of the solute molecule (Fig. **10**) less and less deeply, and the shape of the solvent accessible surface (cavity surface) becomes increasingly smooth.

The PCM model was first developed at the University of Pisa [64, 65] and continuously refined afterwards [66-101]. Extensive reviews of the evolution of the model and the expansion of its applications from its earlier origins to the current state of the art are presented in [41, 42]. It has served as "parent model" of various modified versions differing from each other by the electrostatic expressions describing the ASC density [42, 99]; they include the integral equation formalism PCM (IEP-PCM, [74, 75, 100], the surface and volume polarization for electrostatics (SVPE, [102, 103]), the surface and simulation for volume polarization for electrostatics (SS(V)PE, [104]), and the conductor-like screening model (COSMO, [105, 106]. The IEF-PCM approach is formulated so as to take into account both isotropic systems (like solutions) and anisotropic systems (like liquid crystals), as well as liquid systems having real charges in the bulk of the medium (as is the case, *e.g.*, for ionic solutions). The COSMO approach involves the change in the dielectric constant of the medium from the specific finite value ε, characteristic of each solvent, to ε = ∞, which corresponds to a conductor [42]. The apparent surface charge is then determined by the local value of the electrostatic potential instead of the normal component of its gradient [42], and finally scaled by a function of ε to make it consistent with the fact that the ε of the medium is finite. Since its early stages, the PCM model has been applied also to the study of issues concerning biological systems [66, 68, 69]. An early example is the study of the energetics of the wrapping of DNA around a histone octamer (nucleosome), where the nucleosome and the DNA molecule wrapping around it are viewed as the solute, the cavity surface is built around it and the solvent is regarded as a continuum containing point-like monovalent ions [68].

Discrete Models

In discrete models, solvent molecules are considered individually, *i.e.*, the model considers a solute molecule surrounded by a certain number of solvent molecules. Increasing the number of solvent molecules around the solute molecule corresponds to mimicking a higher number of solvation layers and leads to more detailed information about solvent effects [48, 107]. The higher the number of solvent molecules around the solute molecule, the better the model can mimic the effects of bulk solvation.

The most rigorous way to evaluate the solvent effects on the molecular properties of the solute utilises full quantum mechanical (QM) calculations on the overall system and considers different arrangements of the solvent molecules around the solute molecule. The solute molecule and the solvent molecules surrounding it are treated as a supermolecular structure (adduct). The Schrödinger equation is solved for this supermolecular structure, providing information on its most favourable geometrical arrangements (the most favourable geometrical arrangements of the solvent molecules around the solute molecule). The effect of the solvent on the solute geometry is easily recognized by comparing the geometry parameters of the solute molecule *in vacuo* and in the adduct. However, this approach requires huge computational efforts and may become unaffordable as the number of explicit solvent molecules increases [108]. The fast increase in the computational time of QM calculations as the number of solvent molecules increases permits the inclusion of only a small number of solvent molecules in the adduct.

When an explicit treatment of the solvent requires that many (hundreds or even thousands) solvent molecules are included around the solute molecule, suitable approaches are offered by the "linear scaling" ab initio Density Functional Theory (DFT) methods. These methods have the advantage that their computational and memory requirements scale linearly with the number of atoms (N) in the system [109] (whereas, in the standard methods, computational and memory requirements scale with the cube of the number of atoms). Other apt approaches are the *ab initio* molecular dynamics methods [110–115], which combine a density functional description of electronic structure and finite temperature dynamics, thus being suitable for the study of various chemical processes in the presence of explicit solvent molecules [113]. In practice, however, the treatment of systems with large number of explicit solvent molecules is mostly done at lower levels of theory such as molecular mechanics (MM, [116-118]). In the initial MM versions, electrons are not considered explicitly and atoms are viewed as classical particles interacting through atomic forces determined by a set of parametrized interaction functions (force field), including bonded interactions (chemical bonds), non-bonded van der Waals interactions, and electrostatic interactions based on net atomic charges (*i.e.*, fixed point charge approaches). More recent MM developments attempt to go beyond the fixed point charge approaches; for instance, the polarizable molecular mechanics force fields incorporate multipole electrostatics [119–121].

QM/MM Models

Lower levels of theory such as MM provide no information on electronic effects and other properties that can only be obtained using higher levels of theory. On the other hand, a full QM treatment of a supermolecular structure with a high number of solvent molecules becomes unaffordable, above all for medium-size or larger solute molecules. A combination of the two approaches (QM and MM) may offer a reasonable and realistic compromise. In this combination, known as QM/MM, the solute molecule is treated with a QM approach (either *ab initio* or semi-empirical) while the solvent molecules are treated with an MM approach [122]. Typically, the MM treatment of the solvent molecules replaces their actual electronic distributions (which determine the solute-solvent potential) with partial point charges on the atomic sites, thus accounting only for their electrostatic influence on the solute. In this way, the solute alone is considered polarized, while the polarization of the solvent is neglected.

Continuum models too involve different treatments for the solute and the solvent – a QM treatment for the solute, while the solvent is represented by a continuum dielectric medium, so that the overall approach is of a QM/continuum type. However, unlike in the QM/MM approach, the solvent in the continuum approaches is also polarizable and its effect on the solute is represented by the *reaction potential* part of the Hamiltonian [122-124]).

In both standard QM/MM and QM/continuum models, an effective Schrödinger equation for the solvated system is written as

$$\hat{H}_{eff}\,|\psi\rangle = (H_0 + H_{env})\,|\psi\rangle = \mathrm{E}|\psi\rangle \tag{7}$$

where $\hat{H}_0$ is the hamiltonian of the solute system in the absence of the solvent, the $\hat{H}_{env}$ operator accounts for the effect of the coupling between the solute and the solvent, and ψ is the solute wavefunction. The form of $\hat{H}_{env}$ is different for the two model types:

$$\hat{H}_{env} = \begin{cases} H_{QM/MM} + H_{MM} & \text{QM/MM} \\ V_{cont} & \text{QM/Continuum} \end{cases} \tag{8}$$

The addition of $\hat{H}_{env}$ to the solute hamiltonian modifies the solute wavefunction. The QM/MM hamiltonian can be written as a sum of the hamiltonian ($\hat{H}_{el}$) for the electrostatic interaction between the QM system and the point charges in the MM part of the system and the hamiltonian ($\hat{H}_{pol}$) accounting for molecular polarisabilities at selected points in the solvent molecules (polarization interaction between the induced dipole moments and the electric field from the QM system):

$$\hat{H}_{QM/MM} = \hat{H}_{el} + \hat{H}_{pol} \tag{9}$$

with

$$\hat{H}_{el} = \sum_m q_m(r_m)\hat{V}(r_m)$$

where $\hat{V}(r_m)$ is the electrostatic potential operator for the solute electrons and nuclei at the MM charges q_m.

Molecular Dynamics Models

The situation in a liquid solution is continuously changing with time. A solute molecule is surrounded by solvent molecules, but they are not always the same solvent molecules: they interchange rapidly. This happens even when the solute-solvent interactions are comparatively strong, like in the case of solute-solvent H-bonds.

Molecular dynamics attempts to take into account the time-changing character of liquid solutions. Like in the QM/MM approach, the solute is treated quantum mechanically and, therefore, its chemical properties are well defined. The discrete representation of the solvent molecules is realised through sampling of the degrees of freedom of the solvent, usually using Monte Carlo (MC) or molecular dynamics (MD) techniques to generate a large number of possible configurations of solvent molecules. In a typical study, several QM/MD calculations for the solute's properties are performed and the final description of the solute properties is an average of all the possible outcomes [122, 125]. Because of the large number of possible configurations to be calculated, the solute is often treated at a semi-empirical QM level. Although quite demanding from a computational point of view, MD simulations provide a reasonable description of weak solute-solvent specific interactions which cannot be represented by a single configuration obtained from a QM geometry optimisation [125].

Combinations of explicit and implicit solvation methods consider a certain number of explicit solvent molecules around the solute molecule, and the resulting system is then considered to interact with a continuum solvent. These methods may utilise only QM approaches, as in the *cluster-continuum* model described in the next section, or hybrid QM/MD approaches. Hybrid QM/MD explicit/implicit solvation methods [126-130] enable the inclusion of explicit solvent molecules for a higher number of solvation layers around the solute molecule. The electron density of the solute and of few solvent molecules close to it is described by a localised basis set, whereas the rest of the solvent molecules are described using an MM force field, whose charge distribution adds an electrostatic embedding to the QM Hamiltonian. The interactions of both the QM and the MM parts with the bulk (continuum) solvent are treated by a mean field approach that includes an exact treatment of the electrostatic reaction field [98] and an effective representation of short-range (dispersion and repulsion) interactions, derived in such a way as to minimize edge effects on the solvent density and average energy [127].

Limitations of the Models

Both continuum models and discrete models have limitations. The main limitation of continuum models is their inability to fully take into account specific directional solute-solvent interactions such as H-bonding, exchange repulsions and the unique dielectric characteristics of the first solvation shell [44]. Since the solvent is treated as an isotropic continuum dielectric medium, whose effect on the solute is represented by a perturbation term in the molecular hamiltonian, important interactions such as H-bonding, or phenomena like possible charge-transfers between solute and solvent, are not completely described by the average nature of the continuum averaged reaction potential field [43, 131, 132]. While the electrostatic component of H-bonding may partially be included in the dielectric polarisation terms (so that computational results often show an implicit partial consideration of the effects of H-bonding [133]), the short-range directional components of H-bonding are not taken into account in a uniform-dielectric model [14]. The computational outcomes of the implicit partial consideration of H-bonding suggests the possibility of developing options that might enhance it (*e.g.*, by suitable adaptations in the design of the cavity surface [133]). Other properties that are not completely described by continuum models include the solute's absorption energies and nuclear magnetic shielding properties. Moreover, a rigorous identification of the charge distribution and of physically meaningful size and shape of the solute cavity may not be attainable for some solutes [18].

The limitations of continuum models with respect to specific/directional solute-solvent interactions may be overcome by a combination of discrete and continuum approaches, in what is sometimes called the *cluster-continuum* model. The model considers adducts comprising the solute molecule and the solvent molecules directly interacting with it [16, 43, 133-135]. This selection is justified by the fact that interactions such as H-bonding, or other donor-acceptor interactions, are established between the solute molecule and the solvent molecules that come into "direct contact" with it. When the donor/acceptor sites in the solute molecule are sufficiently close to each other and the solvent molecules are capable of H-bonding to each other (as is the case of water), the adduct should preferably comprise also the solvent molecules bridging those directly bonded to the solute molecule, as they increase the adduct stability (Fig. **2**). The supermolecular structure of the adduct is first calculated in the gas phase (usually quantum-mechanically) and is subsequently considered a "solute" in a cavity embedded in a continuum solvent, to incorporate long-range interactions due to the solvent dielectric properties (Fig. **12**).

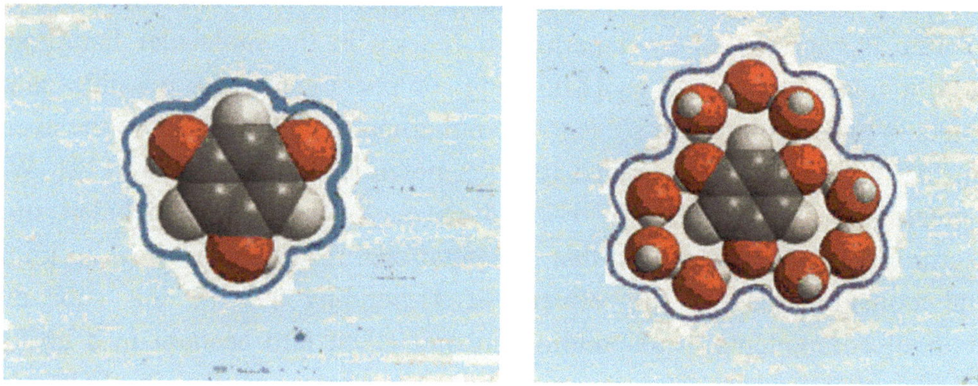

(a) (b)

Figure 12: The combination of discrete and continuum approach (*cluster-continuum* model). The figure considers the case of the lowest energy conformer of phloroglucinol in water [16] as an illustrative example. Fig. (a) illustrates the model for the PCM study of the conformer: the molecule is embedded in a cavity surrounded by a continuum of liquid water. Fig. (b) considers the most stable adduct of phloroglucinol with water molecules as the solute embedded in a cavity surrounded by a continuum of liquid water. Given the spacing of the OH groups in the phloroglucinol molecule, the most stable adduct comprises the water molecules directed H-bonded to the phenol OH and those bridging them.

The main drawback of discrete models is their inability to take into account thermal motions in the solution and their effects [66]. The solute-solvent supermolecular structure is treated as a rigid structure. This description is more apt for systems with highly directional and comparatively strong intermolecular interactions between the explicit solvent molecules and the solute molecule (such as solute-solvent H-bonds). Although, even in this case, the solvent molecules directly H-bonded to the solute molecule interchange rapidly with other solvent molecules, the geometry of the cluster obtained on optimisation is likely to correspond to a time-averaged (or frequently occurring) arrangement of solvent molecules around the solute molecule. Discrete models are less apt to describe situations in which the interactions between the solute molecule and the explicit solvent molecules are intrinsically weak or non-directional. On the other hand, studies of adducts with explicit molecules of non-polar solvents have been conducted and contribute to the understanding of the effects of short-range weak interactions [136, 137]. Another major drawback of discrete models stems from the limitations to the adduct size determined by the fast increase in computational time as the number of solvent molecules increases.

The main drawbacks of QM/MM approaches relate to the fact that the force fields utilised are generated from fully classical force fields. While this is generally suitable for the description of solvent–solvent interactions, it fails for those interactions whose description requires explicit consideration of electrons or other typically quantum mechanical features. For instance, it is not easy to model the van der Waals interactions between the solute and the solvent molecules [108] or to determine reliable force field parameters for an adequate description of the whole range of intermolecular H-bonds [122].

The main limitation for the study of clusters with MD simulations is that, as the size of the cluster increases, the accuracy of the QM level describing the solute has to be reduced. Moreover, an increase in the cluster size also implies that the statistical representativity of the solvent becomes more difficult because of the rapid increase in the number of possible configurations of solvent molecules.

STUDYING A BIOLOGICALLY ACTIVE MOLECULE IN SOLUTION

Major issues for the study of a biologically active molecule in solution are the selection of the solvents to be considered, the selection of the solvation model and the selection of the computational method. They are given specific attention in separate sections.

The Selection of the Most Suitable Solvents

The effects of the solvent on the characteristics of the solute molecule are dominantly determined by the polarity of the solvent molecules, the types of interactions that they can establish with the solute molecule, and the dielectric constant of the solvent as a bulk medium. The same solute molecule will experience different properties-modifications in different solvents. When studying a biologically active molecule, it is important to consider the medium/media in which that molecule may be preferably present within a living organism.

Water is the medium present in the highest proportion in living organisms, and it is a polar solvent with high dielectric constant. The lipid phase constitutes a non-polar medium. Membranes may have intermediate or particular situations. Although, within the organism, each of these media is more complex than a specific pure solvent, it is possible to select solvents that can aptly mimic the medium in which a given compound will preferably exert its biological activity in the organism. If partition coefficients, or other criteria, show a probability that that compound distributes (although not evenly) in more than one medium, it becomes important to study it in solvents mimicking different media. An apt selection would include water, a non-polar solvent and a solvent with intermediate characteristics. If it is sure that a given biologically active molecule prefers only one type of medium, and its presence in the other types is negligible, then the study may be limited to the solvent mimicking that medium. However, a study in water solution is always recommendable because of its dominant abundance in living organisms.

Non-aqueous solvents often play significant roles in the solvation of solute molecules in biological systems or in biotechnology applications [138]. Besides modelling a non-polar medium through non-polar solvents, it is interesting to consider other aspects typical of the complexity of biological systems, such as the interface between polar and non-polar media and the simultaneous presence, in the same medium, of aqueous and non-aqueous solvents mixed together. The most interesting aspect to investigate for aqueous and non-aqueous solvent mixtures is the preferential solvation at the solute-solvent interface. This depends on the characteristics of the solute molecule (*e.g.*, whether it is mostly hydrophilic, or mostly hydrophobic, or has regions that interact preferably with water and regions that tend to be hydrophobic) and on the nature of the other solvent (co-solvent) mixed with water. The role of water may be altered partially or completely, depending on the type of co-solvent. For instance, a non-aqueous medium with amphiphilic character may prefer to interact with the hydrophobic sites or with the polar sites of the solute; in the latter case, it would compete with the water molecules and in some cases might replace some water molecules [139, 140].

The solvent selection may conveniently include also the solvent utilised in the experimental determination of the biological activity of a compound, to know its conformational preferences and other molecular properties in the conditions under which the activity was determined. For instance, ethanol is often utilised for *in vitro* tests of antiradical activities with 2,2-diphenyl-1-picrylhydrazyl (DPPH); if the antiradical activity of a newly discovered compound has been determined in this way, it is important to include ethanol among the solvents selected for the investigation of that compound in solution [141].

The solvent selection depends also on the nature of the objects of a given study. If only one molecule is concerned, the selection will consider the types of media in which that molecule may be preferably present and the medium used in the experimental determination of its activity. For instance, on modelling the antioxidant activity of hyperjovinol A through donor-acceptor maps [141], three solvents were selected: water, because it is always included, as explained previously; ethanol, because the antioxidant activity is determined in ethanol; and pentylethanoate, because it simulates the lipid shell of cell membranes. When a study concerns a large number of molecules of the same class, with different types of biological activity, the correspondence with the solvent used in the experimental determination of the activity, or with the medium in which the molecule may preferably be present in a living organism, may not be possible for all the molecules considered, or even for most of them. Then, the solvent selection will respond to criteria suitable for the whole class, rather than for a specific molecule. These include the importance of covering the polarity-range and dielectric constants of the media present in a living organism and the importance of taking into account the characteristics common to all the molecules of the given class. For instance, in a study of acylphloroglucinols involving more than 120 different molecules [15, 26-28], water, acetonitrile and chloroform were selected. Their different characteristics enable them to mimic different media: water is the most abundant and most polar medium in the organisms, chloroform mimics non-polar media and acetonitrile constitutes a good model for cell membranes.

Furthermore, acylphloroglucinols (Fig. **4**) have several sites capable of forming H-bonds and, therefore, it is important to include solvents with different H-bonding abilities. The three solvents selected respond to this criterion, as:

- Water can be both H-bond donor and H-bond acceptor, and water molecules are also capable of forming H-bonds with each other.
- Acetonitrile can only be H-bond acceptor, and acetonitrile molecules are not capable of forming H-bonds with each other.
- Chloroform is not capable of forming H-bonds.

The selection enabled informative comparisons of solvent effects on acylphloroglucinols' conformational preferences and on various molecular properties, resulting in the identification of trends enabling reasonable predictions for other acylphloroglucinol molecules.

The Selection of the Calculation Method

The computational study of any molecule starts in the gas phase. The computational method – in terms of level of theory and basis set – is therefore selected for the calculations *in vacuo*. The selection aims at attaining optimal balance between results accuracy (which would require higher levels of theory and larger basis sets) and computational costs (which increase for higher levels of theory and for larger basis sets). The "optimal balance" depends largely on the size of the molecule/s under investigation. While the highest levels of theory and large basis sets are affordable for small molecules, reasonable compromises between results accuracy and computational costs become necessary for medium-size or larger molecules. When the highest levels of theory are not affordable, it becomes important to test more than one method, to verify whether there might be significant aspects that one or the other method does not reveal sufficiently. If the study concerns a high number of molecules of the same class, the testing of different methods may be limited to a representative subset of molecules. Several studies test Hartree-Fock (HF, [142, 143]), Density Functional Theory (DFT, [144, 145]) and Møller-Plesset Perturbation Theory (*e.g.*, MP2, [146, 147]) calculations. The results are checked against experimental information, if available. When experimental values are not available, Møller-Plesset Perturbation Theory results constitute a suitable benchmark to assess the performance of the other two methods. Although DFT takes into account part of the electron correlation, whereas HF includes only the limited amount of electron correlation due to the Pauli exclusion principle, several instances have been reported [148-153] in which HF highlights features that do not appear in the DFT results. The comparison of the results of different calculation methods helps ensure that no relevant aspect fails to be recognised.

Calculations in solution are performed on in-vacuo-optimised geometries. They require the selection of the most suitable model for the evaluation of the solvent effects. Since "continuum solvation models are the ideal conceptual framework to describe solvent effects within the QM approach" [101], it is convenient to start with PCM calculations. Depending on the nature of the solute and on the objectives of the study, the use of PCM methods may be complemented by other methods capable of giving the additional desired information.

PCM calculations utilise *in-vacuo*-optimised geometries as their inputs. Therefore, they must be performed at the same level of theory and with the same basis set with which those geometries were obtained, for the comparison between the results *in vacuo* and the results in solution to be meaningful. PCM calculations may be performed with full geometry re-optimization in solution, or as single point (SP) calculations. Full re-optimization is the ideal option, as it shows the effects of the solvent on the molecular geometry and provides better-quality description of the solvation phenomenon [154], including the evaluation of the related thermodynamic quantities. It is also the necessary option if one wants to calculate properties that need to be computed on an equilibrium geometry, such as vibrational frequencies (then, frequencies are calculated on the geometry re-optimised in solution), or if one expects major geometry changes induced by the solvent. However, PCM re-optimisation calculations are computationally demanding, posing affordability problems for medium-size or larger molecules. When no dramatic geometry changes are expected in solution with respect to in vacuo, SP PCM calculations can be viewed as an affordable option. They usually provide reasonable estimation of energetics aspects and enable reasonable identification of trends.

Chemical considerations offer reliable guidance to evaluate when SP PCM calculations are likely to be sufficiently informative. Solute-solvent interactions are generally much weaker than intramolecular forces (with exceptions such as the case of acids, for which solute-solvent interactions are strong enough to dissociate the solute molecule into ions). Therefore, for molecules that do not undergo dissociation in a given solvent (as is the large majority of biologically active molecules), it is reasonable to expect that the geometry of the solute molecule does not undergo important changes on going from the gas phase to solution. Then, SP PCM calculations can be utilised to determine the influence of the solvent on the relative energies and other molecular properties (*e.g.*, dipole moments) and to estimate important quantities of the solution process, such as the bulk solvent effect (ΔG_{solv}) and its G_{el} and $G_{non\text{-}el}$ components.

It is also convenient to verify the reliability of SP PCM calculations for a given molecular system or a given class of compounds by performing full re-optimisation PCM calculations on the lowest energy conformers of the given molecule, or on selected smaller molecules of the given class. For instance, in the study of acylphloroglucinols [25-28], both full re-optimization and SP PCM calculations were performed for all the conformers of a considerable number of molecules (preferably selecting the smaller ones) in all the three solvents considered. The number of calculations with full re-optimisation was sufficiently high to enable a reliable estimation of the performance of SP PCM calculations by comparing their results with the full re-optimisation ones. The comparisons showed a good degree of consistency for individual values and close similarities of the identifiable trends. Table **5** compares the results for selected molecules, whose structures are shown in Fig. **13**.

On the other hand, given the importance of geometry aspects for a molecule's biological activity [4], it is advisable – whenever affordable – to perform full re-optimisation PCM calculations for the conformers that might be involved in the activity. This is particularly important when the biologically active molecule contains IHBs, because of their frequent roles in the biological activity mechanisms [155, 156]. For instance, at least one comparatively strong IHB is present in all acylphloroglucinol molecules; therefore, in the study of this class of compounds [25, 26, 35], it was opted to perform full re-optimisation PCM calculations for all the conformers with relative energy below the 3.5 kcal/mol threshold, to investigate the solvent influence on the characteristics of the IHB.

Adducts with Explicit Solvent Molecules

In non-polar solvents, the solvent effect on the energies of organic compounds is often reasonably well related to the solvent dielectric constant and may have minimal influence on the conformers' relative energies; therefore, PCM calculations can provide all the information that is relevant to understand the solvent effect. However, when comparatively strong and directional solute-solvent interactions are possible, a better understanding of the situation in solution is obtained by utilising also other approaches in addition to PCM calculations. A combination of discrete and continuum solvation models like the one illustrated in (Fig. **12**) offers a compromise capable of providing valuable information at reasonable computational costs.

The main features in the study of an adduct are the arrangement of the solvent molecules around the solute molecule, the solute-solvent distances in the sites of directional interactions (*e.g.*, the length of solute-solvent H-bonds) and the interaction energy between the solute molecule and the solvent molecules. In its general form, the interaction energy (ΔE_{adduct}) is given [157] by the difference between the energy of the adduct (E_{adduct}) and the energies of its constituting units, *i.e.*, the energy of the isolated solute molecule ($E_{solute\text{-}(isolated)}$) and the energy of the n solvent molecules surrounding it ($E_{solvent\text{-}molecules}$):

$$\Delta E_{adduct} = E_{adduct} - E_{solute\text{-}(isolated)} - E_{solvent\text{-}molecules} \quad (10)$$

The evaluation of $E_{solvent\text{-}molecules}$ depends on the presence or absence of interactions between the solvent molecules. If these interactions are negligible, $E_{solvent\text{-}molecules}$ can be approximated by the sum of the energies of the n separated solvent molecules

$$E_{solvent\text{-}molecules} = n\, E_{solvent\text{-}(isolated)} \quad (11)$$

where $E_{solvent\text{-}(isolated)}$ is the energy of an isolated solvent molecule. If the interactions between solvent molecules are not negligible, it is necessary to consider their contribution to the energy of the adduct. For instance, in the case of an adduct of a certain solute with n interacting explicit water molecules (water molecules bonded by water-water H-bond, as in the examples shown in Fig. **2**), the contribution of the water-water interactions ($E_{aq\text{-}aq}$) is estimated as the difference between the energy of the interacting water molecules ($E_{aq\text{-}(interacting)}$) and the total energy of n isolated water molecules (n $E_{aq\text{-}(isolated)}$):

$$E_{aq\text{-}aq} = E_{aq\text{-}(interacting)} - n\, E_{aq\text{-}(isolated)} \quad (12)$$

where $E_{aq\text{-}(isolated)}$ is the energy of an isolated water molecule. $E_{aq\text{-}(interacting)}$ is evaluated through a single point calculation (at the same level of theory as the adduct calculation) of a system consisting of the n interacting water molecules arranged exactly as in the adduct, but without the solute molecule (Fig. **14**). Then, the interaction energy (ΔE_{adduct}) between the solute molecule and the water molecules in the adduct is estimated as:

$$\Delta E_{adduct} = E_{adduct} - (E_{solute\text{-}(isolated)} + n\, E_{aq\text{-}(isolated)}) - E_{aq\text{-}aq} \quad (13)$$

Comparison of equations (12) and (13) leads to

$$\Delta E_{adduct} = E_{adduct} - E_{solute\text{-}(isolated)} - E_{aq\text{-}(interacting)} \quad (14)$$

Both E_{adduct} and $E_{aq\text{-}(interacting)}$ should be corrected for basis set superposition error, BSSE [160], usually done with the counterpoise method [158]. ΔE_{adduct} can be viewed as a good approximation to the total solute-solvent interaction energy in the adduct, resulting from the competition between the intermolecular solute-water and water-water interactions, including intermolecular H-bonds (when the solute molecule can form them), and electrostatic, exchange-repulsion, dispersion and polarization (induction) contributions.

The selection of the number of solvent molecules apt to provide a sufficiently informative description depends on the characteristics of both the solute molecule and the solvent molecules. In the case of water, the ability of water molecules to H-bond to each other plays important roles. The study of solutes as different as polyhydroxybenzenes [31], acylphloroglucinols [26-28] or alkaloids [158] shows that the presence of additional water molecules bridging those directly H-bonded to the solute molecule has a stabilising effect. For instance, the interaction energy between the solute molecule and the solvent molecules in the adduct shown in Fig. (**2**) is 10.051 kcal/mol stronger than for the adduct in Fig. (**1**). The two adducts have the same number of water molecules attached to the same sites of the solute molecule, but these water molecules are bridged by other water molecules in the adduct in Fig. (**2**) and not in the adduct in Fig. (**1**). Similarly, in the case of phloroglucinol, the interaction energy for the adduct with 6 water molecules directly attached to the OH groups of the phloroglucinol molecule (Fig. **15**-a) is 25.737 kcal/mol and the interaction energy for the adduct with additional water molecules bridging those directly attached to the OH of phloroglucinol (Fig. **15**-b) is 39.804 kcal/mol.

The case of acylphloroglucinols (Fig. **4**) is particularly apt to illustrate the relevant aspects that can be investigated by calculating adducts with explicit water molecules, because their molecules contain several H-bond donor or acceptor sites and at least one IHB. The issues that were investigated [15] can be summarised as follows:

- The strength of the interaction of each site of the solute molecule with a water molecule. This is investigated by considering adducts with one water molecule, attached in turn to different sites of the solute molecule. An illustrative example is shown in (Fig. **16**).
- The definition of *first solvation layer* that is more suitable for the given class of compounds. The distribution and spacing of the H-bond donor/acceptor sites in the acylphloroglucinol molecules enable arrangements of water molecules in which the ones directly H-bonded to the solute molecule are bridged by one water molecule (by two in the vicinity of the first IHB, to remain sufficiently far away from the IHB). Therefore, in the case of acylphloroglucinols, it is convenient to consider that the "first solvation layer" concept is better approximated by ensembles of water molecules like the one shown in (Fig. **2**) rather than by ensembles of water molecules like the one shown in (Fig. **1**).

- The best arrangement/s of water molecules around each conformer of a given molecule. This is identified by considering different geometrical arrangements of water molecules. Illustrative examples are shown in (Figs. **17** and **18**).

- The identification of preferences in the arrangement of water molecules around specific parts of the solute molecule. For instance, the best arrangement of water molecules around each phenol OH is the one enabling a square of O atoms (a known tendency with phenol OH [159]); and the best arrangement of water molecules around the first IHB in acylphloroglucinols corresponds to a pentagon of O atoms (Figs. **17** and **18**).

- The effect of structural features on the arrangement of water molecules around the solute molecule. For instance, when there is no substituent at C3 (atom numbering shown in Fig. (**4**), the arrangement of water molecules around the solute molecule is continuous, whereas the presence of a methyl or a bigger substituent at C3 introduces an interruption in this continuity (Fig. **18**). The presence of additional H-bond donor/acceptor sites in a substituent chain influences the distribution of water molecules around the solute molecule (as shown in (Fig. **2**) for the case of caespitate).

- The outcome of the competition between IHBs and intermolecular solute-solvent H-bonds. For instance, the study of adducts of caespitate shows that the first IHB does not break in water solution (Figs. **1**, **2**), whereas the second IHB (involving a phenol OH and one of the O of the ester function at the end of the prenyl chain at C3) is broken in favour of intermolecular solute-water H-bonds.

- The importance of a good, chemically-based initial guess of a reasonable arrangement of water molecules around the solute molecule. Water molecules tend to cluster together, and computational algorithms account for it. This may lead to a shift of water molecules on optimization, moving away from the site to which they were attached (or placed) in the input, to H-bond to other water molecules. An illustrative example is shown in (Fig. **19**). Experience has shown that this occurs more rarely if the arrangement of water molecules in the inputs takes into adequate account aspects like the appropriate directionality of all the solute-water H-bonds.

The information on the preferred arrangement of the water molecules around the solute molecule, or the strength with which each site of the solute molecule can bind a water molecule, may be relevant when trying to understand the mechanism of action of a given biologically active molecule within a living organism.

Compound AA

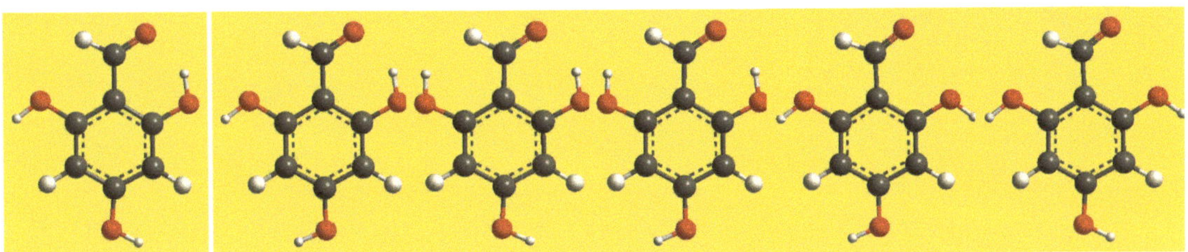

d-r 0.000 d-w 1.391 d-r-u 4.481 d-w-u 4.800 w-x 13.117 r-x 13.305

Compound A

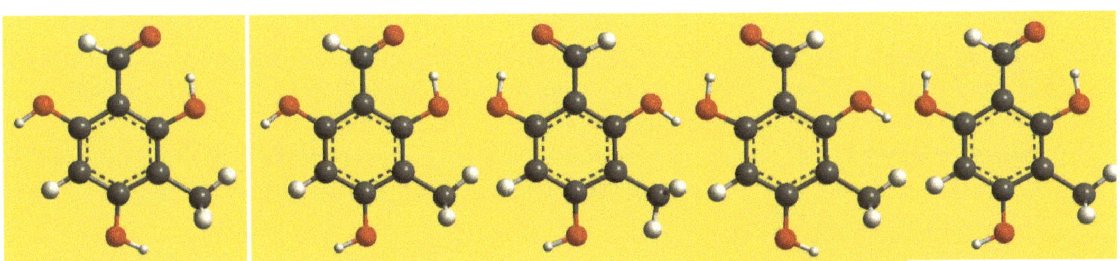

d-r 0.000 d-w 1.016 s-w 1.139 s-r 3.695 d-w-u 4.430

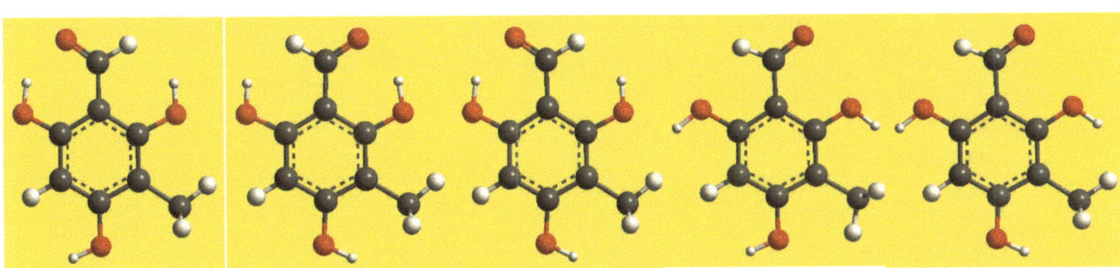

s-w-u 4.520 d-r-u 4.522 s-r-u 5.352 w-x 14.474 r-x 15.800

Compound BB

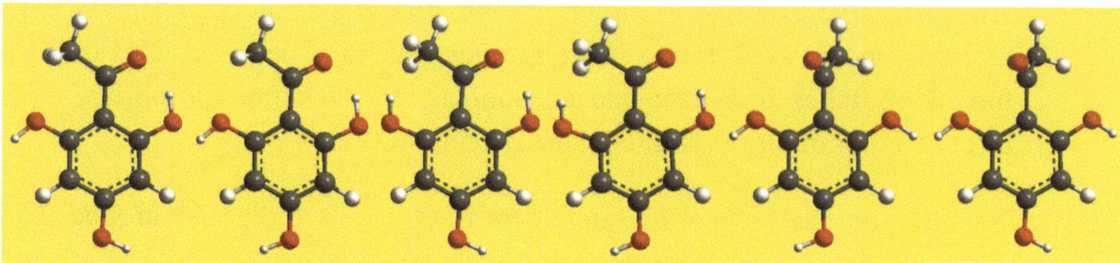

d-r 0.000 d-w 1.519 d-r-u 4.665 d-w-u 5.109 w-y 13.140 r-y 13.239

Compound B

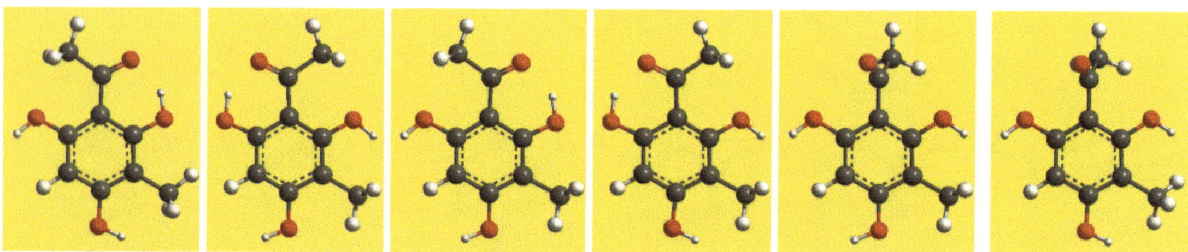

d-r 0.000 s-w 0.380 d-w 0.880 s-r 3.857 w-y 12.970 r-y 14.690

Figure 13: Geometries of the conformers of the molecules considered in Table **5**.
For each structure, the acronyms denoting the conformers are reported under each image on the left, and the relative energies (kcal/mol, from HF/6-31G(d,p) calculations) on the right.

Table 5: Comparison of the solvent effect (ΔG_{solv}) and its electrostatic (G_{el}) and non-electrostatic (G_{non-el}) components for representative acylphloroglucinols in the HF/6-31G(d,p) results from full-optimization and single point PCM calculations [15].
All the values are in kcal/mol. The molecules considered have R = H and R′ = H (AA), R = H and R′ = CH_3 (A), R = CH_3 and R′ = H (BB) and R = CH_3 and R′ = CH_3 (B). The geometries of their conformers are shown in Fig. **13**.

Molecule	**Conformer**	**Results from Full Optimisation PCM Calculations**			**Results from Single Point PCM Calculations**		
		ΔG_{solv}	G_{el}	G_{non-el}	ΔG_{solv}	G_{el}	G_{non-el}
AA	d-r	-14.70	-16.65	1.95	-13.80	-15.72	1.91
	d-w	-15.95	-17.90	1.95	-15.00	-16.92	1.92
	d-r-u	-14.88	-16.86	1.98	-13.90	-15.84	1.94
	d-w-u	-15.23	-17.22	1.98	-14.20	-16.15	1.95
	r-x	-22.91	-24.97	2.06	-21.54	-23.55	2.01
	w-x	-23.08	-25.15	2.06	-21.30	-23.41	4.02
A	d-r	-11.63	-13.92	2.29	-13.78	-16.03	2.25

Table 5: contd…

	d-w	-14.76	-17.04	2.28	-10.71	-12.97	2.26
	s-w	-11.74	-14.02	2.27	-10.78	-13.03	2.25
	s-r	-12.05	-14.33	2.28	-13.04	-15.32	2.28
	d-w-u	-14.08	-16.39	2.31	-12.65	-14.90	2.25
	s-w-u	-13.67	-15.94	2.27	-10.98	-13.23	2.26
	d-r-u	-11.91	-14.23	2.32	-10.88	-13.16	2.28
	s-r-u	-12.27	-14.56	2.28	-11.13	-13.38	2.26
	w-y	-19.84	-22.22	2.38	-18.43	-20.77	2.34
	r-x	-18.88	-21.25	2.38	-17.71	-20.06	2.35
BB	d-r	-14.03	-16.05	2.02	-13.11	-15.10	1.99
	d-w	-15.42	-17.43	2.02	-14.43	-16.43	1.99
	d-r-u	-14.20	-16.40	2.21	-11.70	-13.84	2.14
	d-w-u	-14.69	-16.90	2.21	-11.94	-14.08	2.14
	w-y	-20.66	-23.80	3.14	-19.39	-22.52	3.13
	r-y	-20.77	-23.91	3.14	-19.44	-22.61	3.16
B	d-r	-10.93	-13.29	2.35	-13.06	-15.39	2.32
	s-w	-11.01	-13.36	2.35	-10.02	-12.34	2.32
	d-w	-14.01	-16.36	2.35	-10.06	-12.38	2.33
	s-r	-11.48	-13.85	2.37	-10.41	-12.77	2.35
	w-y	-18.23	-21.85	3.62	-16.40	-19.95	3.55
	r-y	-16.76	-20.50	3.74	-15.36	-19.06	3.70

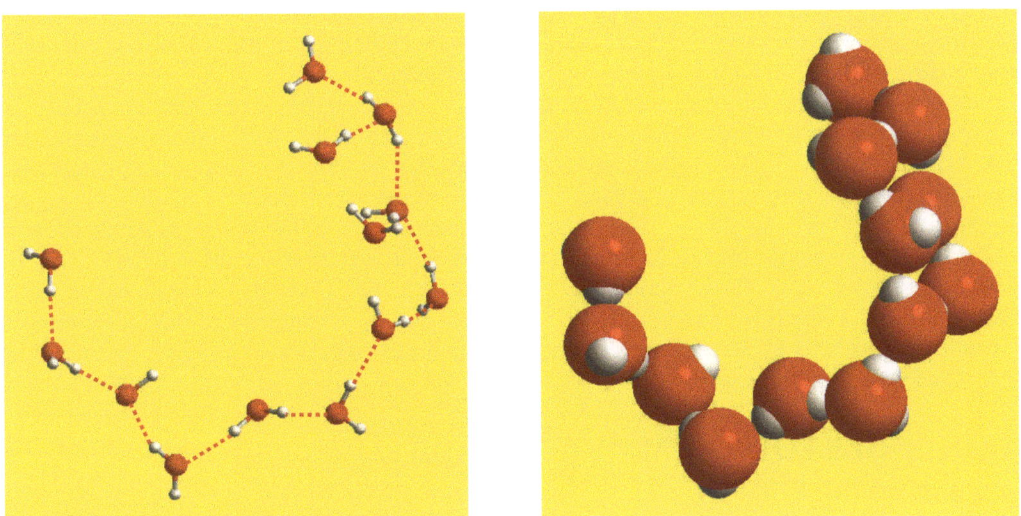

Figure 14: Water molecules in the adduct shown in (Fig. **2**), after removing the solute molecule. This adduct, constituted only by the water molecules present in the original adduct, is utilised to calculate $E_{aq\ (interacting)}$ (eqn. 12).

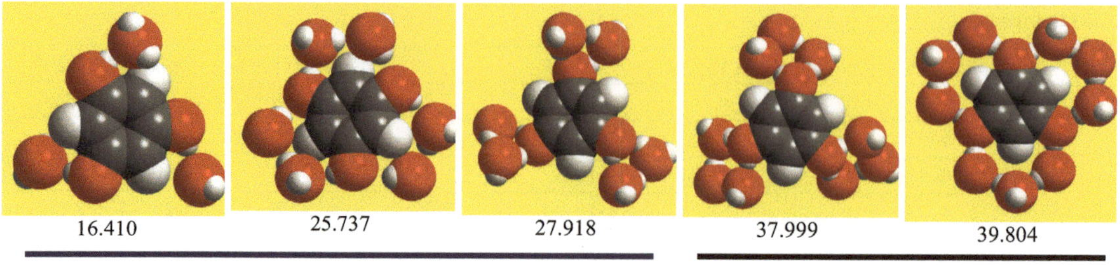

(a) (b)

Figure 15: Adducts of the lowest-energy conformer of phloroglucinol with only the water molecules directly attached to the solute molecule (a) and with other water molecules bridging them (b) [16].
The interaction energy between the phloroglucinol molecule and the water molecules (kcal/mol, from HF/6-31G(d,p) calculations and corrected for BSSE) is shown under each image.

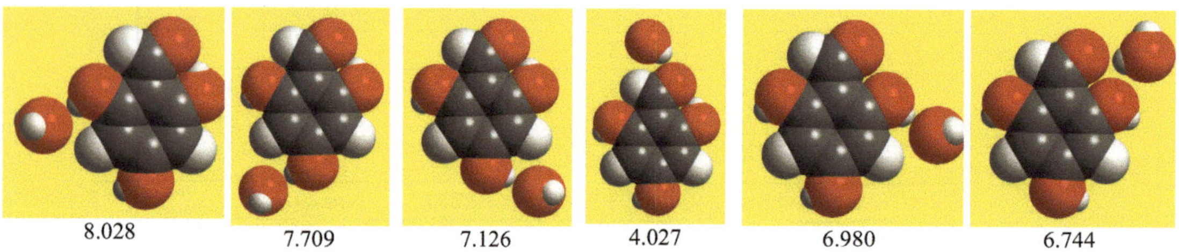

(a) (b) (c) (d) (e) (f)

Figure 16: Effects of the nature of the H-bond donor or acceptor site in the solute molecule on the characteristics of the acylphloroglucinol-water intermolecular interaction [15].
The figure considers the simplest acylphloroglucinol – the aldehyde of phloroglucinol (structure AA in Fig. **13**) – as an illustrative example. The interaction energy of each adducts (kcal/mol, from HF/6-31G(d,p) calculations, corrected for BSSE) is reported under the corresponding image. Their comparison corresponds to a comparison of the strength of the solute-water H-bond.
Adducts (a), (b), (c) and (d) refer to the lowest energy conformers of the solute molecule (in which the first IHB is present), considering the interaction of a water molecule with H17 (a), H16 oriented "to the left" (b), H16 oriented "to the right" (c), and O14 (d; the atom numbering is shown in Fig. **4**). Adducts (e) and (f) refer to the higher energy conformers of the solute molecule (in which the first IHB is absent), considering the interaction of a water molecule with H15 and with O8 and O14 simultaneously.

CHALLENGES FOR THE WAY FORWARD

The awareness of the importance of considering solvent effects when studying biologically active molecules is continuously increasing. The main challenges concern a general aspect – the description of solvent effects – and an aspect typical of the study of the action of biologically active molecules – the description of solvation and desolvation processes.

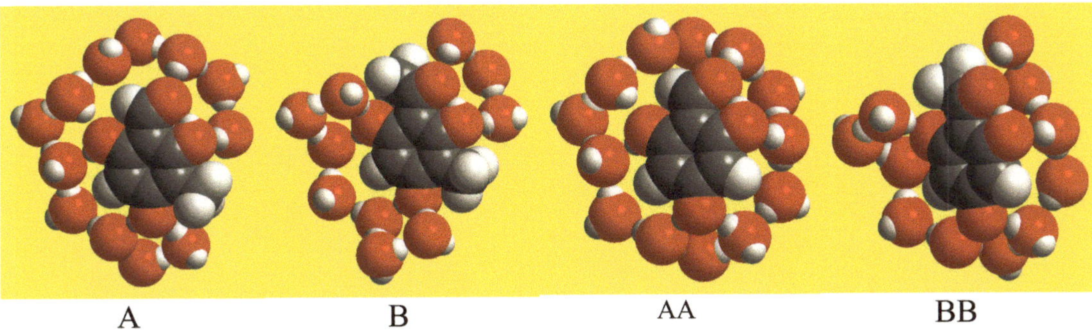

Figure 17: Effect of the nature of R, and of the presence or absence of a substituent at C3, on the arrangement of water molecules around an acylphloroglucinol molecule. HF/6-31G(d,p) results [15]. The images are denoted with the symbols used for the corresponding acylphloroglucinol molecules (Fig. **13**). Molecules A and AA have R = H, molecules B and BB have R = CH_3; the presence of R ≠ H in B and BB interrupts the continuity of water molecules in the region around the acyl chain. Molecules A and B have R′ = CH_3 (mimicking any R′ substituent that might be present at C3); molecules AA and BB have R′ = H; the presence of a substituent at C3 interrupts the continuity of water molecules in the region around C3.

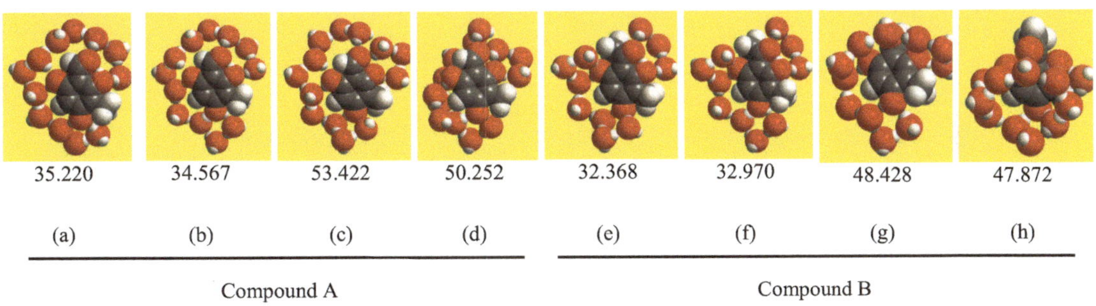

Figure 18: Combined effects of the nature of the acyl chain and of the conformers' geometries on the arrangement of water molecules around an acylphloroglucinol molecule. [15].
The figure considers adducts of the conformers of compounds A and B in Fig. 13. They both have a methyl group at C3, and differ by the acyl group (CHO in A, $COCH_3$ in B). The interaction energy of each adduct (kcal/mol, from HF/6-31G(d,p) calculations, corrected for BSSE) is reported under the corresponding image.
Adducts (a), (b), (e) and (f) refer to lowest energy conformers of the two compounds (conformers in which the first IHB is present); the conformers differ by the orientation of H16 (to the left in (a) and (e), to the right in (b) and (f)). Adducts (c), (d), (g) and (h) refer to higher energy conformers of the solute molecule (in which the first IHB is absent); the conformers differ by the orientation of H16 (to the left in (c) and (g), to the right in (d) and (h)). The presence of the methyl at C3 causes an "interruption" in the continuity of water molecules directly bonded to the solute

molecule. The presence of the methyl in the acyl chain in B causes another interruption in this continuity for the conformers with the first IHB (e and f), whereas the continuity is possible for the conformers without the first IHB (g and h), as the off-plane orientation of O14 makes it available for H-bonds enabling a continuous arrangement of water molecules.

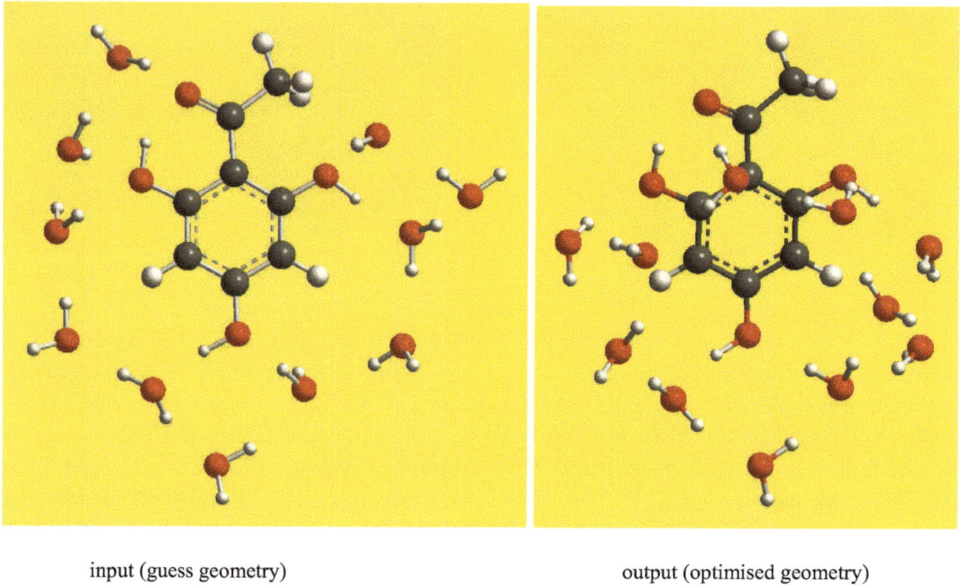

Figure 19: Illustration of how PCM optimization may take into account the tendency of water molecules to cluster together.

Challenges for the Description of Solute-Solvent Interactions

The challenges for the description of solute-solvent interactions are related to the nature of liquids as systems without a regular structure (*e.g.*, without periodicity in the arrangement of molecules) and continuously changing with time, as the molecules move within the system. In liquid solutions, the solvent molecules surrounding a solute molecule interchange continuously. The information obtainable from continuum solvation models like PCM concerns thermodynamic quantities such ΔG_{solv} and its components. The preferred arrangements of solvent molecules around a solute molecule (solute-solvent configurations), identified through the study of adducts with explicit solvent molecules, can be viewed as somehow time-averaged, because the actual arrangement is neither rigid nor constant through time. Adducts are more informative when the solute-solvent interactions are strong and directional (*e.g.*, solute-solvent H-bonds), and less close to what might be the actual situation when the solute-solvent interactions are weaker and non-directional.

Quantum mechanical (electronic structure) calculations provide the best descriptions of molecular systems. However, they allow the consideration of either a continuum solvent or a limited number of solvent molecules. When it is important to consider a large number of solvent molecules explicitly, one has to resort to less powerful levels of theory.

None of the existing models is capable of taking into account all the aspects that would be interesting for a complete understanding of what happens to a certain molecule in a certain medium. The combination of more than one approach enables the obtainment of relevant information from different perspectives; however, the information does not yet provide a complete picture. Meeting this challenge entails increasing the descriptive power of the models for the dissolution process and the solvent effects.

Challenges for the Description of Solvation-Desolvation Phenomena

The interaction of the drug molecule with a receptor's active site implies desolvation of the part of the molecule that gets into the active site and of the active site itself. Thus, desolvation phenomena play fundamental roles in the interaction between a drug and the receptor and may be viewed as part of the recognition between them [69, 160, 161]. The desolvation extent depends largely on the characteristics of the active site in the receptor [69]. The active site has different shapes, depending on the receptor itself: a shallow indentation, a deep pocket, and a variety of intermediate shapes. As long as the drug molecule and the receptor are not interacting, the drug molecule is completely solvated (surrounded by solvent molecules). The active site of the receptor is "filled" by solvent molecules and its surface interacts with them (it is solvated). When the drug molecule comes sufficiently close to the active site for short range interactions between them (charge transfers, attractions, repulsions, H-bonding, *etc*) to be activated, the drug molecule ends up binding to the active site. For the drug molecule to attach itself to the active site, the solvent molecules surrounding the active site and the solvent molecules surrounding the part of the drug molecule that binds to the active site get "squeezed out". The deeper the active site pocket, the larger the portion of the drug molecule which enters into it; then, a large portion of its surface gets desolvated, and this may also modify the solvation pattern around the part of the molecule which does not enter the pocket. If the pocket is completely inserted into the receptor, the drug molecule gets completely desolvated.

Desolvation phenomena condition the binding affinity of a ligand for its receptor, expressed in terms of the free energy change $\Delta G_{binding}$. Its value depends on the interaction free energy of the two molecules relative to their free energies in solution [162, 163]:

$$\Delta G_{binding} = \Delta G_{interaction} - \Delta G_{solv,ligand} - \Delta G_{solv,receptor} \quad (15)$$

where $\Delta G_{interaction}$ is the interaction free energy of the ligand-receptor complex, $\Delta G_{solv,ligand}$ is the free energy of the ligand desolvation and $\Delta G_{solv,receptor}$ is the free energy for barring the solvent from the receptor site. The main difficulty relates to the fact that the right hand side of this equation involves a small difference of large terms, which extensively affects the accuracy of the calculated difference.

Desolvation in the contact region between the ligand and the active site is often complete. However, in some cases, one or more solvent molecules remain bonded to the drug molecule and play a role in its interaction with the active site of the receptor. The occurrence of this permanence may be determined experimentally and through calculations. The use of more than one calculation method may be advisable to ensure that such occurrence does not remain undetected. For instance, in a case reported in [148], the permanence of a water molecule attached to the solute molecule when the solute had already entered the active site was highlighted by HF calculations and by experimental determinations, but not by DFT calculations.

Considering solvation and desolvation phenomena is important for all forms of drug design [164-172], including drug-design techniques based on geometry-complementarity, such as docking. In these studies, potentially active molecules are designed so that they fit the structurally-known active site of a relevant receptor [162, 172]. What will actually happen between the designed molecule and the active site of that receptor within a living organism is largely conditioned by its interactions with the molecules of the medium in which it dissolves (and, therefore, also by its solvation characteristics). These interactions may determine aspects such as whether the expectedly active molecule reaches its biological target within the organism. If it reaches the intended site of the receptor, desolvation phenomena become relevant in the establishing of the molecule-receptor interactions. Difficulties in incorporating the modelling of solvation-desolvation phenomena may be at least partially responsible for the lower-than-expectation success of docking techniques. An in-vacuo-only study risks to miss determining aspects of the actual modes in which the biological activity is exerted, even if the geometrical mutual "fitting" of a designed molecule and its receptor may appear ideal in vacuo ("when the energy of the solvated state is not considered. the ligands that are selected often bear high formal charge or are larger than expected" [173]). Although some modelling approaches have been successful in accounting for desolvation phenomena in the interaction between a ligand and the active site [173, 174], their complete description is still a challenge [164-171, 175].

CONCLUSION

Computational approaches can provide a wealth of information in the design of new drugs, including the possibility of modelling a molecule's ability for a certain activity, or predicting whether a new structure may have enhanced activity (which, in turn, enables a pre-selection for the more costly experimental studies, so that they are performed only on potentially promising structures). The fact that the activity of a drug is exerted in a medium within a living organism requires that the computational study of a biologically active molecule considers also its properties in solution, selecting the solvents that more closely mimic the media within which that molecule is more likely to be present in a living organism. Similarly, studies of the interaction between a drug and the active site of its receptor need to consider solvent-related phenomena, such as the desolvation of the drug molecule and the receptor's active site. Improving the descriptive abilities of the models for solute-solvent interactions and for solvation-desolvation phenomena is a major challenge to improve the predictive abilities of computational studies about the fate of a drug molecule, once introduced into a living organism.

ACKNOWLEDGEMENTS

Declared None.

CONFLICT OF INTEREST

The authors confirm that this chapter content have no conflict of interest.

REFERENCES

[1] Emilien, G.; Ponchon, M.; Caldas, C.; Isacson, O.; Maloteaux, J.M. Impact of genomics on drug discovery and clinical medicine. *Q. J. Med.,* **2000**; *93*, 391–423.

[2] Lin, J.H.; Lu, A.Y.H. Role of pharmacokinetics and metabolism in drug discovery and development. *Pharm. Rev*., **1997**, *49*, 403-449.

[3] Cumming, J.G.; Davis, A.M.; Muresan, S.; Haeberlein, M.; Chen, H. Chemical predictive modelling to improve compound quality. *Nat. Rev. Drug Discov.,* **2013**, *12*, 948-962.

[4] Bushelyev S.N.; Stepanov N.F. *Elektronnaya Struktura y Biologhicheskaya Aktivnost Molecul*. Khimiya, Snanye: Moscow, **1989**.

[5] Prabhu, N.; Sharp, K. Protein–solvent interactions. *Chem. Rev.,* **2006**, *106*, 1616–1623.

[6] Levy, Y.; Onuchic, J.N. Water and proteins: a love–hate relationship. *Proc. Natl. Acad. Sci. U.S.A.,* **2004**, *101*, 3325–3326.

[7] Guo, Z.Y.; Thirumalai, D.; Honeycutt, J.D. Folding kinetics of proteins – a model study. *J. Chem. Phys.*, **1992**, *97*, 525–535.

[8] Dill, K.A.; Fiebig, K.M.; Chan, H.S. Cooperativity in protein – folding kinetics. *Proc. Natl. Acad. Sci. U.S.A.,* **1993**, *90*, 1942–1946.

[9] Avbelj, F.; Baldwin, R.L. Role of backbone solvation and electrostatics in generating preferred peptide backbone conformations: distributions of phi. *Proc. Natl. Acad. Sci. U.S.A.,* **2003**, *100*, 5742–5747.

[10] Imai, T.; Kovalenko, A.; Hirata, F. Hydration structure, thermodynamics, and functions of protein studied by the 3D-RISM theory. *Mol. Simul.,* **2006**, *32*, 817–824.

[11] Feig, M. *Modeling solvent environments. Applications to simulations of biomolecules*. WILEY-VCH Verlag GmbH & Co. KGaA, Weinheim, **2010**.

[12] Melandri, S.; Sanz, M.E.; Caminati, W.; Favero, P.G.; Kisiel, Z. The hydrogen bond between water and aromatic bases of biological interest: An experimental and theoretical study of the 1:1 complex of pyrimidine with water. *J. Am. Chem. Soc.,* **1998**, *120,* 11504-11509.

[13] Nogrady, T.; Weaver, D.F. *Medicinal Chemistry*. Oxford University Press: Oxford, **2005**.

[14] Cramer C.J.; Truhlar D.G. Continuum solvation models: Classical and quantum mechanical implementations. In: Lipkowitz, K.B.; Boys, D.B (Eds.), Reviews in computational chemistry, VCH publishers, New York, **1995**, *6*, 1-75.

[15] Mammino, L.; Kabanda, M.M. Adducts of acylphloroglucinols with explicit water molecules: similarities and differences across a sufficiently representative number of structures. *Int. J. Quantum Chem.,* **2010**, *110*, **2378-2390**.

[16] Mammino, L.; Kabanda, M.M. A Computational study of the interactions of the phloroglucinol molecule with water. *J. Mol. Struct.* (*Theochem.)* **2008**, *852*, 36-45.

[17] Alagona, G.; Ghio, C.; Igual, J.; Tomasi, J. Theoretical study in vacuo of the first step of the reversible aldol cleavage catalyzed by aldolase from rabbit muscle. *J. Mol. Struct.* (*Theochem)* **1990**, *204*, 253-259.

[18] Barone, V.; Cossi, M.; Tomasi, J. A New definition of cavities for the computation of solvation free energies by the polarizable continuum model. *J. Chem. Phys.,* **1997**, *107*, 3210-3221.

[19] Cossi, M.; Scalmani, G.; Rega, N.; Barone, V. New developments in the polarisable continuum model for quantum mechanical and classical calculations on molecules in solution. *J. Chem. Phys.,* **2002**, *117*, 43-54.

[20] Olivares del Valle, F.; Tomasi, J. A simple model for molecular vibrations in solution: Application to hydrogen fluoride and its dimer in polar and non-polar solvents. *J. Chem. Phys.,* **1987**, *114*, 231-239.

[21] Bonaccorsi, R.; Cimiraglia, R.; Tomasi, On the free energy changes of a solution in light absorption or emission processes. J. *Chem. Phys. Lett.,* 1983, *99*, 77-82.

[22] Bonaccorsi, R.; Cimiraglia, R.; Tomasi, J. *Ab initio* evaluation of absorption and emission transitions for molecular solutes, including separate consideration of orientational and inductive solvent effects. *J. Comput. Chem.,* **1983**, *4*, 567−577.

[23] Bonaccorsi, R.; Cimiraglia, R.; Tomasi, J. The effect of the solvent in electronic transitions: some recent developments in the continuum model. *J.Mol. Struct.,* 1984, *107*, 197−209.

[24] Naray-Szabo, G.; Surjan, P. R.; Angyan, J.G., *Applied Quantum Chemistry*; Akademiai Kiado: Budapest, **1987**.

[25] Bonaccorsi, R.; Cimiraglia, R.; Tomasi, J.; Miertus, S. The mechanism of carbonyl reduction by $LiBH_4$: An ab initio investigation with inclusion of solvent effects. *J. Mol. Struct.,* **1983**, *94*, 11−23.

[26] Kabanda, M.M.; Mammino, L. The conformational preferences of acylphloroglucinols—a promising class of biologically active compounds. *Int. J. Quantum Chem.,* **2012**, *112*, 3691–3702.

[27] Mammino, L.; Kabanda, M.M. A Computational study of the effects of different solvents on the characteristics of the intramolecular hydrogen bond in acylphloroglucinols. *J. Phys. Chem. A*, **2009,** *113,* 15064-15077.

[28] Mammino, L.; Kabanda, M.M. The role of additional O–H···O intramolecular hydrogen bonds for acylphloroglucinols' conformational preferences *in vacuo* and in solution. *Mol. Simul.,* **2013**, *39*, 1–13.

[29] Nandini, G.; Sathyanarayana D.N. *Ab initio* studies of solvent effect on molecular conformation and vibrational spectra of diacetamide. *Spectrochim. Acta Part A,* **2004**, *60*, 1115–1126.

[30] Mammino, L.; Kabanda, M.M. A Computational study of the carboxylic acid of phloroglucinol in vacuo and in water solution. *Int. J. Quantum Chem.,* **2010**, *110*, **595-623**.

[31] De la Paz, M.L.; Vicent, C. Hydrogen bonding and cooperativity effects on the assembly of carbohydrates. *Chem. Commun*., **1998**, *4*, 465–466.

[32] Qian, X. The effect of cooperativity on hydrogen bonding interactions in native cellulose Ib from ab initio molecular dynamics simulations. *Mol. Simul.*, **2008**, *34*, 183–191.

[33] Mammino, L.; Kabanda, M.M. Interplay of intramolecular hydrogen bonds, OH orientations and symmetry factors in the stabilization of polyhydroxybenzenes. *Int. J. Quant. Chem.*, **2011**, *111*, 3701–3716.

[34] Mó, O.; Yañez, M.; Elguero, J. Cooperative (nonpairwise) effects in water trimers: An ab initio molecular orbital study. *J. Chem. Phys.*, **1992**, *97*, 6628–6638.

[35] Mammino, L.; Kabanda, M.M. A Study of the interactions of the caespitate molecule with water. *Int. J. Quantum Chem.,* **2008**, *108*, 1772-1791.

[36] Mammino, L.; Kabanda, M.M. The geometric isomers of caespitate: a computational study *in vacuo* and in solution. *Int. J. Biol. Biomed. Eng.,* **2012**, *6*, 114-133.

[37] http://srdata.nist.gov/cccbdb/vibscale.asp, 2006.

[38] Cramer, C. *Essentials of Computational Chemistry: Theories and Models*. Wiley: **2004**.

[39] Orozco, M.; Luque, F.J. Theoretical methods for the description of the solvent effect in biomolecular systems. *Chem. Rev.*, **2000**, *100*, 4187– 4225.

[40] Cossi, M.; Rega, N.; Soteras, I.; Blanco, D.; Huertas, O.; Bidon-Chanal, A.; Luque, F.J.; Truhlar, D.G.; Pliego, J.R.; Ladanyi, B.M.; Newton, M.D.; Domcke, W.; Sobolewski, A.L.; Laage, D.; Burghardt, I.; Hynes, J.T.; Persico, M.; Granucci, G.; Huxter, V.M.; Scholes, G.D.; Curutchet, C. In *Continuum solvation models in chemical physics*. John Wiley & Sons, Ltd: **2007**, pp. 313-485.

[41] Tomasi, J.; Persico, M. Molecular interactions in solution: An overview of methods based on continuous distributions of the solvent. *Chem. Rev.*, **1994**, *94*, 2027-2094.

[42] Tomasi, J.; Mennucci, B.; Cammi, R. Quantum mechanical continuum solvation models. *Chem. Rev.,* **2005**, *105*, 2999–3094.

[43] Sunoj, R.G.; Anand, M. Microsolvated transition state models for improved insight into chemical properties and reaction mechanisms. *Phys. Chem. Chem. Phys.*, **2012**, *14*, 12715–12736.

[44] Marenich, A.V; Cramer, C.J; Truhlar, D.G. Generalized Born Solvation Model SM12. *J. Chem. Theory Comput.* **2013**, *9*, 609−620.

[45] Andreussi, O; Dabo, I; Marzari N. Revised self-consistent continuum solvation in electronic structure calculations. *J. Chem. Phys.,* **2012**, *136*, 064102–064121.

[46] Dziedzic, J; Fox, S.J; Fox, T; Tautermann, C.S; Skylaris, C. Large-Scale DFT Calculations in Implicit Solvent—A Case Study on the T4 Lysozyme L99A/M102Q Protein. *Int. J. Quantum Chem.* **2013**, *113,* 771–785.

[47] Pomogaeva, A; Chipman, D.M. Hydration energy from a composite method for implicit representation of solvent. *J. Chem. Theory Comput.* **2014**, *10*, 211−219.

[48] Cammi, R.; Mennucci, B.; Tomasi, J. Nuclear magnetic shieldings in solution: Gauge invariant atomic orbital calculation using the polarizable continuum model. *J. Chem. Phys.,* **1999**, 110, 7627−7638.

[49] Amovilli, C.; Barone, V.; Cammi, R.; Cancès, E.; Cossi, M.; Mennucci, C.; Pomelli, C.S.; Tomasi, J. Recent advances in the description of solvent effects with the polarisable continuum model. *Adv. Quantum Chem.,* **1999**, *32*, 227-259.

[50] Soteras, I.; Curutchet, C.; Bidon-Chanal, A.; Orozco, M.; Luque, F.J. Extension of the MST model to the IEF formalism: HF and B3LYP parametrizations. *J. Mol. Struct,* (*Theochem.*) **2005**, *727*, 29–40.

[51] Cramer, C.J.; Truhlar, D.G. General parameterized SCF model for free energies of solvation in aqueous solution. *J. Am. Chem. Soc.*, **1991**, *113*, 8305–8311.

[52] Cramer, C.J.; Truhlar, D.G. An SCF Solvation model for the hydrophobic effect and absolute free energies of aqueous solvation. *Sci.,* **1992**, *256*, 213–217.

[53] Cramer, C.J.; Truhlar, D.G. AM1-SM2 and PM3-SM3 parameterized SCF solvation models for free energies in aqueous solution. *J. Computer-Aided Mol. Des.*, **1992**, *6*, 629–666.

[54] Cortis, C; Langlois, J.M; Beachy, M; Friesner, R. Quantum mechanical geometry optimization in solution using a finite element continuum electrostatics method. *J. Chem. Phys*. **1996**, 105, 5472–5484.

[55] Klapper, I; Hagstrom, R; Fine,R; Sharp, K; Honig, B. Focusing of electric fields in the active site of Cu-Zn superoxide dismutase: Effects of ionic strength and amino-acid modification. *Proteins: Struct., Funct. Genet.* **1986**, 1, 47–59

[56] Pascual-Ahuir Giner J.L. *GEPOL: un metodo para el calculo de superficies moleculares.* Universitat de Valencia, **1988**.

[57] Flower, D.R. SERF: A program for accessible surface area calculations. *J. Mol. Graphics and Modeling,* **1997**, *15*, 238-244.

[58] Nilsson, O.; Pascual-Ahuir, J.L.; Tapia, O. Molecular volumes and surfaces of biomacromolecules *via* GEPOL: a fast and efficient algorithm**.** *J. Mol. Graphics,* 1990, *8*, 168-172.

[59] Pascual-Ahuir, J.L.; Silla, E. GEPOL: An improved description of molecular surfaces. I. Building the spherical surface set. *J. Comp. Chem.*, **1990**, *11*, 1047–1060.

[60] Silla, E.; Pascual-Ahuir, J.L.; Tunon, I. GEPOL: An improved description of molecular surfaces. II. Computing the molecular area and volume. *J. Comp. Chem.*, **1991**, *12*, 1077–1088.

[61] Pascual-Ahuir, J.L.; Silla, E.; Tunon, I. GEPOL: An improved description of molecular surfaces. III. A new algorithm for the computation of a solvent-excluding surface. *J. Comp. Chem.*, **1994**, *15*, 1127–1138.

[62] Lee, B.; Richards, F.M. The interpretation of protein structures: estimation of static accessibility. *J. Mol. Biol.*, **1971**, *55*, 379-400.

[63] Banerjee, T.; Singh, M.K.; Sahoo, R.K.; Khanna, A. Volume, surface and UNIQUAC interaction parameters for imidazolium based ionic liquids *via* polarizable continuum model. *Fluid Phase Equilibria*, **2005**, *234*, 64–76.

[64] Miertus, S.; Scrocco, E.; Tomasi, J. Electrostatic interaction of a solute with a continuum. A direct utilization of *ab initio* molecular potentials for the prevision of solvent effects. *Chem. Phys.,* **1981**, *55*, 117-129.

[65] Miertus, S.; Tomasi, J. Approximate evaluations of the electrostatic free-energy and internal energy changes in solution processes. *Chem. Phys.*, **1982**, *65*, 239-256.

[66] Bonaccorsi, R.; Scrocco, E.; Tomasi, J. Simple theoretical models for biochemical systems, with applications to DNA. *Proc. Int. Symp. Biomol. Struct. Interactions, Suppl. J. Biosci.,* **1985**, *8*, 627-634.

[67] Pascual-Ahuir, J.L.; Silla, E.; Tomasi, J.; Bonaccorsi, R. Electrostatic interaction of a solute with a continuum. Improved description of the cavity and of the surface cavity bound charge distribution. *J. Comp. Chem.,* **1987**, *8*, 778-787.

[68] Tomasi, J. Effective and practical ways of introducing the effect of the solvent in the theoretical evaluation of conformational properties of biomolecules. In *QSAR in drug design and toxicology (Proceedings of the sixth European symposium on quantitative structure activity relationships, Portorož-Portorose 22-26 September 1986)*, Elsevier: **1987**.

[69] Bonaccorsi, R.; Hodošček, N.; Tomasi, J. Introduction of solvent effects in the electrostatic recognition of biological receptors. *J. Mol. Struct. (Theochem)*, **1988**, *164*, 105-119.

[70] Tomasi, J.; Bonaccorsi, R.; Cammi, R.; Olivares del Valle, F.J. Theoretical chemistry in solution. Some results and perspectives of the continuum methods and in particular of the polarizable continuum model. *J. Mol. Struct. (Theochem)*, **1991**, *234*, 401-424.

[71] Cossi, M.; Barone, V.; Cammi, R.; Tomasi, J. *Ab-initio* study of solvated molecules: a new implementation of the polarizable continuum model. *Chem. Phys. Lett.*, **1996**, *255*, 327–335.

[72] Barone, V.; Cossi, M.; Tomasi, J. A new definition of cavities for the computation of solvation free energies by the polarizable continuum model. *J. Chem. Phys.*, **1997**, *107*, 3210-3221.

[73] Mennucci, B.; Tomasi, J. Continuum solvation models: A new approach to the problem of solute's charge distribution and cavity boundaries. *J. Chem. Phys.*, **1997**, *106*, 5151-5158.

[74] Mennucci, B.; Cancès, E.; Tomasi, J. Evaluation of solvent effects in isotropic and anisotropic dielectrics and in ionic solutions with a unified integral equation method: Theoretical bases, computational implementation, and numerical applications. *J. Phys. Chem. B*, **1997**, *101*, 10506-10517.

[75] Cancès, E.; Mennucci, B.; Tomasi, J. A new integral equation formalism for the polarizable continuum model: Theoretical background and applications to isotropic and anisotropic dielectrics. *J. Chem. Phys.*, **1997**, *107*, 3032–3041.

[76] Barone, V.; Cossi, M.; Tomasi, J. Geometry optimization of molecular structures in solution by the polarizable continuum model. *J. Comput. Chem.*, **1998**, *19*, **404–417.**

[77] Cossi, M.; Barone, V.; Mennucci, B.; Tomasi, J. *Ab-initio* study of ionic solutions by a polarizable continuum dielectric model. *Chem. Phys. Lett.*, **1998**, 286, 253-260.

[78] Cossi, M.; Barone, V. Analytical second derivatives of the free energy in solution by polarizable continuum models. *J. Chem. Phys.*, **1998**, *109*, 6246-6254.

[79] Cossi, M.; Barone, V. Quantum calculation of molecular energies and energy gradients in solution by a conductor solvent model. *J. Phys. Chem. A*, **1998**, *102*, 1995-2001.

[80] Adamo, C.; Barone, V. Exchange functionals with improved long-range behavior and adiabatic connection methods without adjustable parameters: The Mpw and Mpw1pw Models. *J. Chem. Phys.*, **1998**, *108*, 664-675.

[81] Tomasi, J.; Mennucci, B. In Schleyer, P. v. R. (editor-in-chief), *The Encyclopaedia of Computational Chemistry*, John Wiley & Sons Ltd.: Athens, USA, **1998**, *4*, 2547-2560.

[82] Rega, N.; Cossi, M.; Barone, V. Towards linear scaling in continuum solvent models. A new iterative procedure for energies and geometry optimizations. *Chem. Phys. Lett.*, **1998**, *293*, 221-229.

[83] Scalmani, G.; Barone, V. Use of molecular symmetry in the computation of solvation energies and their analytical derivatives by the polarizable continuum model. *Chem. Phys. Lett.*, **1999**, *301*, 263-269.

[84] Pomelli, C.S.; Tomasi, J.; Cossi, M.; Barone, V. Effective generation of molecular cavities in the polarizable continuum model by the DefPol procedure. *J. Comp. Chem.*, **1999**, *20*, **1693–1701.**

[85] Tomasi, J.; Mennucci, B.; Cancès, E. The IEF version of the PCM solvation method: an Overview of a new method addressed to study molecular solutes at the QM *Ab initio* level. *J. Mol. Struct. (Theochem)*, **1999**, *464*, 211-226.

[86] Cammi, R.; Mennucci, B.; Tomasi, J. Second-order Moller-Plesset analytical derivatives for the polarizable continuum model using the relaxed density approach. *J. Phys. Chem. A,* **1999**, *103*, 9100-9108.

[87] Cossi, M.; Barone, V.; Robb, M.A. A Direct procedure for the evaluation of solvent effects in MC-SCF Calculations. *J. Chem. Phys.,* **1999**, *111*, 5295-5302.

[88] Amovilli, C.; Barone, V.; Cammi, R.; Cancès, E.; Cossi, M.; Mennucci, B.; Pomelli, C.S.; Tomasi, J. Recent advances in the description of solvent effects with the polarisable continuum model. *Adv. Quantum Chem.,* **1999**, *32*, 227-259.

[89] Cammi, R.; Mennucci, B.; Tomasi, J. Fast evaluation of geometries and properties of excited molecules in solution: A Tamm-Dancoff model with application to 4-dimethylaminobenzonitrile. *J. Phys. Chem.* A, **2000**, *104*, 5631-5637.

[90] Cossi, M.; Barone, V. Solvent effect on vertical electronic transitions by the polarizable continuum model. *J. Chem. Phys.,* **2000**, *112*, 2427-2435.

[91] Cossi, M.; Barone, V. Time-dependent density functional theory for molecules in liquid solutions. *J. Chem. Phys.,* **2001**, *115*, 4708-4717.

[92] Cossi, M.; Rega, N.; Scalmani, G.; Barone, V. Polarizable dielectric model of solvation with inclusion of charge penetration effects. *J. Chem. Phys.,* **2001**, *114*, 5691-5701.

[93] Cossi, M.; Scalmani, G.; Rega, N.; Barone, V. New Developments in the polarisable continuum model for quantum mechanical and classical calculations on molecules in solution. *J. Chem. Phys.,* **2002**, *117*, 43-54.

[94] Mennucci, B.; Tomasi, J.; Cammi, R.; Cheeseman, J.R.; Frisch, M.J.; Devlin, F.J.; Gabriel, S.; Stephens, P.J. Polarizable continuum model (PCM) calculations of solvent effects on optical rotations of chiral molecules. *J. Phys. Chem.* A, **2002**, *106*, 6102-6113.

[95] Cappelli, C.; Corni, S.; Mennucci, B.; Cammi, R.; Tomasi, J. Vibrational circular dichroism within the polarizable continuum model: A theoretical evidence of conformation effects and hydrogen bonding for (*S*)-(-)-3-Butyn-2-ol in CCl_4 solution. *J. Phys. Chem.* A, **2002**, *106*, 12331-12339.

[96] Tomasi, J.; Cammi, R.; Mennucci, B.; Cappelli, C.; Corni, S. Molecular properties in solution described with a continuum solvation model. *Phys. Chem. Chem. Phys.,* **2002**, *4*, 5697-5712.

[97] Tomasi, J. Cavity and reaction field: "robust" concepts. Perspective on "electric moments of molecules in liquids". *Theor. Chem. Acc.,* **2000**, *103*, 196-199.

[98] Cossi, M.; Rega, N.; Scalmani, G.; Barone, V. Energies, structures and electronic properties of molecules in solution with the C-PCM solvation model. *J. Comp. Chem.,* **2003**, *24*, 669-681.

[99] Scalmani, G.; Barone, V.; Kudin, K.N.; Pomelli, C.S.; Scuseria, G.E.; Frisch, M.J. Achieving linear-scaling computational cost for the polarizable continuum model of solvation. *Theor. Chem. Acc.,* **2004**, *111*, 90-100.

[100] Soteras, I.; Curutchet, C.; Bidon-Chanal, A.; Orozco, M.; Luque, F.J. Extension of the MST model to the IEF formalism: HF and B3LYP parametrizations. *J. Mol. Struct. (Theochem),* **2005**, *727*, 29–40.

[101] Mennucci, B. Continuum solvation models: What else can we learn from them? *J. Phys. Chem. Lett.*, **2010**, *1*, 1666–1674.

[102] Zhan, C.G.; Bentley, J.; Chipman, D.M. Volume polarization in reaction field theory. *J. Chem. Phys.*, **1998**, *108*, 177−192.

[103] Vilkas, M.J.; Zhan, C.G. An efficient implementation for determining volume polarization self-consistent reaction field theory. *Chem. Phys.*, **2008**, *129*, 194109–194115.

[104] Chipman, D.M. Reaction field treatment of charge penetration *J. Chem. Phys.*, **2000**, *112*, 5558-5565.

[105] Klamt, A.; Schüüman, G.J. COSMO: A New approach to dielectric screening in solvents with explicit expressions for the screening energy and its gradient. *J. Chem. Soc. Perkin Trans.*, **1993**, *2*, 799–805.

[106] Klamt, A.; Jonas, V. **Treatment of the outlying charge in continuum solvation models.** *J. Chem. Phys.*, **1996**, *92*, 9972-9981.

[107] Goldblum, A.; Perahia, C.; Pullman, A. Hydration scheme of complementary base-pairs of DNA. *F. E. B. S. L.,* **1979**, *91*, 213-215.

[108] Tunega, D.; Aquino, A.J.A.; Haberhauer, G.; Gerzabek M.H.; Lischka H. *Hydrogen bonds and solvent effects in soil processes: a theoretical view*. Springer Science+Business Media B.V: **2008**, pp. 321–347.

[109] Bowler. D.R; Miyazaki, T. *O*(*N*) methods in electronic structure calculations. *Rep. Prog. Phys.*, **2012**, *75*, 036503–036545.

[110] Titantah, J.T.; Karttunen, M. Water dynamics: Relation between hydrogen bond bifurcations, molecular jumps, local density & hydrophobicity. *Scientific Reports*, **2013**, *3*, 2991–2999.

[111] Su, J.T.; Xu, X.; Goddard III, W.A. Accurate energies and structures for large water clusters using the X3LYP hybrid density functional. *J. Phys. Chem. A*, **2004,** *108,* 10518–10526.

[112] Titantah, J.T.; Karttunen, M. Long-time correlations and hydrophobe modified hydrogen bonding dynamics in hydrophobic hydration. *J. Am. Chem. Soc.*, **2012**, *134*, 9362–9368.

[113] Tuckerman, M.; Laasonen, K.; Sprik, M.; Parrinello, M. *Ab Initio* Molecular Dynamics Simulation of the solvation and transport of H_3O^+ and OH^- ions in water. *J. Phys. Chem.*, **1995**, *99*, 5749–5752.

[114] Laasonen, K.; Pasquarello, A.; Car, R.; Lee, C.; Vanderbilt, D. Car-Parrinello molecular dynamics with Vanderbilt ultrasoft pseudopotentials. *Phys. Rev. B*, **1993**, *47*, 10142

[115] Car, R.; Parrinello, M. Unified approach for molecular dynamics and density functional theory. *Phys. Rev. Lett.*, **1985**, *55*, 2471–2474.

[116] Burkert, U.; Allinger, N.L. *Molecular Mechanics*, ACS Monograph, *American Chemical Society*: Washington, DC, **1982**.

[117] Rappe´, A.K.; Casewit, C.J. *Molecular Mechanics across Chemistry*, University Science Books: Sausalito, CA, **1997**.

[118] Halgren, T.A. **Merck molecular force field. 1. Basis, form, scope, parameterization, and performance of MMFF94.** *J. Comp. Chem.*, **1996**, *17*, 490–519.

[119] Ren, P.; Ponder, J.W. Polarizable Atomic Multipole water model for molecular mechanics simulation. *J. Phys. Chem. B*, **2003**, 107, *24*, 5933–5947.

[120] Shi, Y.; Xia, Z.; Zhang, J.; Best, R.; Wu, C.; Ponder, J.W.; Ren, P. Polarizable atomic multipole-based AMOEBA force field for proteins. *J. Chem. Theory Comput.*, **2013**, 9, *9*, 4046–4063.

[121] Kaminski, G.A.; Stern, H.A.; Berne, B.J.; Friesner, R.A. Development of an accurate and robust polarizable molecular mechanics force field from ab initio quantum chemistry. *J. Phys.Chem. A*, **2003**, *108*, 621–627.

[122] Mennucci, B. Solvation models for molecular properties: Continuum *versus* discrete approaches. In Canuto, S. *Solvation effects on molecules and biomolecules*. Springer Science+Business Media B.V: **2008**, pp. 1–22.

[123] Schwörer, M.; Breitenfeld, B.; Troster, P.; Bauer, S.; Lorenzen, K.; Tavan P.; Mathias G. Coupling density functional theory to polarizable force fields for efficient and accurate Hamiltonian molecular dynamics simulations. *J Chem Phys*., **2013**, *138, 24*, 244103.

[124] Lipparini, F.; Cappelli, C.; Barone, V. A gauge invariant multiscale approach to magnetic spectroscopies in condensed phase: General three-layer model, computational implementation and pilot applications. *J. Chem. Phys*., **2013**, *138*, 234108–234117.

[125] Van Duijnen, P.T.; Swart M.; Jensen L. The discrete reaction field approach for calculating solvent effects. in Canuto, S. Solvation effects on molecules and biomolecules. Springer Science+Business Media B.V: **2008**, pp. 39–102.

[126] Brancato, G.; Rega, N.; Barone, V. A hybrid explicit/implicit solvation method for first-principle molecular dynamics simulations. *J. Chem. Phys*., **2008**, *128*, 144501–144510.

[127] Brancato, G.; Di Nola, A.; Barone, V.; Amadei, A. A mean field approach for molecular simulations of fluid systems. *J. Chem. Phys*., **2005**, *122*, 154109.

[128] Brancato, G.; Rega, N.; Barone, V. Reliable molecular simulations of solute-solvent systems with a minimum number of solvent shells. *J. Chem. Phys*., **2006**, *124*, 214505.

[129] Rega, N.; Brancato, G.; Barone, V. Non-periodic boundary conditions for ab initio molecular dynamics in condensed phase using localized basis functions. *Chem. Phys. Lett*., **2006**, *422*, 367–371.

[130] Brancato, G.; Barone, V.; Rega, N. Theoretical modeling of spectroscopic properties of molecules in solution: toward an effective dynamical discrete/continuum approach. *Theor. Chim. Acta*, **2007**, *117*, 1001–1015.

[131] Nadig, G.; van Zant, L.C.; Dixon S.L.; Merz Jr., K.M. Charge-transfer interactions in macromolecular systems: A new view of the protein/water interface. *J. Am. Chem. Soc*., **1998**, *120*, 5593–5594.

[132] van der Vaart, A.; Merz Jr., K.M. Charge transfer in small hydrogen bonded clusters. *J. Chem. Phys*., **2002**, *116*, 7380–7388.

[133] Mammino, L. Could geometry considerations help take into account solute–solvent hydrogen bonding in continuum solvation models? *Chem. Phys. Lett*., **2009**, *473*, 354–357.

[134] Asthagiri, D.; Pratt, L.R. Quasi-chemical study of Be^{2+}(aq) speciation. *Chem. Phys. Lett*., **2003**, *371*, 613–619.

[135] Asthagiri, D.; Pratt, L.R.; Kress J.D.; Gomez, M.A. The hydration state of HO^{-}(aq). *Chem. Phys. Lett*., **2003**, *380*, 530–535.

[136] Elmer, S.P.; Park, S.; Pande, V.S. Foldamer dynamics expressed *via* Markov state models. I. Explicit solvent molecular-dynamics simulations in acetonitrile, chloroform, methanol, and water. *J. Chem. Phys*., **2005**, *115*, 114902.

[137] Poopari, M.R.; Dezhahang, Z.; Xu, Y.A. Comparative VCD study of methyl mandelate in methanol, dimethyl sulfoxide, and chloroform: explicit and implicit solvation models. ***Phys. Chem. Chem. Phys*.**, **2013**, ***15***, 1655–1665.

[138] Roccatano, D. Computer simulations study of biomolecules in non - aqueous or cosolvent/water mixture solutions. *Curr. Protein Pept. Sci.,* **2008**, *9*, 407–426.

[139] Bennion, B.J.; Daggett, V. The molecular basis for the chemical denaturation of proteins by urea. *Proc. Natl. Acad. Sci. U.S.A*., **2003**, *100*, 5142–5147.

[140] Lee, M.E.; van der Vegt, N.F.A. Does urea denature hydrophobic interactions? *J. Am. Chem. Soc.,* **2006**, *128*, 4948–4949.

[141] Delgado Alfaro, R.A.; Gomez-Sandoval, Z.; Mammino, L. Evaluation of the antiradical activity of hyperjovinol A utilizing donor-acceptor maps. *J. Mol. Model.*, **2014**, 20(7), 2337.

[142] Hehre, W.J.; Ditchfield, R.; Pople, J.A. Self-consistent molecular orbital methods. XII. Further extensions of gaussian-type basis sets for use in molecular orbital studies of organic molecules. *J. Chem. Phys.,* **1972**, *56*, 2257-2261.

[143] Hariharan, P.C.; Pople, J.A. The Influence of polarization functions on molecular orbital hydrogenation energies. *Theor. Chem. Acta,* **1973**, *28*, 213-222.

[144] Korth, H.G.; de Heer, M.I.; Mulder, P.A. DFT study on intramolecular hydrogen bonding in 2-substituted phenols: conformations, enthalpies, and correlation with solute parameters. *J. Phys. Chem. A,* **2002**, *106*, 8779-8789.

[145] Tiwari, A.K.; Sathyamurthy, N. Structure and stability of salicylic acid-water complexes and the effect of molecular hydration on the spectral properties of salicylic acid. *J. Phys. Chem. A,* **2006**, *110*, 5960-5964.

[146] Tsuzuki, S.; Uchimaru, T.; Matsumura, K.; Mikami, M.; Tanabe, K. Effects of the higher electron correlation correction on the calculated intermolecular interaction energies of benzene and naphthalene dimers: comparison between MP2 and CCSD(T) calculations. *Chem. Phys. Lett.,* **2000**, *319*, 547–554.

[147] Bartlett R.J.; Purvis, G.D. Many-body perturbation theory, coupled-pair many-electron theory, and the importance of quadruple excitations for the correlation problem. *Int. J. Quantum Chem.,* **1978**, *14*, 561−581.

[148] Hannongbua, S. *EuAsC$_2$S-11 Conference*, The Dead Sea, October **2010**.

[149] Xanides, D.; Randolf, B.R.; Rode, B.M. Structure and ultrafast dynamics of liquid water. A quantum mechanics / molecular mechanics molecular dynamics simulation study. *J. Chem. Phys*., **2005**, *122*, 174506.

[150] Xanides, D.; Randolf, B.R.; Rode, B.M. Hydrogen bonding in liquid water. An *ab initio* QM/MM MD simulation study. *J. Mol. Liquids*, **2006**, *123*, 61-67.

[151] Oliveira, A.A.; Nagurniak, G.R.; da Silva, A.B.F. Theoretical approach to differentiate set of steric conformations of octahydroquinolizine with μ-opioid activity antagonism. *XXXVIII International Conference of Theoretical Chemists of Latin Expression*, Natal (Brazil), December **2012.**

[152] Pereira, E.B.; Nagurniak, G.R.; da Silva, A.B.F. A DFT and HF study of trans-3,4-Dimethyl-4-(3-carboxamidophenyl)piperidines with μ-opioid activity. *XXXVIII International Conference of Theoretical Chemists of Latin Expression*, Natal (Brazil), December **2012.**

[153] Mammino, L.; Bilonda, M.K. Computational study of antimalarial pyrazole alkaloids from *newbouldia laevis*. *Congress of Theoretical Chemists of Latin Expression*, Granada, July **2013**.

[154] Saracino, G.A.A.; Improta, R.; Barone, V. Absolute pKa determination for carboxylic acids using density functional theory and the polarizable continuum model. *Chem. Phys. Lett.,* **2003**, *373*, 411-415.

[155] Wahl, M.C., Sundaralingam, M. C-H···O hydrogen bonding in biology. *Trends Biochem. Sci.,* **1997**, *22*, 97–102.

[156] Scheiner, S.; Maksic, Z.B. (Eds.), *Theoretical Models of Chemical Bonding*, vol.4, Springer Verlag: Berlin, **1991**.

[157] Alagona, G.; Ghio, C. 5-fluorouracil dimers in aqueous solution: molecular dynamics in water and continuum solvation. *Int. J. Quantum Chem.*, **2002**, *88*, 133-146.

[158] Boys, S.F.; Bernardi, F. The calculation of small molecular interactions by the differences of separate total energies. Some procedures with reduced errors. *Molec. Phys.*, **1970**, *19*, 553-566.

[159] Ahn, D-S.; Jeon, I-S.; Jang, S-H.; Park, S-W.; Lee, S.; Cheong, W. Hydrogen bonding in aromatic alcohol-water clusters: A brief review. *Bull. Korean Chem. Soc.,* **2003**, *24*, 695-702.

[160] Davies, T.G.; Tame, J.R.; Hubbard, R.E. Generating consistent sets of thermodynamic and structural data for analysis of protein-ligand interactions. *Persp. Drug Discov. Des*., **2000**, *20*, 29–42.

[161] Baum, B.; Muley, L.; Smolinski, M.; Heine, A.; Hangauer, D.; Klebe, G. Non-additivity of functional group contributions in protein–ligand binding: A comprehensive study by crystallography and isothermal titration calorimetry. *J. Mol. Biol.,* **2010**, *397*, 1042–1054.

[162] Sousa, S.P.; Fernandes, P.A.; Ramos, M.J. Protein–ligand docking: Current status and future challenges. *Proteins: Structure, Function, and Bioinformatics,* **2006**, *65,* 15–26.

[163] Majeux, N.; Scarsi, M.; Caflisch, A. Efficient electrostatic solvation model for protein-fragment docking. *PROTEINS: Structure, Function, and Genetics*, **2001**, *42*, 256–268.

[164] Young, T.; Abel, R.; Kim, B.; Berne, B.J.; Friesner, R.A. Motifs for molecular recognition exploiting hydrophobic enclosure in protein-ligand binding., *Proc Natl. Acad. Sci. USA.*, **2007**, *104*, 808−813.

[165] Klebe, G. Virtual ligand screening: strategies, perspectives and limitations. *Drug Discov. Today*, **2006**, *11*, 580−594.

[166] Schneider, G.; Bohm, H.J. Virtual screening and fast automated docking methods. *Drug. Discov. Today*, **2002**, *7*, 64–70.

[167] Muegge, I.; Rarey, M. Small molecule docking and scoring. Vol. 17. In: Lipkowitz K.B.; Boyd, D.B. eds. *Reviews in Computational Chemistry*. John Wiley & Sons Inc: **2001**, pp. 1–60.

[168] Waszkowycz, B.; Perkins, T.D.J.; Sykes, R.A.; Li, J. Large-scale v.irtual screening for discovering leads in the postgenomic era. *Ibm. Syst. J.*, **2001**, *40*, 360–376.

[169] Böhm, H.; Stahl, M. The use of scoring functions in drug discovery applications. In: Lipkowitz, K.; Boyd, D, eds. *Reviews in Computational Chemistry*. John Wiley & Sons Inc.: **2002**, pp. 41–86.

[170] Klebe, G.; Grädler, U.; Grüneberg, S.; Krämer, O.; Gohlke, H. Understanding receptor ligand interactions as a prerequisite for virtual screening. In: Böhm, H.; Schneider, G. eds. *Virtual Screening for Bioactive Molecules*. Wiley-VCH: Weinheim, Germany, **2000**, pp. 207–227.

[171] Walters, W.P.; Stahl, M.T.; Murcko, M.A. Virtual screening—an overview. *Drug. Discov. Today*, **1998**; *3*, 160–178.

[172] Shoichet, B.K.; Leach, A.R.; Kuntz, I.D. Ligand solvation in molecular docking. *PROTEINS: Structure, Function, and Genetics*, **1999**, *34*, 4–16.

[173] Grosdidier, A.; Zoete, V.; Michielin, O. EADock: docking of small molecules into protein active sites with a multiobjective evolutionary optimization. *Proteins: Structure, Function, and Bioinformatics,* **2007**, *67,* 1010−1026.

[174] Verdonk, M.L.; Chessari, G.; Cole, J.C.; Hartshorn, M.J.; Murray, C.W.; Nissink, J.W.M.; Taylor, R.D.; Taylor, R. Modeling water molecules in protein-ligand docking using GOLD. *J. Med. Chem.,* **2005**, *48*, 6504–6515.

[175] Tirado-Rives, J.; Jorgensen, W.L. Contribution of conformer focusing to the uncertainty in predicting free energies for protein-ligand binding. *J. Med. Chem.,* **2006**, *49*, 5880–5884.

CHAPTER 7

A Novel Coarse-Grained Description of Protein Structure and Folding by UNRES Force Field and Discrete Nonlinear Schrödinger Equation

Adam Liwo[1], Antti Niemi[2,3], Xubiao Peng[2] and Adam K. Sieradzan[1,2,*]

[1]Faculty of Chemistry, University of Gdańsk, Wita Stwosza 63, 80-952 Gdańsk, Poland; [2]Department of Physics and Astronomy and Science for Life Laboratory, Uppsala University, P.O. Box 803, S-75108 Uppsala, Sweden and [3]Laboratoire de Mathematiques et Physique Theorique CNRS UMR 6083, Fédération Denis Poisson, Université de Tours, Parc de Grandmont, F37200 Tours, France and Department of Physics, Beijing Institute of Technology, Haidian District, Beijing 100081, People's Republic of China

Abstract: The UNited RESidue (UNRES) force field has been developed for over two decades. This force field has been derived carefully as a potential of mean force of the system studied, which is further expressed in terms of the Kubo cluster-cumulant functions. New terms in the energy function to improve loop structures have been introduced recently. On the other hand, new concept was developed, in which wave-analysis physics is applied to the protein folding problem. At present, the energy function is based on the Landau Hamiltonian, the minima of which are stable conformations of protein fragments; these minima are obtained as kink solutions of the Discrete Nonlinear Schrödinger Equation. The parameters of the Hamiltonian have been obtained by statistical analysis of known protein structures. The unique feature of this approach is that the curvature description is sufficient for protein folding without any long-distance interactions other than the excluded-volume interactions. The combination of those two methodologies - molecular dynamics with the use of physics-base UNRES force field and the kink approach have been applied to study the flexibility and movement of the kinks as well as their formation and disappearance in the folding process.

Keywords: UNRES, force-field, Davydov soliton, dark soliton, molecular dynamics, energy landscape, Landau Hamiltonian, gauge inverse, physics-based, cumulant-cluster expansion, mean-force potentials, loop structures, wave-analysis physics, protein flexibility, kink formation, kink disappearance, kink movement, folding pathways, local interactions.

[*]**Corresponding author Adam K. Sieradzan:** Faculty of Chemistry, University of Gdańsk, Wita Stwosza 63, 80-952 Gdańsk, Poland; Tel: +48585235351; Fax: +48585235350; E-mail: adasko@sun1.chem.univ.gda.pl

Zaheer Ul-Haq and Jeffry D. Madura (Eds). **10.1016/B978-1-60805-865-5.50007-1**

INTRODUCTION

Over the last two decades there has been a huge progress in the protein structure prediction field [1-3]. The most successful method, which is applied routinely nowadays, is homology modeling. The huge advantage of this method is high accuracy [3-5] and considerably low time required to achieve a meaningful result. However, when the template is not available for homology modeling the accuracy is questionable [6]. The homology modeling methods are constantly developed. Introduction of four body interactions in homology modeling [7] is one of the examples. Despite high accuracy obtained in the last Community Wide Experiment on the Critical Assessment of Techniques for Protein Structure Prediction (CASP10) the homology modeling reached plateau. Moreover, not only single static structure is currently required but also the protein flexibility plays an important role in its activity [8].

The molecular dynamics [9](MD) and Monte Carlo [10] (MC) simulations are alternative methods to homology modeling that do not require the template for getting high accuracy [11-13]. Both methods give valuable insight into crucial movements of the proteins. Usually the MC and MD are used with all-atom force fields. The use of all-atom level of description requires great computational effort. Nevertheless there has been a tremendous advance in the computational techniques. One of the examples is implementation of all-atom molecular dynamics (MD) programs on graphical processor units (GPUs) [18]. Another example is use of world-distributed computing (the FOLDING@HOME project) [14]. Moreover, very efficient load-balanced parallel codes such as GROMACS [15], NAMD [16], or DESMOND [17] have been introduced. Finally dedicated machines [19] have been constructed. However, even for very small (50 amino-acid residue) systems and with the use of special purpose computers, time scales are restricted to 200 - 500μs [13].

The alternative for all-atom force fields are coarse-grained ones. In coarse grained models groups of atoms are united into one interacting center (Fig. **1**).

By coarse-graining, the number of degrees of freedom is reduced as well as fast movements, high frequency movements connected with all-atom structure, are averaged out. This may leads to speed up by 3-6 orders of magnitude with respect to all-atom simulation in explicit solvent [20-22].

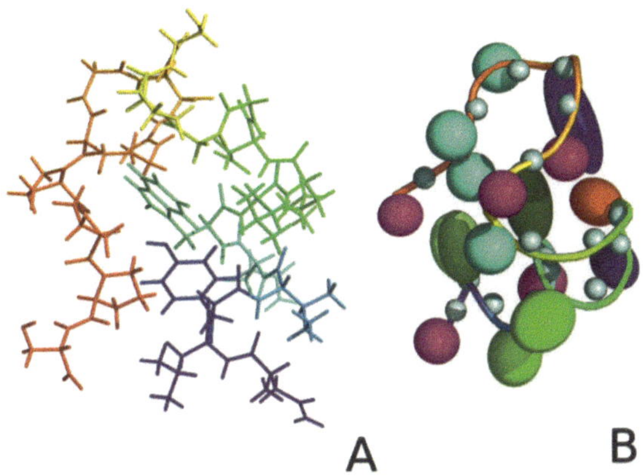

Figure 1: The comparison between all atom representation of tryptophan cage (A) with coarse-grain representation of this peptide (B).

There are two types of force-fields: "universal", such as C^{α} C^{β} Side Groups (CABS)[23], UNited Residue (UNRES)[24], CHARMM [25] or AMBER [26] and "structure based" like Gō model [27] or kink description [28]. The "universal" force field is the one in which Hamiltonian does not require the knowledge of the tertiary structure *a priori*. The "universal" coarse-grained force field is usually able to predict the overall fold of a protein but the details of the structure are not reproduced accurately. The structure-based force field, despite having high accuracy, requires the native structure and parameterization for each protein separately.

Advances in Coarse-Grained Models

Recently many important and new concepts appeared in the field of coarse graining. There has been great advance in the development of coarse-grained force fields, which have more than one residue per center of interaction [29] and conceptual "ultra"-coarse-grained force fields development [30]. Moreover, the "ultra" coarse-grained force fields have been optimized for memory and parallelized [31]. There has been also improvement in energy function parameterization. For example, in the MARTINI force field [32,33] new parameters with regards to side-chain properties have been introduced [34]. In the

ROSETTA force field [35] a lot of emphasis has been put into improving the energy function with respect to unfolded state [36], which is important for obtaining the correct folding pathway. Moreover, the side-chain rotamer library has been improved [36]. In the OPEP force field [37], where backbone is in all-atom representation and side-chains are coarse-grained, new energy function for side-chain side-chain interactions has been introduced improving the resolution of the force field [38-40].

UNRES FORCE FIELD

The United RESidue (UNRES) force field [24] is a coarse grained force field in which the polypeptide chain is represented by a sequence of C^{α} atoms with side-chains (SC) attached to the C^{α} atoms (Fig. **2**).

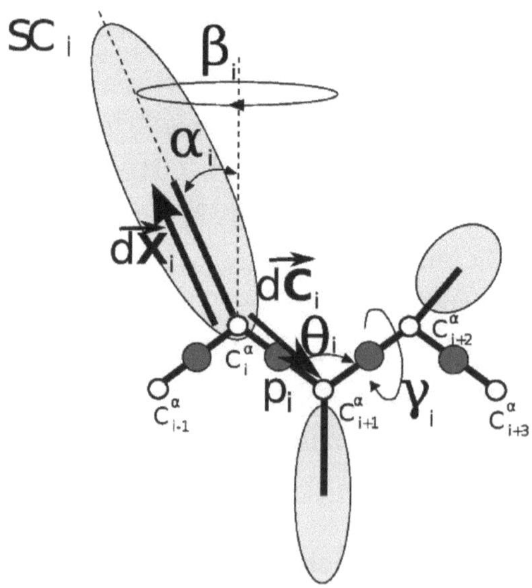

Figure 2: The UNRES model of polypeptide chains. The interaction sites are peptide-group centers (p) represented as dark grey circles, and side-chain centers (SC) represented as light grey ellipsoids. The side chains are attached to the corresponding α-carbon atoms with different $C^{\alpha\cdots}SC$ bond lengths, d_{SC}. The α-carbon atoms are represented by small open circles and are not an interacting site. The geometry of the chain can be described either by virtual-bond lengths, backbone virtual-bond angles θ_i, i=1,2,…,n-2, backbone virtual-bond-dihedral angles γ_i, i=1,2,…,n-3, and the angles α_i and β_i i=2,…,n-1 that describe the location of a side chain with respect to the coordinate frame defined by C^{α}_{i-1}, C^{α}_i, and C^{α}_i, C^{α}_{i+1}, or in terms of the virtual-bond vectors dC_i (from C^{α}_i to C^{α}_{i+1}), i=1,2,…,n-1 and dX_i (from C^{α}_i to SC_i), i=2,…,n-1, represented by thick lines, where n is the number of residues.

The peptide groups are located half-way between two consecutive C^{α} atoms and are represented by spheres (Fig. **2**). The side-chains are represented by spheroids with the axis of revolution along the $C^{\alpha}\cdots$SC virtual bond (Fig. **2**). The geometry of the backbone is defined either by the sequence of vectors $\mathbf{dC}_i$, i=1,2,...,n or the internal coordinates: virtual-bond valence angles (θ) between three consecutive C^{α} atoms and virtual-bond torsional angles (γ) between four consecutive C^{α} atoms. The location of a side-chain with respect to the backbone is defined either by the **dX** virtual-bond vector [41] or by the angle α between the center of the side-chain and C^{α} plane and the angle β describing the rotation of the $C^{\alpha}\cdots$ SC axis about the bisector of the virtual-bond angle θ (Fig. **3**).

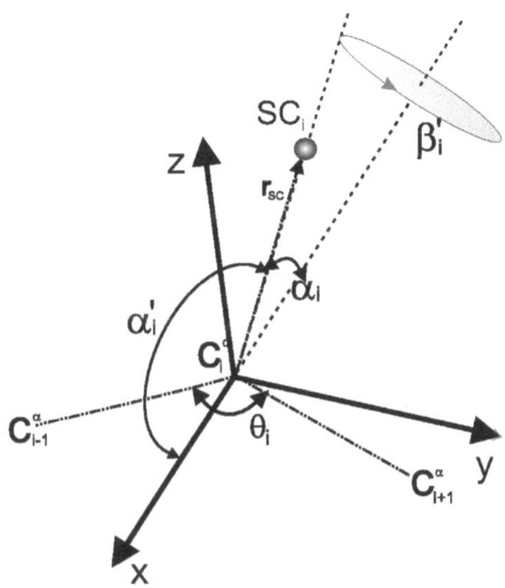

Figure 3: Local coordinate for rotameric potentials of side-chains [42]. θ_i is virtual valence angle between three consecutive C^{α} atoms. The *x* axis is on the bisection of the angle θ_i. The *y* is perpendicular to *x* in the direction of C^{α}_{i+1} and is on the C^{α}_{i-1}, C^{α}_{i}, C^{α}_{i+1} plane. The *z* axis is orthogonal to *x* and *y* axis and create right handed coordinate system. The α is a conjugate angle to α'. The α' angle is the angle between the side-chain vector (SC) and the *x* axis. The β angle is the revolution angle around *x* axis.

"In the UNRES force field, the effective energy function is defined as the restricted free energy (RFE) or the potential of mean force (PMF) of the chain constrained to a given coarse-grained conformation along with the surrounding solvent [43-45]." This effective energy function, including the new terms that improve loop representation introduced in our earlier work [12], is expressed by eq 1.

$$U = w_{sc}\sum_{i<j} U_{SCiSCj} + w_{scp}\sum_{i\neq j} U_{SCipj} + w_{pp}^{VDW}\sum_{i<j-1} U_{pipj}^{VDW} + w_{pp}^{el} f_2(T)\sum_{i<j-1} U_{pipj}^{VDW}$$
$$+ w_{tor} f_2(T)\sum_i U_{tor}(\gamma_i) + w_{tord} f_3(T)\sum_i U_{tord}(\gamma_i, \gamma_{i+1}) + w_b \sum_i U_b(\theta_i)$$
$$+ w_{rot}\sum_i U_{rot}(\alpha_{SCi}, \beta_{SCi}) + w_{bond}\sum_i U_{bond}(d_i) + w_{corr}^{(3)} f_3(T)\sum_i U_{corr}^{(3)} \quad (1)$$
$$+ w_{corr}^{(4)} f_4(T)\sum_i U_{corr}^{(4)} + + w_{turn}^{(3)} f_3(T)\sum_i U_{turn}^{(3)} + w_{turn}^{(4)} f_4(T)\sum_i U_{turn}^{(4)}$$
$$+ w_{SC-corr} f_2(T)\sum_{m=1}^{3}\sum_i U_{SC-corr}(\tau_i^{(m)})$$

"where the U's are energy terms, θ_i is the backbone virtual-bond angle, γ_i is the backbone virtual-bond-dihedral angle, α_i and β_i are the angles defining the location of the center of the united side chain of residue *I* (Fig. **2**), and d_i is the length of the *i*th virtual bond, which is either a $C^{\alpha}\cdots C^{\alpha}$ virtual bond or $C^{\alpha}\cdots$ SC virtual bond." Each energy term is multiplied by an appropriate weight, w_x, and the terms corresponding to the Kubo cumulant-cluster expansion [46] "factors of order higher than 1 are additionally multiplied by the respective temperature factors which were introduced in our work [47] and which reflect the dependence of the first generalized-cumulant term in those factors on temperature, as discussed in refs [47] and [48]." The factors f_n are defined by eq 2.

$$f_n(T) = \frac{\ln[\exp(1) + \exp(-1)]}{\ln\{\exp[(T/T_0)^{n-1}] + \exp[-(T/T_0)^{n-1}]\}} \quad (2)$$

where T_0=300K.

"The term U_{SCiSCj} represents the mean free energy of the hydrophobic (hydrophilic) interactions between the side chains, which implicitly contains the contributions from the interactions of the side chain with the solvent. The term U_{SCipj} denotes the excluded-volume potential of the side-chain - peptide-group interactions. The peptide-group interaction potential is split into two parts: the Lennard-Jones interaction energy between peptide-group centers ($U^{VDW}{}_{pipj}$) and the average electrostatic energy between peptide-group dipoles ($U^{el}{}_{pipj}$); the second of these terms accounts for the tendency to form backbone hydrogen bonds between peptide groups p_i and p_j. The terms U_{tor}, U_{tord}, U_b, U_{rot} and U_{bond} describe the local properties of the backbone and are the virtual-bond-dihedral angle torsional terms, virtual-bond dihedral angle double-torsional terms, virtual-bond angle bending terms, side-chain rotamer, and virtual-bond-deformation terms. The terms $U^{(m)}_{corr}$ represent correlation or multibody contributions from the coupling between backbone-local and backbone-electrostatic interactions (dipole moment alignment), and the terms $U^{(m)}_{turn}$ are correlation contributions involving m consecutive peptide groups; they are, therefore, termed turn contributions. The multibody terms are indispensable for reproduction of regular α-helical and β-sheet structures [43, 44, 49]. The $U_{SC\text{-}corr}$ terms are newly introduced side-chain backbone correlation potentials; they are expressed as Fourier series in the $SC\cdots C^{\alpha}\cdots C^{\alpha}\cdots C^{\alpha}$ ($\tau^{(1)}$), $C^{\alpha}\cdots C^{\alpha}\cdots C^{\alpha}\cdots SC$ ($\tau^{(2)}$), and $SC\cdots C^{\alpha}\cdots C^{\alpha}\cdots SC$ ($\tau^{(3)}$) virtual-bond-dihedral angles [12]."

There are two types of force field weights set which are frequently used in the UNRES force field. The energy-term weights of the first set were determined [47] by force-field calibration to reproduce the structure and folding thermodynamics of the GA (protein G-related albumin-binding) module (an α protein; PDB code: 1GAB) [50], while the energy-term weights were determined [51] by global search of optimal force-field parameters to reproduce the structure and folding thermodynamics of the tryptophan cage (PDB code: 1L2Y)[52] and the tryptophan zipper 2 (PDB code: 1LE1)[53].

Derivation of Potentials of Mean Force

As mentioned in the previous section the energy components are expressed as potentials of mean force (PMF). In the UNRES force field three different methodologies to derive PMF are used. The side-chain - side-chain interaction potentials have been determined recently [54-56] by all-atom molecular dynamics in water of the model system. Earlier [57] these potentials were determined from PDB statistics by applying the inversion of the Boltzmann law, according to which the dimensionless potentials of mean force were calculated from eq (3):

$$f_{XY}(\tau_i) = \ln(N_{XY,\max}) - \ln(N_{XY,i}) \qquad (3)$$

where $N_{XY,i}$ is the number of counts of type for residue types X and Y in the *i*th bin for variable (τ), and $N_{XY,max}$ is the largest number of counts over all bins, for a given type of variable (τ) and given types of residues.

The torsional [56], the double torsional potentials [58], the valence bending potentials [57], the bond stretching potentials [22, 41, 42] and the side-chain rotation potentials [41,42] were determined by Boltzmann integration over the potential-energy surfaces of model systems calculated with the *ab initio* method of molecular quantum mechanics at the MP2 (Møller-Plesset) theory level or with use of semi-empirical calculation with use of the AM1 method of the model system. The respective integrals corresponded to the terms of the Kubo cluster-cumulant expansion [46] of the total potential of mean force of polypeptide chains (for more details of deriving potentials see [32]). As an example, the equation for derivation of the torsional potentials is given below:

$$\begin{aligned}U_{XY}(\gamma) = &-\beta^{-1}\ln\{\frac{1}{(2\pi)^3}\int_{-\pi}^{\pi}\int_{-\pi}^{\pi}\int_{-\pi}^{\pi}[\det H^*(\lambda_1,\gamma-\pi-\lambda_2)\times\det H^*(\lambda_2,\lambda_3)]^{-1/2}\\ &\exp[-\beta[e_X(\lambda_1,\gamma-\pi-\lambda_2)+e_Y(\lambda_2,\lambda_3)]]d\lambda_1 d\lambda_2 d\lambda_3\}\\ &+\beta^{-1}\ln\{\frac{1}{(2\pi)^3}\int_{-\pi}^{\pi}\int_{-\pi}^{\pi}[\det H^*(\lambda_1,\lambda_2)]^{-1/2}\exp[-\beta e_X(\lambda_1,\lambda_2)]d\lambda_1 d\lambda_2\}\\ &+\beta^{-1}\ln\{\frac{1}{(2\pi)^3}\int_{-\pi}^{\pi}\int_{-\pi}^{\pi}[\det H^*(\lambda_2,\lambda_3)]^{-1/2}\exp[-\beta e_X(\lambda_2,\lambda_3)]d\lambda_2 d\lambda_3\}\end{aligned} \quad (4)$$

"where e_x and e_y are the energy surfaces for the terminally-blocked residues of type X and Y respectively, *H** is the energy Hessian computed over all variables except for the angles λ_i and λ_j (the terms with det**H*** account for the harmonic-entropy contribution to the PMF), where β=1/RT where R is the universal gas constant and T is the absolute temperature. The angles λ_i and λ_j are the variable which are the averaged out degrees of freedom for the equation 4."

The $U_{SC\text{-}corr}$ and correlation potentials [44] were derived from statistical analysis of structures of the known proteins obtained from Protein Data Bank [12], by applying the Boltzmann inversion method (equation 3).

After obtaining the potentials the analytical function is fitted to obtained PMF. For the torsional, double torsional, valence bending potentials the Fourier expansion series is used [58,59]. For the side-chain rotation potential the multiple Gaussian-like function is fitted [41,42], for other potentials the more complicated functions are used [55,56].

Simulations with Use of the UNRES Force Field

The UNRES force field is routinely used in three variants: conformational space annealing [60], molecular dynamics [61] and replica exchange [62].

Conformational Space Annealing

The Conformational Space Annealing (CSA) method is an algorithm for searching local minimum in multidimensional space [62-65]. "It combines the buildup and genetic algorithm. As in the genetic algorithm, it is started with a random population of conformations whose energies are then minimized. These local minima constitute the "bank" (n groups)." The typical size of the bank for UNRES CSA runs is 50-100 conformations. "At the beginning, conformations in the bank are distributed randomly *i.e.*, minimized from random conformations in the conformational space of local minima." Conformations in the bank are kept as diverse as possible, in terms of the difference between their virtual-bond dihedral angles γ. A newly generated conformation replaces the highest-energy conformation of the bank if it is more diverse from each of the conformations of the bank than the selected cut-off. If the newly-generated conformation is similar to one existing in the bank, it replaces it if its energy is lower. The cut-off is shrunk during the progress of the procedure, which results in gradual focusing of the search in the low-energy region(s). "Schematically, each conformation in the bank is considered as a representative of a group of local minima within a certain distance of each other in the conformational space."

Molecular Dynamics

In molecular dynamics, Newton's equations of motion are solved numerically to determine the time evolution of a system. The Newton equations are shown (eq. 5):

$$m\vec{a}(t) = \vec{F} = -\vec{\nabla}U[x(t)] \tag{5}$$

where m is a mass of interacting center, $\vec{a}$ is a center's acceleration, U is a potentials energy (determined by coordinates of centers of interaction x), t is a time and $\vec{\nabla}$ is a derivative operator.

In the UNRES model, the equations of motion are more complicated because the variables are the $C^{\alpha}\cdots C^{\alpha}$ and the $C^{\alpha}\cdots$SC virtual-bond vectors and not centers of interactions - peptide group (p) or side-chain (SC). This is due to the fact that C^{α} atoms are not the centers of interaction whereas the peptide groups are located in between two consecutive C^{α} atoms. Equations of motion are expressed as non-diagonal, conformation independent matrix of inertia. For the full description of equation of motion and their derivation see the [51].

Due to practical reasons the numerical version of equation of motion is used. In the UNRES force field the Verlet algorithm [9,61] is used. The simplified version of this algorithm is described with the equations (eq 6):

$$\begin{aligned} m\vec{a}(t+\partial t) &= -\vec{\nabla}U(t+\partial t) \\ \vec{v}(t+\partial t) &= \vec{v}(t) + \frac{1}{2}[\vec{a}(t) + \vec{a}(t+\partial t)]\partial t \\ \vec{x}(t+\partial t) &= \vec{x}(t) + \vec{v}(t)\partial t + \frac{1}{2}[\vec{a}(t) + \vec{a}(t+\partial t)]\partial t^2 \end{aligned} \tag{6}$$

where δt is a time step, $\vec{\mathbf{v}}$ is a velocity vector of a center of interaction, $\vec{x}$ is a position vector of center of interaction. For the full version of the Verlet algorithm in the UNRES force field see the [61].

Replica Exchange

The replica exchange molecular dynamics (REMD) method [66,67] is a sampling improving method. This idea is based on running parallel molecular dynamics simulations, each at a different temperature, and allow to exchange temperatures between trajectories every pre-assigned number of steps (Fig. **4**).

Figure 4: The schematic illustration of replica exchange molecular dynamics method. The four trajectories marked by four different colors are in four different temperatures T_1,T_2,T_3,T_4.

In UNRES force field the exchange between replicas is usually every 10000 or 20000 time steps. The criterion of acceptance of exchange is modified Metropolis criterion [47,62,68] (eq 7):

$$p(x) = \min[1; \exp\{-[\beta_j U(X_j,\beta_j) - \beta_i U(X_j,\beta_i)] - [\beta_j U(X_i,\beta_j) - \beta_i U(X_i,\beta_i)] \quad (7)$$

where p(x) is probability of exchange of temperature between *i*-th and *j*-th replica, U is effective energy of given conformation in given temperature (as mentioned before the U is potential of mean force and is temperature dependent), β=1/RT, R is universal gas constant, T is absolute temperature.

Examples of Application of the UNRES Force Field

The UNRES force field has been applied with success to study kinetics and folding pathways for various systems [69-72]. The UNRES force field has high predictive power for the overall fold of the protein and the domain packing. In the 10th Community Wide Experiment on the Critical Assessment of Techniques for Protein Structure Prediction (CASP10) UNRES force field as the best force field predicted domain packing of target T0663 [73].

Apart from the prediction of the protein folding the UNRES force field can be applied to study the protein association process [72,74,75]. The UNRES force field has a high predictive power for the alignment of monomers in oligomers (Fig. **5**).

The UNRES force field has been applied also for large molecular systems. The molecular studies of the chaperon closing and opening [76] or studies of the PDZ binding to the BAR domain of PICK1[77] give insight into functional mechanism at atomic level.

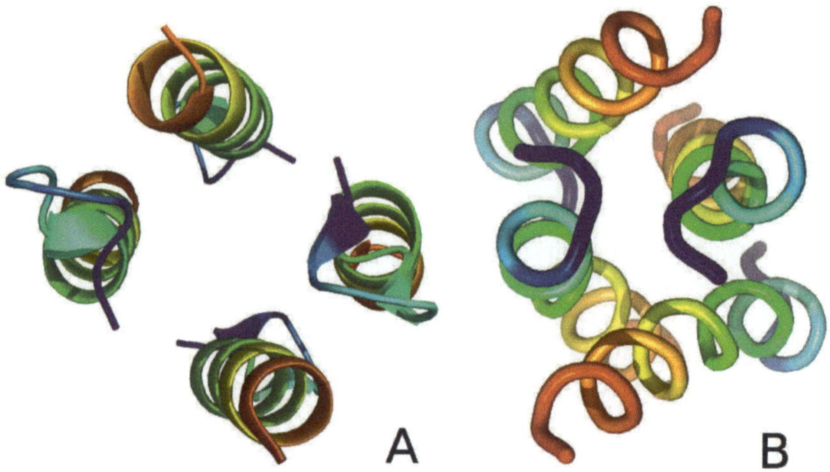

Figure 5: The comparison between the native structure (A) of ββα tetramer (PDB code:1SN9) with the representative structure of the dominant cluster (B). Both native and simulated reveal a pseudo C_{4s} symmetry.

KINK DESCRIPTION IN PROTEINS

Kink Model and Energy Function

80,000 structures have been determined and deposited in the Protein Data Bank (PDB) [78] but only 1393 unique folds have been identified by SCOP [79]. Even fewer topologies (1282) have been identified by CATH classification [80]. This leads to straightforward conclusion that many proteins share the same fold and bear structural similarities.

The study of string-like objects and their properties, both continuous and discrete, is similarly pivotal to several apparently disparate sub-fields of physics. Examples include polymers [81], Kirchoff-type elastic rods [82], vortexes in fluid dynamics [83], turbulence [84], superconductors [85], super-fluids [86], cosmic [87] and fundamental [88] strings in high energy physics, and numerous other applications. Based on the limited number of protein folds the theory of continuous curves in three dimensional space [89] should be of high applicability to protein-folding problem.

The further analysis has shown that only few region are occupied in θ, γ space (Fig. **6**). Moreover, the loop formation is connected with "circulating path" (Fig. **6B**).

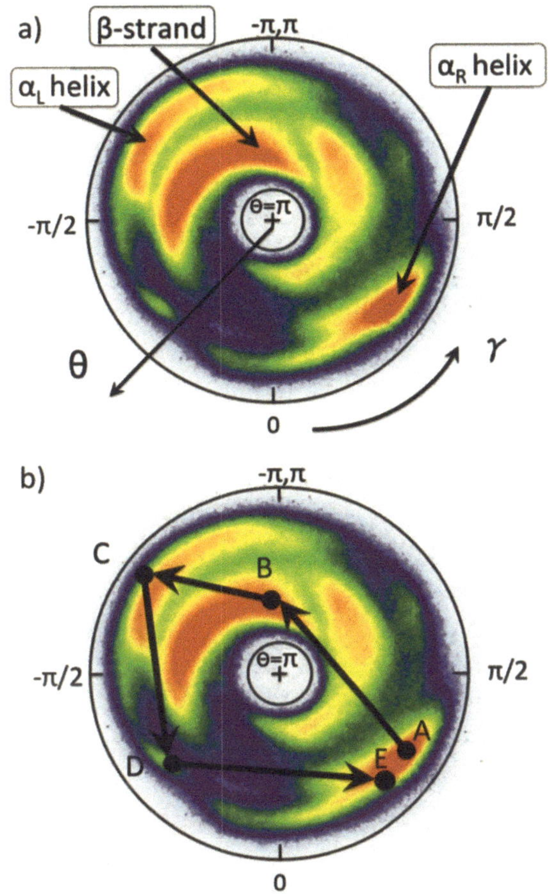

Figure 6: (a) The distribution of bond and torsion angles on the stereographically projected two-sphere (θ,γ); The color intensity is (logarithmically) proportional to the number of PDB entries (red > yellow > green > blue > white). (b) An example of a circular path (corresponding to a loop structure), as an oriented trajectory on the stereographically projected two-sphere. The circular path starts from the right-handed α-helical region (A), proceeds to the β-strand region (B), to the left-handed region (C), followed by steps (D) and (E), and terminates in region of the right-handed α-helical region (A). Reprinted with permission from Krokhotin, A.; Liwo, A.; Maisuradze, G. G.; Niemi, A. J.; Scheraga, H. A. The Journal of Chemical Physics 2014, 140, 025101. Copyright 2014, AIP Publishing LLC.

With the use of the Discrete Nonlinear Schrödinger Equation (DNLSE)[90] the derived energy function [91] for the kink is expressed by equation (eq 8):

$$E = -\sum_{i=1}^{N-1} 2\theta_{i+1}\theta_i + \sum_{i=1}^{N}[2\theta_i^2 + c(\theta_i^2 - m^2)^2] + \sum_{i=1}^{N}[b\theta_i^2\gamma_i^2 + d\gamma_i + e\gamma_i^2 + q\theta_i^2\gamma_i] \qquad (8)$$

where E is energy, θ is the angle between three consecutive C^{α} atoms, γ is torsional angle between four consecutive C^{α} atoms (see Fig. **2**), N is number of residues in kink, *b, c, d, e, m, q* are fitted parameters to reflect the regular and loop (kink) structure behavior. For the full derivation of equation 8 see the [91].

A single kink describes two regular structures (α-helix or β-sheet) with a loop (or other unorganized fragment) in-between. The graphical interpretation of the kink connected with structural changes is shown in the (Fig. **7**).

In the kink the system undergoes the gauge inversion after the kink transition:

$$\begin{aligned} &\gamma_i \to \gamma_i - \pi \\ &\theta_k \to -\theta_k \quad for \quad all \quad k \geq i \end{aligned} \tag{9}$$

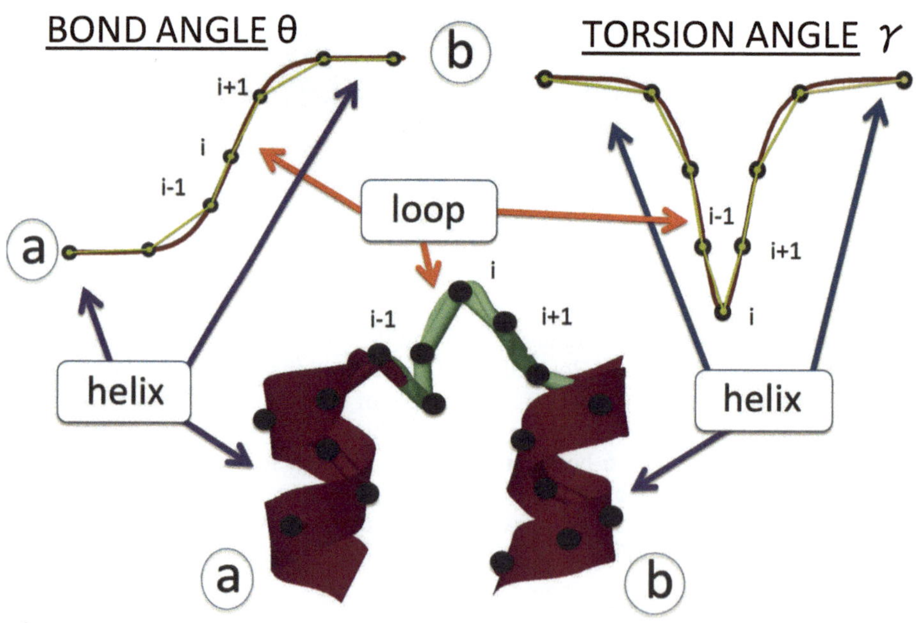

Figure 7: Top: Schematic sketches of the profiles of angles θ (left) and γ (right) along the chain. Bottom: The solutions of the generalized DNLS equation are the modular building blocks of folded proteins. They correspond to super-secondary structures such as right-handed-α-helix-loop-right-handed-α-helix (strand-loop-strand). Reprinted with permission from Krokhotin, A.; Liwo, A.; Maisuradze, G. G.; Niemi, A. J.; Scheraga, H. A. The Journal of Chemical Physics 2014, 140, 025101. Copyright 2014, AIP Publishing LLC.

It must be noted that the current model only uses C^{α} trace (Fig. **6**) and the position of the side-chain is not taken into account in Hamiltonian explicitly. Nevertheless, the statistical analysis results of the side-chain positions [92] with respect to backbone gives the possibility that side-chain position will be taken into account in this model.

APPLICATIONS OF THE KINK MODEL

Currently the parameters, *b, c, d, e, m, q* in the equation 8 are optimized to minimize the root-mean-square division (RMSD) of the structure corresponding to energy minimum [28]. Despite the similarity with Gō model [27] and the need for reparametrization for each system separately, the number of parameters used is significantly smaller then in case of Gō model. For 33 residue chicken villin headpiece (PDB code: 1YFR) the 14 parameters are sufficient for obtaining the structure with 0.51 Å RMSD from the native structure (Fig. **8**).

Figure 8: The superposition of the villin headpiece (PDB code: 1YFR) native structure (red) with the simulated structure (blue).

Apart from high precision in structure reproduction, the kink model can be applied to determining the folding pathway [93]. The simulated annealing with Monte Carlo was applied to determine the key steps in unfolding and re-folding process. With kink model the first step of folding for villin headpiece is formation of C-terminal helix followed by formation of the middle helix. After the formation of the two helices the next step in folding pathway is formation of loop structures with the final stabilization of the helices after obtaining the proper topology (Fig. **9**). Therefore the diffusion-collision [94,95] seem to fit the obtained data.

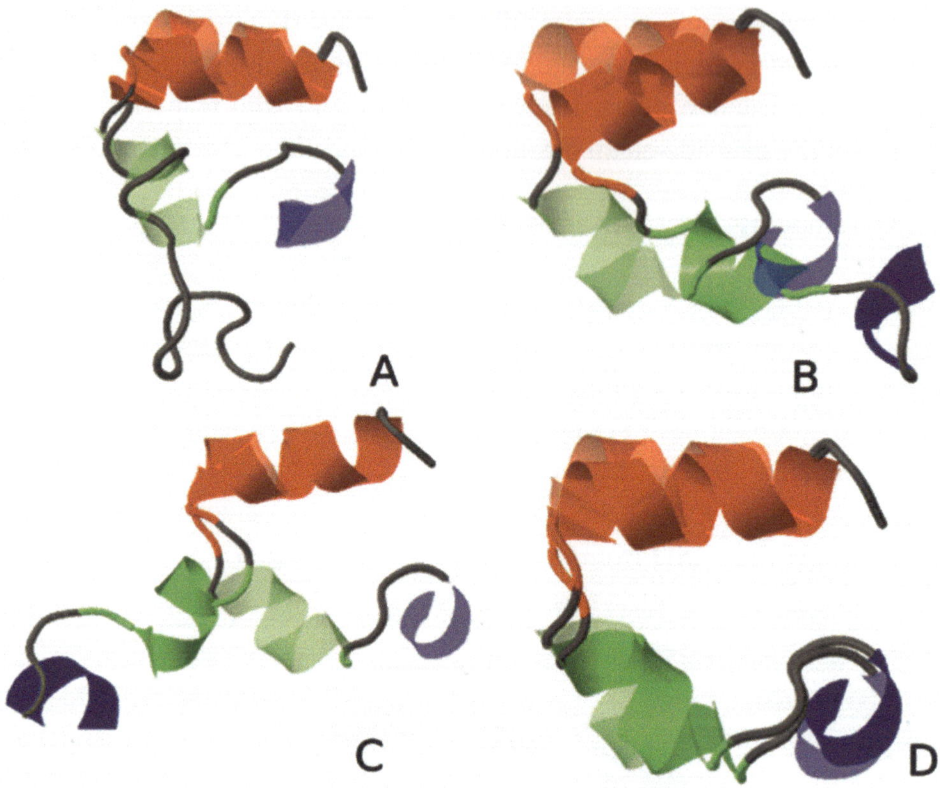

Figure 9: "A set of snap-shots of a generic folding pathway in simulation, over the shadow of the native structure. Color coding shows how the three helices are formed. (a) The folding starts with formation of helix 3, colored red. (b) After this, there is the formation of the other two helices and loops. (c) The folding proceeds with stabilization of the middle helix (green) and loops. (d)" The final fold and the PDB structure of 1YRF coincide with RMSD about 0.5 Å.

The kink description is not only restricted to small proteins. The λ-repressor 92-residue protein, can be folded with sub-atomic accuracy of 0.5Å [96]. With the use of the kink description the folding nucleus has been identified leading to the partially folded and the collapsed structure. Therefore the nucleation-condensation mechanism [97] seem to fit the obtained results. The mechanism of folding λ-repressor is different than in case of villin headpiece showing that the kink model is not biased toward one folding model. The interesting feature of the kink model is simplicity in analyzing the flexible part of the protein. In case of the λ -repressor the flexible N-terminal part has been identified [96], showing that it is the last to form in folding process.

Beyond the Single Protein

Despite the fact that, currently the parameters are fitted to the experimental structure [28,96], the kink description seem to be universal feature of the proteins. Due to the fact that the kink is very sensitive to experimental noises the universal analysis was done only on the protein with resolution of 2.0Å and better [98]. The statistical analysis of 3.027 proteins leading to the analysis of 193.640 loop fragments was performed. The clustering with use of single-link clustering method [99] with RMSD cutoff at 0.5 Å level revealed occurrence of only 200 loop clusters [98]. This proves that the kink description has a universal character and at least 92% of deposited protein structures can be described by kinks [98].

The interesting conclusion emerging from analysis of the kink description [98] was the occurrence of the kink in the DNA-binding protein which was not similar with any other [96]. That kink was describing the loop fragment binding to DNA. This suggest that the kink description may also be used for a structure-activity studies.

COMPARISON OF THE MODELS

Before the application of those two methodologies is shown the differences and similarities of those two model have to be presented. Both models are coarse-grain models which use the C^{α} trace to define the geometry of the protein. The kink model is deprived of the side-chain but statistical analysis has been done to determine the behavior of the side-chains in the regular and loop parts of the protein. The UNRES force field Hamiltonian in contrast to the kink description Hamiltonian does not require reparametrization for every protein separately. Very important difference is the accuracy of the model. In case of the UNRES force field a protein with RMSD within 4Å can be treated as the one with the native structure, whereas in case of the kink description the RMSD has to be at the experimental level. As the simulations in further part of this section will be conducted with the use of the UNRES force field and only the analysis will be done with the kink model the criterion of 4Å will be valid.

ENERGY CHANGES IN KINK FORMATION

As mentioned before, the UNRES force field is a powerful tool for the molecular dynamics simulation of peptides and proteins. The recent improvement of the representation of the loop structures [12] enabled to study in detailed and more reliable way very important parts of proteins - loops. The kink description on the other hands is very potent tool for the loop changes and the other local conformational changes analysis. As the loop fragments have been shown experimentally that play a key role in protein folding [100,101] the UNRES force field along with the kink description have been applied to study this problem.

Those two tools have been applied to study protein A [92] (PD code:1BDD). During the simulation of the protein, the kink emerged, disappear, propagate or annihilate.

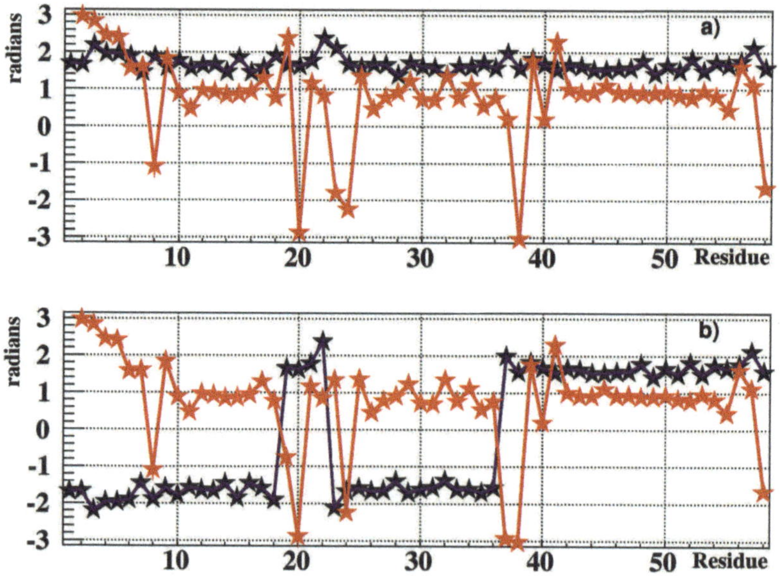

Figure 10: The virtual-bond θ_i(black) and torsion γ_i (red) angle spectra of 1BDD. Fig. **10(a)** uses the convention that the bond angle is positive. In Fig. **10(b)**, Z2 transformation (eq. 9) have been introduced to reveal the kink content at the peaks. Reprinted with permission from Krokhotin, A.; Liwo, A.; Maisuradze, G. G.; Niemi, A. J.; Scheraga, H. A. The Journal of Chemical Physics 2014, 140, 025101. Copyright 2014, AIP Publishing LLC.

The study has shown the energy barrier from the kink formation (Fig. **11**).

The kink formation can be observed with the changes of γ and θ values (Fig. **10**) This study have shown that the energy barrier (Peierls-Nabarro barrier, NPB)[102,103] of the kink formation is equal to 7 kcal/mol (Fig. **11**). This value is similar to dissociation of a phosphate group of ATP to ADP [104].

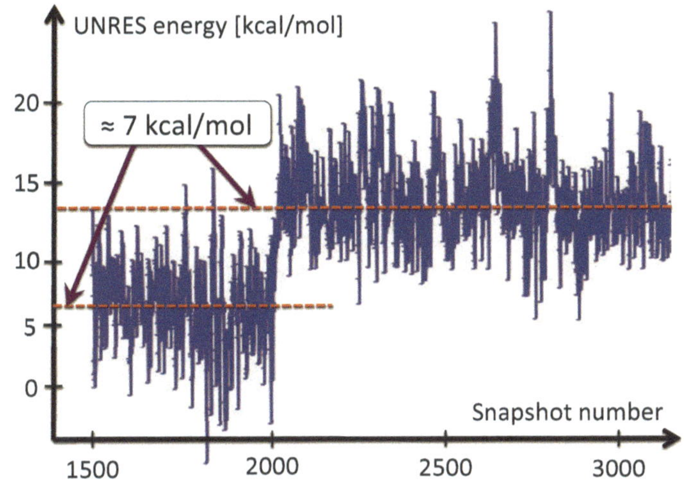

Figure 11: The energy changes of the kink in the protein A during the kink formation process. Reprinted with permission from Krokhotin, A.; Liwo, A.; Maisuradze, G. G.; Niemi, A. J.; Scheraga, H. A. The Journal of Chemical Physics 2014, 140, 025101. Copyright 2014, AIP Publishing LLC.

CHANGES IN SIDE-CHAIN ORIENTATION IN KINK FORMATION

Despite the fact that currently in the kink description there are no side-chains expicite, the statistical analysis of the side-chains positions with respect to backbone has been done [92]. During the kink formation the geometry of the side-chain position with respect to the geometry of the backbone undergo a dramatic changes. The side-chains in the loop fragments act as domain wall (Fig. **12**).

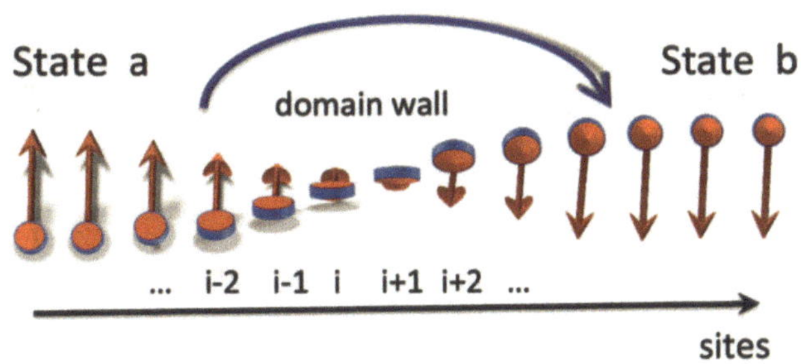

Figure 12: The kink is the boundary between the two minima a and b. It can be interpreted as a continuum limit of a magnetic Bloch-type domain wall that interpolates between spin-up state (a) and spin-down state (b). Here, we show, as an example, the Bloch wall in the transversal O(2) spin model. Reprinted with permission from Krokhotin, A.; Liwo, A.; Maisuradze, G. G.; Niemi, A. J.; Scheraga, H. A. The Journal of Chemical Physics 2014, 140, 025101. Copyright 2014, AIP Publishing LLC.

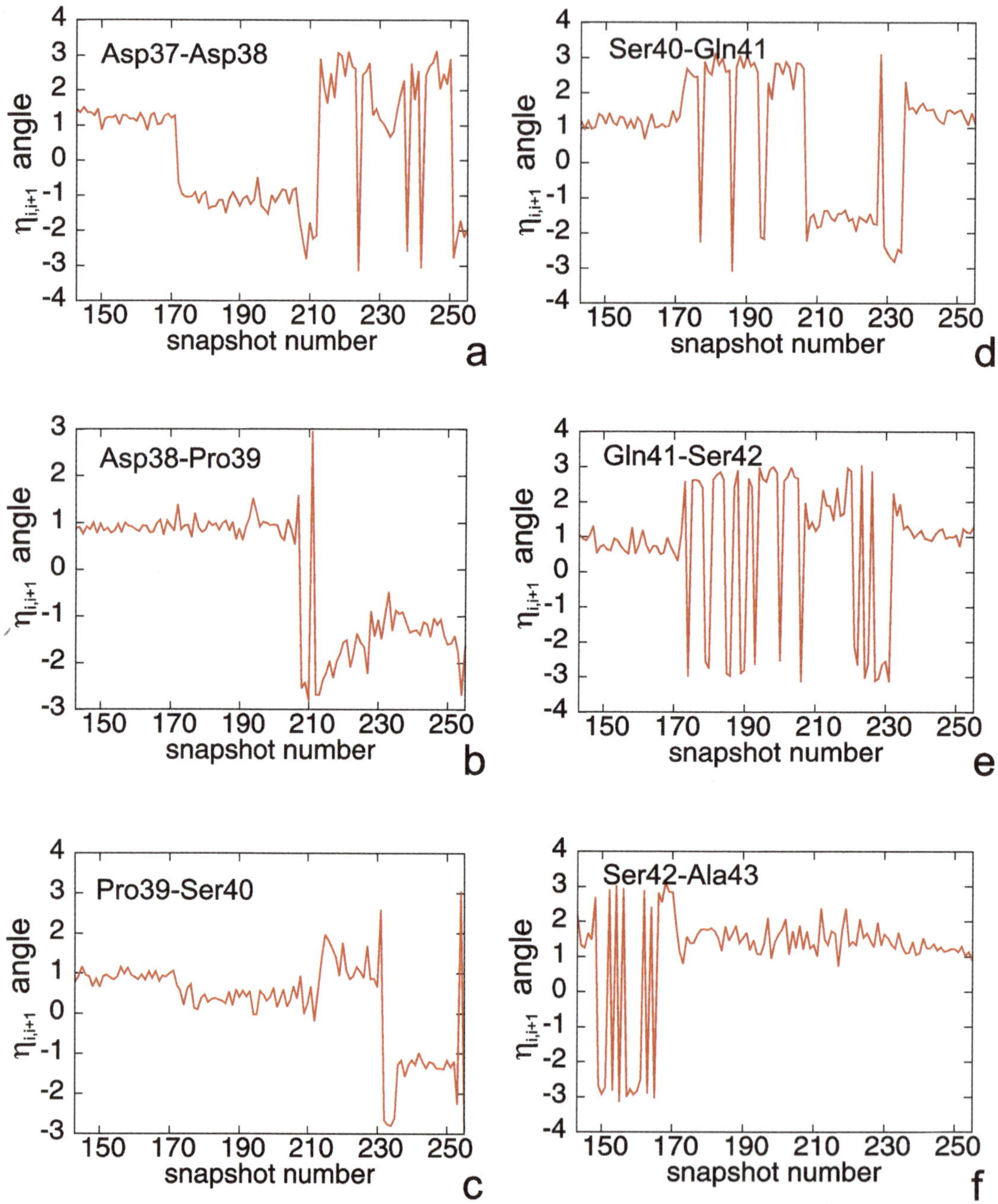

Figure 13: a-f Time evolution of the angles η_i, during a generic UNRES simulation, over the location of the kink.Reprinted with permission from Krokhotin, A.; Liwo, A.; Maisuradze, G. G.; Niemi, A. J.; Scheraga, H. A. The Journal of Chemical Physics 2014, 140, 025101. Copyright 2014, AIP Publishing LLC.

The changes of η angles (angles describing the twist of consecutive side-chains with respect to backbone) during the kink formation are shown in the Fig. **13**. Where the angle η is defined by:

$$\eta_{i,i+1} = \mathrm{sgn}[t_i \circ (t_i \times t_{i+1})] \arccos(u_i \circ v_i) \tag{10}$$

where the t_i is the unit length vector pointing from C^{α}_i to C^{α}_{i+1} the u_i vector is orthogonalized, unit vector pointing from C^{α}_i to SC_i (s_i), the v_i vector is orthogonalized unit vector pointing from C^{α}_{i+1} to SC_{i+1} (s_i). The u_i and v_i are given by the equations:

$$\begin{aligned} u_i &= \frac{s_i - (s_i \circ t_i)t_i}{\sqrt{[1-(s_i \circ t_i)^2]}} \\ v_i &= \frac{s_{i+1} - (s_{i+1} \circ t_i)t_i}{\sqrt{[1-(s_{i+1} \circ t_i)^2]}} \end{aligned} \tag{11}$$

As can be seen from the plot the η values change from value 1 (corresponding to the α-helical structure) to value η corresponding to the loop structure or the unstructured parts of the protein.

The changes in γ, θ and η angles made a kink formation analysis quantitative and able for distinction.

Kink Disappearance

When the temperature is too low the protein cannot pass through the NPB (Fig. **14**).

It goes through conformational changes and finally leading to disappearance of the kink. These mechanism may be similar to process which proteins undergo during cold denaturation. In the low temperature the energetic penalty associated with exposing the hydrophobic side-chains is lowering [105], therefore the formation of the most stable secondary structure becomes the structure determining factor. As can be seen in the Fig. **15** the kink disappeared and the loop fragment merged into one α-helix.

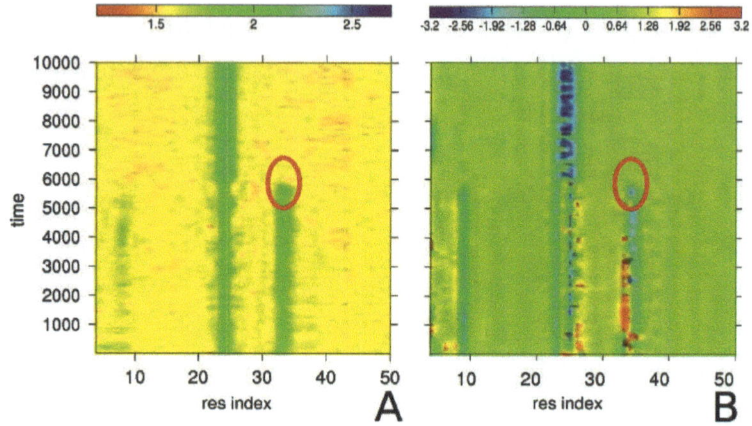

Figure 14: The changes of θ angle (A) and γ angle (B) in a function of a time and a residue number, when the protein cannot pass NPB. With the red ellipse the kink disappearance is marked.

Figure 15: The structure of the simulated protein (PDB code: 1GAB) after the disappearance of the kink. One of the loop fragments merged into helix.

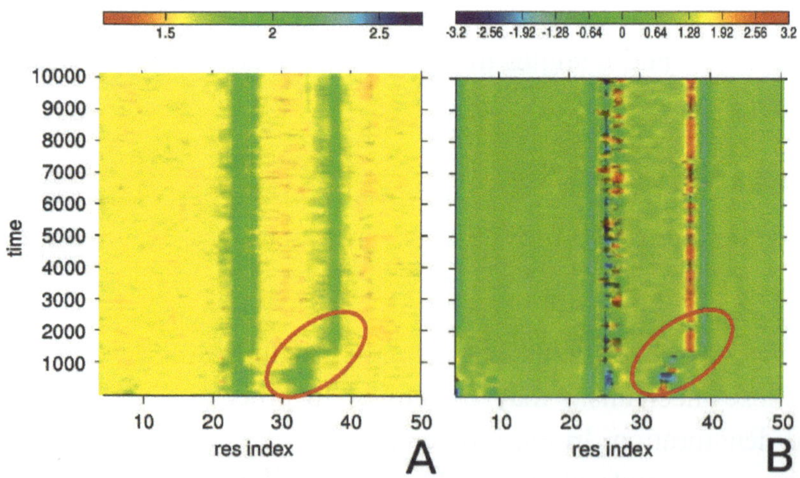

Figure 16: The changes of θ angle (A) and γ angle (B) in a function of time and residue number, when the kink movement undergoes simple transition without second soliton interaction. With the red ellipse the kink movement is marked.

Kink Movement

When the temperature is high enough at least two mechanisms of the kink movement can be identified. Both mechanisms despite having different transition state lead to the same final (native) structure.

The first one (Fig. **16**) is the simple kink movement toward the native conformation, in which energy changes of the kink movement are connected with NPB (Fig. **17**).

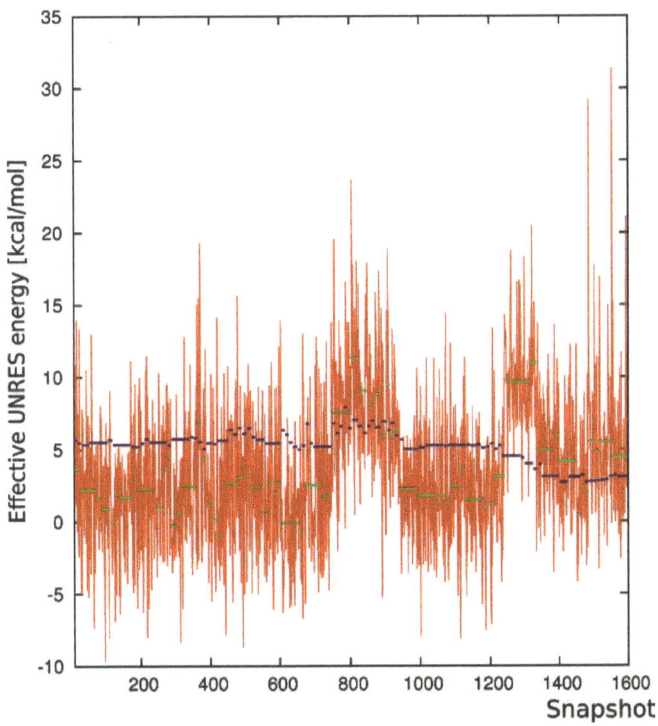

Figure 17: The energy changes (kcal/mol) throughout the simulation (snapshots) of G-related albumin-binding protein. Red color represents the energy at a given time, green color is the grouped energy by the minimizied standard deviation criterion, blue color represents the average RMSD of the clustered structures.

As can be seen from the Fig. **17** the two transition occur. Nevertheless, the first transition (around 800) does not lead to any dramatic changes in the RMSD to native structure. The second transition around 1300 is also connected with the similar as in case of the kink creation barrier of 7kcal/mol (compare with Fig. **10**). The second transition is connected with the drop of average RMSD from ~5Å to ~3Å. It means that the structure in the second transition goes from partially folded state to native structure. The interesting observation is that initial and the final energy of the kink fragment remains almost the same. On the other hand the total potential energy of the whole protein (Fig. **18**) changes with the kink movement process.

As can be seen from the Fig. **18** there is a great energy benefit connected with kink movement (~17 kcal/mol). Such a huge gap between those two states (before and after kink movements) come from the long range interactions and not the local energy of the loop (kink center). Another interesting observation comes from the difference of the transition state potential energy. In case of the kink movement NPB for local energy is ~7kcal/mol, however for the whole protein this barrier is much lower and is equal to ~3kcal/mol. This indicates that the collective motion has a lower energy barrier then the local atom movements.

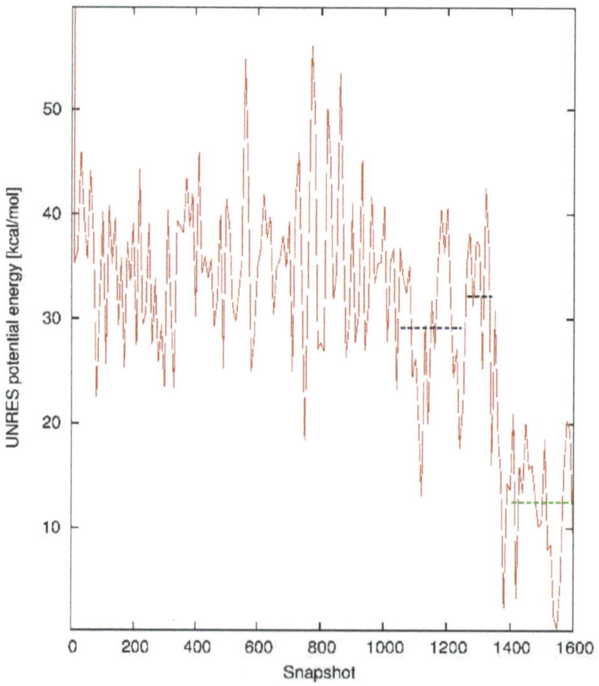

Figure 18: The potential energy changes of the whole protein in the kink movement process. The red line is the energy of a given snapshot, green dashed lines indicates the average potential energy after the kink transition, the blue dashed line indicates the average potential energy before the transition, the black dashed line indicates the average potential energy during the kink transition.

Apart from the mechanism described above the second mechanism can be distinguished (Fig. **19**). The second mechanism involve the interaction between two kinks.

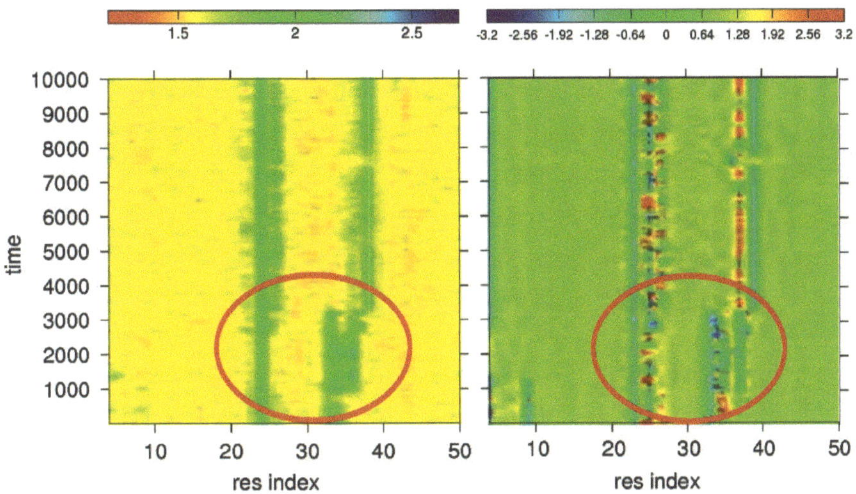

Figure 19: The changes of θ angle (A) and γ angle (B) in a function of time and residue number, when the kink movement undergoes transition with second soliton interaction. With the red ellipse the kink movement is marked.

As can be noticed from Figs. **16** and **19** both of the lead to the same final θ and γ profile. However, in the first case (Fig. **16**) the profile resembles shifting toward final positioning whereas the second one is much more complicated. In the second one a kink widening in the 30-37 residue region can be noticed with at the same time kink shortening in the 22-24 residue region. As the both mechanism occurred at the same temperature this indicate that more than one kink movement mechanism is possible. This also lead to the conclusion that the problem of multiple folding pathway [106-108] is much more common phenomenon than expected.

CONCLUDING REMARKS

In this chapter, two very distinct approaches to study the protein structure and folding have been presented. The first is the physics-based coarse-grained force field UNRES, which has high applicability for determining the loop structure and overall fold of the protein but lacks in high resolution details. The second method is application of discrete nonlinear Schrödinger equation to the kink description of the proteins. This method is able to fold protein with sub-atomic resolution which is comparable with the experimental accuracy. The downside of the latter method is quite tedious process of parameters fitting and requirement for experimental structure. In this chapter the combination of the two techniques has been shown. The UNRES force field was used for simulating the system whereas the kink description was used for the analysis for the elementary process occurring during the protein folding. The initial results from combining those two methodologies are promising, giving solution to the problem of when the loop fragments are formed, their shift and their disappearance leading a step closer to understanding the complicated process of the protein folding.

ACKNOWLEDGEMENTS

This book chapter was supported by Polish Science Foundation (FNP START 100.2014 and Mistrz7./201) and from Polish National Science Center (DEC-2012/06/A/ST4/00376). AKS research at Uppsala University was supported by Swedish Institute scholarship. AN acknowledge support from the program "Recherche d'Initiative Académique" of Region Centre, France, from the Sino-French Cai Yuan-pei Exchange Program, from a Qian Ren Grant at Beijing Institute of Technology, China, from Vetenskapsrådet, Sweden, and from Carr Tryggers Stiftelse.

CONFLICT OF INTEREST

The authors confirm that this chapter content have no conflict of interest.

REFERENCES

[1] Moult, J.; Hubbard, T.; Fidelis, K.; Pedersen, J. T. Critical assessment of methods of protein structure prediction (CASP): round III *Proteins* **1999**, 3, 2-6.

[2] Moult, J.; Fidelis, K.; Kryshtafovych, A.; Rost, B.; Tramontano, A. Critical assessment of methods of protein structure prediction—Round VIII *Proteins: Structure, Function, and Bioinformatics* **2009**, 77, 1-4.

[3] Zhang, Y. I-TASSER: Fully automated protein structure prediction in CASP8 *Proteins: Structure, Function, and Bioinformatics* **2009**, 77, 100-113.

[4] Moult, J. A decade of CASP: progress, bottlenecks and prognosis in protein structure prediction *Current Opinion in Structural Biology* **2005**, 15, 285 - 289.

[5] Moult, J.; Fidelis, K.; Kryshtafovych, A.; Rost, B.; Hubbard, T.; Tramontano, A. Critical assessment of methods of protein structure prediction—Round VII *Proteins: Structure, Function, and Bioinformatics* **2007**, 69, 3-9.

[6] Marti-Renom, M.; Stuart, A.; Fiser, A.; Sanchez, R.; Melo, F.; Sali, A. Comparative protein structure modeling of genes and genomes *Annual Review of Biophysics and Biomolecular Structure* **2000**, 29, 291-325.

[7] Faraggi, E.; Kloczkowski, A. A global machine learning based scoring function for protein structure prediction *Proteins: Structure, Function, and Bioinformatics* **2013**, in press

[8] Hamelberg, D.; McCammon, J. A Fast Peptidyl cis−trans Isomerization within the Flexible Gly-Rich Flaps of HIV-1 Protease. *J. Am. Chem. Soc.* **2005**, 127, 13778-13779.

[9] Verlet, L. Computer" experiments" on classical fluids. I. Thermodynamical properties of Lennard-Jones molecules *Phys. Rev.* **1967**, 159, 98.

[10] Fermi E., Pasta J.R., Ulam S. M. Parallel libraries In Studies of Nonlinear Problems, 1st ed.; Los Alamos Scientific Laboratory of the University of California, 1955; Chapter 1, pp 2-20.

[11] Skolnick, J.; Kolinski, A. Simulations of the folding of a globular protein *Science* **1990**, 250, 1121-1125.

[12] Krupa, P.; Sieradzan, A. K.; Rackovsky, S.; Baranowski, M.; Ołdziej, S.; Scheraga, H. A.; Liwo, A.; Czaplewski, C. Improvement of the Treatment of Loop Structures in the UNRES Force Field by Inclusion of Coupling between Backbone- and Side-Chain-Local Conformational States *J. Chem. Theory Comput.* **2013**, 9, 4620-4632.

[13] Lindorff-Larsen, K.; Trbovic, N.; Maragakis, P.; Piana, S.; Shaw, D. E. Structure and Dynamics of an Unfolded Protein Examined by Molecular Dynamics *Simulation J. Am. Chem. Soc.* **2012**, 134, 3787-3791.

[14] Pande, V. S.; Baker, I.; Chapman, J.; Elmer, S.; Kaliq, S.; Larson, S. M.; Rhee, Y. M.; Shirts, M. R.; Snow, C. D.; Sorin, E. J.; Zagrovic, B. Atomistic protein folding simulations on the submillisecond timescale using worldwide distributed computing *Biopolymers* **2003**, 68, 91-109.

[15] Hess, B.; Kutzner, C.; van der Spoel, D.; Lindahl, E. GROMACS 4: Algorithms for highly efficient, load-balanced, and scalable molecular simulation *J. Chem. Theory Comput.* **2008**, 4, 435-447.

[16] Phillips, J. C.; Braun, R.; Wang, W.; Gumbart, J.; Tajkhorshid, E.; Villa, E.; Chipot, C.; Skeel, R. D.; Kalé, L.; Schulten, K. Scalable molecular dynamics with NAMD *J. Comput. Chem.* **2005**, 26, 1781-1802.

[17] Bowers, K. J.; Chow, E.; Xu, H.; Dror, R. O.; Eastwood, M. P.; Gregersen, B. A.; Klepeis, J. L.; Kolossvary, I.; Moraes, M. A.; Sacerdoti, F. D.; Salmon, J. K.; Shan, Y.; E.Shaw, D. Scalable algorithms for molecular dynamics simulations on commodity clusters ACM/IEEE SC **2006** Conference (SC.06) 2006, 43-43.

[18] Friedrichs, M. S.; Eastman, P.; Vaidyanathan, V.; Houston, M.; Legrand, S.; Beberg, A. L.; Ensign, D. L.; Bruns, C. M.; Pande, V. S. Accelerating molecular dynamic simulation on graphics processing units *J. Comput. Chem.* **2009**, 30, 864-872.

[19] Shaw, D. E., Deneroff M. M., Dror R.O., Kuskin J.S., Larson R. H., Salmon J.K., Young C., Batson B., Bowers K.J., Chao J. C., Eastwood M. P., Gagliargo J., Grossman J. P., Ho C. R., Ierardi Dj. J., Kolossvary I., Klepeis J.L., Layman T., Mcleavy C., Moraes M. A., Mueller R., Priest E. C., Shan Y., Spengler J., Theobald M., Towels B., Wang S. C., Anton,

a special-purpose machine for molecular dynamics simulation *Commun. ACM* **2008**, 51, 91-97.

[20] Khalili, M.; Liwo, A.; Jagielska, A.; Scheraga, H. Molecular dynamics with the united-residue model of polypeptide chains. II. Langevin and Berendsen-bath dynamics and tests on model a-helical systems *J. Phys. Chem. B* **2005**, 109, 13798-13810.

[21] Liwo, A.; Khalili, M.; Scheraga, H. A. Ab initio simulations of protein-folding pathways by molecular dynamics with the united-residue model of polypeptide chains *Proc. Natl. Acad. Sci. U.S.A.* **2005**, 102, 2362-2367.

[22] Sieradzan, A. K.; Scheraga, H. A.; Liwo, A. Determination of Effective Potentials for the Stretching of Cα···Cα Virtual Bonds in Polypeptide Chains for Coarse-Grained Simulations of Proteins from ab Initio Energy Surfaces of N-Methylacetamide and N-Acetylpyrrolidine *J. Chem. Theory Comput.* **2012**, 8, 1334-1343.

[23] Kolinski, A.; Skolnick, J. Reduced models of proteins and their applications *Polymer* **2004**, 45, 511-524.

[24] Liwo, A.; Lee, J.; Ripoll, D. R.; Pillardy, J.; Scheraga, H. A. Protein structure prediction by global optimization of a potential energy function *Proc. Natl. Acad. Sci., U. S. A.* **1999**, 96, 5482-5485.

[25] Brooks, B. R.; Brooks, C. L.; Mackerell, A. D.; Nilsson, L.; Petrella, R. J.; Roux, B.; Won, Y.; Archontis, G.; Bartels, C.; Boresch, S. CHARMM: the biomolecular simulation program *J. Comput. Chem.* **2009**, 30, 1545-1614.

[26] Wang, J.; Wolf, R. M.; Caldwell, J. W.; Kollman, P. A.; Case, D. A. Development and testing of a general amber force field *J. Comput. Chem.* **2004**, 25, 1157-1174.

[27] Koga, N.; Takada, S. Roles of native topology and chain-length scaling in protein folding: a simulation study with a Go-like model *J. Mol. Biol.* **2001**, 313, 171-180.

[28] Molkenthin, N.; Hu, S.; Niemi, A. Discrete Nonlinear Schrodinger Equation and Polygonal Solitons with Applications to Collapsed Proteins *J. Phys. Rev. Lett.* **2011**, 106, 078102.

[29] Zhang, Z., & Voth, G. A. Coarse-grained representations of large biomolecular complexes from low-resolution structural data *J. Chem. Theory*, **2010,** 6(9), 2990-3002.

[30] Dama, J. F., Sinitskiy, A. V., McCullagh, M., Weare, J., Roux, B., Dinner, A. R., & Voth, G. A. The theory of ultra-coarse-graining. 1. General principles. *J. Chem. Theory*, **2013,** 9(5), 2466-248

[31] Grime, John MA, and Gregory A. Voth. (2014) "Highly Scalable and Memory Efficient Ultra-coarse-grained Molecular Dynamics Simulations." *J. Chem. Theory,* **2014,** 10 (1), pp 423–431

[32] Marrink, S. J., Risselada, H. J., Yefimov, S., Tieleman, D. P., & de Vries, A. H. The MARTINI force field: coarse grained model for biomolecular simulations. *J. Phys.Chem. B,* **2007** 111(27), 7812-7824.

[33] Monticelli, L., Kandasamy, S. K., Periole, X., Larson, R. G., Tieleman, D. P., & Marrink, S. J. The MARTINI coarse-grained force field: extension to proteins. *J. Chem. Theory*, **2008,** 4(5), 819-834.

[34] de Jong, D. H., Singh, G., Bennett, W. D., Arnarez, C., Wassenaar, T. A., Schäfer, L. V., Periole X., Tieleman P.D., Marrink, S. J. Improved parameters for the martini coarse-grained protein force field. *J. Chem. Theory*, **2013** 9(1), 687-697.

[35] Rohl, C. A., Strauss, C. E., Misura, K. M., & Baker, D. Protein structure prediction using Rosetta. *Methods in enzymology,* **2004** 383, 66-93.

[36] Leaver-Fay, A., O'Meara, M. J., Tyka, M., Jacak, R., Song, Y., Kellogg, E. H, Thompson J., Davis I.W., Pache R. A., Lyskov S., Gray J. J., Kortemme T., Richardson J. S., Havranek J. J., Snoeyink J., Baker D.,Kuhlman, B. Scientific benchmarks for guiding macromolecular energy function improvement. *Methods in enzymology,* **2013,** 523, 109.

[37] Maupetit, J., Tuffery, P., Derreumaux, P. A coarse-grained protein force field for folding and structure prediction. *Proteins: Structure, Function, and Bioinformatics,* **2013**, 69(2), 394-408.

[38] Chebaro, Y., Pasquali, S., Derreumaux, P. The coarse-grained OPEP force field for non-amyloid and amyloid proteins. J. Phys. Chem. B, 2012, 116(30), 8741-8752.

[39] Sterpone, F.; Nguyen, P.H; Kalimeri, M.; Derreumaux, P. Importance of the ion-pair Interactions in the OPEP coarse-grained force field: parametrization and validation. J. Chem. Theory Comput. **2013**, 9, 4574-4584.

[40] Sterpone F, Melchionna S, Tuffery P, Pasquali S, Mousseau N, Cragnolini T, Chebaro Y, St-Pierre JF, Kalimeri M, Barducci A, Laurin Y, Tek A, Baaden M, Nguyen PH, Derreumaux P. The OPEP protein model: from single molecules, amyloid formation, crowding and hydrodynamics to DNA/RNA systems. Chem. Soc. Rev., 2014, DOI: 10.1039/C4CS00048J.

[41] Kozłowska, U.; Liwo, A.; Scheraga, H. A. Determination of side-chain-rotamer and side-chain and backbone virtual-bond-stretching potentials of mean force from AM1 energy surfaces of terminally-blocked amino-acid residues, for coarse-grained simulations of protein structure and folding. 1. The method. *J. Comput. Chem.* **2010**, 31, 1143-1153.

[42] Kozłowska, U.; Maisuradze, G. G.; Liwo, A.; Scheraga, H. A. Determination of side-chain-rotamer and side-chain and backbone virtual-bond-stretching potentials of mean force from AM1 energy surfaces of terminally-blocked amino-acid residues, for coarse-grained simulations of protein structure and folding. 2. Results, comparison with statistical potentials, and implementation in the {UNRES} force field. *J. Comput. Chem.* **2010**, 31, 1154-1167.

[43] Liwo, A.; Kazmierkiewicz, R.; Czaplewski, C.; Groth, M.; Ołdziej, S.; Wawak, R. J.; Rackovsky, S.; Pincus, M. R.; Scheraga, H. A. United-residue force field for off-lattice protein-structure simulations; III. Origin of backbone hydrogen-bonding cooperativity in united-residue potentials *J. Comput. Chem.* **1998**, 19, 259-276.

[44] Liwo, A.; Czaplewski, C.; Pillardy, J.; Scheraga, H. A. Cumulant-based expressions for the multibody terms for the correlation between local and electrostatic interactions in the united-residue force field *J. Chem. Phys.* **2001**, 115, 2323-2347.

[45] Liwo, A.; Czaplewski, C.; Ołdziej, S.; Rojas, A. V.; Kazmierkiewicz, R.; Makowski, M.; Murarka, R. K.; Scheraga, H. A. Simulation of protein structure and dynamics with the coarse-grained UNRES force field In Coarse-Graining of Condensed Phase and Biomolecular Systems; Voth, G., Ed.; CRC Press, 2008, Chapter 8, pp 1391-1411.

[46] Kubo, R. Generalized cumulant expansion method *J. Phys. Soc. Japan* **1962**, 17, 1100-1120.

[47] Liwo, A.; Khalili, M.; Czaplewski, C.; Kalinowski, S.; Ołdziej, S.; Wachucik, K.; Scheraga, H Modification and optimization of the united-residue (UNRES) potential energy function for canonical simulations. I. Temperature dependence of the effective energy function and tests of the optimization method with single training proteins. *J. Phys. Chem. B* **2007**, 111, 260-285.

[48] Shen, H.; Liwo, A.; Scheraga, H. A. An improved functional form for the temperature scaling factors of the components of the mesoscopic UNRES force field for simulations of protein structure and dynamics *J. Phys. Chem. B* **2009**, 113, 8738-8744.

[49] Kolinski, A.; Skolnick, J. Discretized model of proteins. I. Monte Carlo study of cooperativity in homopolypeptides *J. Chem. Phys.* **1992**, 97, 9412-9426.

[50] Johansson, M. U.; de Chăteau, M.; Wikström, M.; Forsön, S.; Drakenberg, T.; Björck, L. Solution structure of the albumin-binding GA module: a versatile bacterial protein domain *J. Mol. Biol.***1997**, 266, 859 - 865.

[51] He, Y.; Xiao, Y.; Liwo, A.; Scheraga, H. A. Exploring the parameter space of the coarse-grained UNRES force field by random search: Selecting a transferable medium-resolution force field *J. Comput. Chem.* **2009**, 30, 2127-2135.

[52] Neidigh, J. W.; Fesinmeyer, R. M.; Andersen, N. H. Designing a 20-residue protein *Nat. Struct. Biol.* **2002**, 9, 425-430.

[53] Cochran, A. G.; Skelton, N. J.; Starovasnik, M. A. Tryptophan zippers: Stable, monomeric β-hairpins *Proc. Natl. Acad. Sci. USA* **2001**, 98, 5578-5583.

[54] Makowski, M.; Sobolewski, E.; Czaplewski, C.; Liwo, A.; Ołdziej, S.; No, J. H.; Scheraga, H. A. Simple Physics-Based Analytical Formulas for the Potentials of Mean Force for the Interaction of Amino Acid Side Chains in Water. 3. Calculation and Parameterization of the Potentials of Mean Force of Pairs of Identical Hydrophobic Side Chains *J. Phys. Chem. B* **2007**, 111, 2925-2931.

[55] Makowski, M.; Sobolewski, E.; Czaplewski, C.; Ołdziej, S.; Liwo, A.; Scheraga, H. A. Simple Physics-Based Analytical Formulas for the Potentials of Mean Force for the Interaction of Amino Acid Side Chains in Water. IV. Pairs of Different Hydrophobic Side Chains *J. Phys. Chem. B* **2008**, 112, 11385-11395.

[56] Makowski, M.; Liwo, A.; Scheraga, H. A. Simple Physics-Based Analytical Formulas for the Potentials of Mean Force of the Interaction of Amino-Acid Side Chains in Water. VI. Oppositely Charged Side Chains *J. Phys. Chem. B* **2011**, 115, 6130-6137.

[57] Liwo, A.; Ołdziej, S.; Pincus, M. R.; Wawak, R. J.; Rackovsky, S.; Scheraga, H. A. {A united-residue force field for off-lattice protein-structure simulations. I. Functional forms and parameters of long-range side-chain interaction potentials from protein crystal data *J. Comput. Chem.* **1997**, 18, 849-873.

[58] Sieradzan, A. K.; Hansmann, U. H. E.; Scheraga, H. A.; Liwo, A. Extension of UNRES Force Field to Treat Polypeptide Chains with d-Amino Acid Residues *J. Chem. Theory Comput.* **2012**, 8, 4746-4757.

[59] Kozłowska, U.; Liwo, A.; Scheraga, H. A. {Determination of virtual-bond-angle potentials of mean force for coarse-grained simulations of protein structure and folding from ab initio energy surfaces of terminally-blocked glycine, alanine, and proline *J. Phys.: Cond. Matter* **2007**, 19, 285203.

[60] Czaplewski, C.; Liwo, A.; Pillardy, J.; Ołdziej, S.; Scheraga, H. Improved Conformational Space Annealing method to treat β-structure with the UNRES force-field and to enhance scalability of parallel implementation *Polymer* **2004**, 45, 677-686.

[61] Khalili, M.; Liwo, A.; Rakowski, F.; Grochowski, P.; Scheraga, H. Molecular dynamics with the united-residue model of polypeptide chains. I. Lagrange equations of motion and tests of numerical stability in the microcanonical mode *J. Phys. Chem. B* **2005**, 109, 13785-13797.

[62] Czaplewski, C.; Kalinowski, S.; Liwo, A.; Scheraga, H. A. Application of multiplexing replica exchange molecular dynamics method to the UNRES force field: tests with α and α +β proteins *J. Chem. Theor. Comput.* **2009**, 5, 627-640.

[63] Lee, J.; Scheraga, H. A.; Rackovsky, S. New optimization method for conformational energy calculations on polypeptides: Conformational space annealing *J. Comput. Chem.* **1997**, 18, 1222-1232.

[64] Lee, J.; Scheraga, H. A.; Rackovsky, S. Conformational analysis of the 20-residue membrane-bound portion of melittin by Conformational space annealing *Biopolymers* **1998**, 46, 103-115.

[65] Lee, J.; Scheraga, H. A. Conformational space annealing by parallel computations: extensive conformational search of Met-enkephalin and of the 20-residue membrane-bound portion of melittin *Int. J. Quant. Chem.* **1999**, 75, 255-265.

[66] Hansmann, U. H. E.; Okamoto, Y. Monte Carlo simulations in generalized ensemble: Multicanonical algorithm *versus* simulated tempering *Phys. Rev. E* **1996**, 54, 5863-5865.

[67] Sugita, Y.; Okamoto, Y. Replica-exchange multicanonical algorithm and multicanonical replica-exchange method for simulating systems with rough energy landscape *Phys. Rev. Lett.* **2000**, 329, 261-270.

[68] Metropolis, N.; Rosenbluth, A. W.; Rosenbluth, M. N.; Teller, A. H.; Teller, E. Equation of state calculations by fast computing machines *J. Chem. Phys.* **1953**, 21, 1087-1092.

[69] Khalili, M.; Liwo, A.; Scheraga, H. Kinetic studies of folding of the B-domain of staphylococcal protein A with molecular dynamics and a united-residue (UNRES) model of polypeptide chains *J. Mol. Biol.* **2006**, 355, 536-547.

[70] He, Y.; Chen, C.; Xiao, Y. United-residue (UNRES) langevin dynamics simulations of trpzip2 folding *J. Comput. Biol.* **2009**, 16, 1719-1730.

[71] Maisuradze, G. G.; Senet, P.; Czaplewski, C.; Liwo, A.; Scheraga, H. A. Investigation of protein folding by coarse-grained molecular dynamics with the UNRES force field *J. Phys. Chem. A* **2010**, 114, 4471-4485.

[72] Sieradzan, A. K.; Liwo, A.; Hansmann, U. H. E. Folding and Self-Assembly of a Small Protein Complex *J. Chem. Theory Comput.* **2012**, 8, 3416-3422.

[73] He, Y.; Mozolewska, M. A.; Krupa, P.; Sieradzan, A. K.; Wirecki, T. K.; Liwo, A.; Kachlishvili, K.; Rackovsky, S.; Jagieła, D.; Ślusarz, R.; Czaplewski, C. R.; Ołdziej, S.; Scheraga, H. A. Lessons from application of the UNRES force field to predictions of structures of CASP10 targets *Proc. Natl. Acad. Sci. USA* **2013**, 110, 14936-14941.

[74] Rojas, A.; Liwo, A.; Browne, D.; Scheraga, H. A. Mechanism of fiber assembly; treatment of Aβ peptide peptide aggregation with a coarse-grained united-residue force field *J. Mol. Biol.* **2010**, 404, 537-552.

[75] Rojas, A.; Liwo, A.; Scheraga, H. A study of the α -helical intermediate preceding the aggregation of the amino-terminal fragment of the Aβ-amyloid peptide (1-28) *J. Phys. Chem. B* **2011**, 115, 12978-12983.

[76] Golas, E. I.; Maisuradze, G. G.; Senet, P.; Ołdziej, S.; Czaplewski, C.; Scheraga, H. A.; Liwo, A. Simulation of the opening and closing of Hsp70 chaperones by coarse-grained molecular dynamics *J. Chem. Theor. Comput.* **2012**, 8, 1334-1343.

[77] Y. He, A. L.; Weinstein, H.; Scheraga, H. PDZ Binding to the BAR Domain of PICK1 is Elucidated by Coarse-grained Molecular Dynamics *J. Mol. Biol.* **2011**, 405, 298-314.

[78] Berman, H. M. The protein data bank: a historical perspective *Acta Crystallographica Section A: Foundations of Crystallography* **2007**, 64, 88-95.

[79] Murzin, A. G.; Brenner, S. E.; Hubbard, T.; Chothia, C. SCOP: a structural classification of proteins database for the investigation of sequences and structures *J. Mol. Biol* **1995**, 247, 536-540.

[80] Greene, L. H.; Lewis, T. E.; Addou, S.; Cuff, A.; Dallman, T.; Dibley, M.; Redfern, O.; Pearl, F.; Nambudiry, R.; Reid, A. he CATH domain structure database: new protocols and classification levels give a more comprehensive resource for exploring evolution. *Nucleic acids research* **2007**, 35, D291-D297.

[81] Kratky, O.; Porod, G. Röntgenuntersuchung gelöster Fadenmoleküle *Rec Trav Chim 1949* **1949,** 68, 1106.

[82] Langer, J.; Singer, D. A. Lagrangian aspects of the Kirchhoff elastic rod *SIAM review* **1996**, 38, 605-618.

[83] Ricca, R. L. Applications of knot theory in fluid mechanics *Banach Center Publications* **1998**, 42, 321-346.

[84] Saffman, P. G. Vortex dynamics; Cambridge university press, 1992.

[85] Abrikosov, A. The magnetic properties of superconducting alloys *Journal of Physics and Chemistry of Solids* **1957**, 2, 199-208.

[86] Volovik, G. E.; Volovik, G. The universe in a helium droplet; Oxford University Press, 2009.

[87] Kibble, T. W. Topology of cosmic domains and strings *Journal of Physics A: Mathematical and General* **1976**, 9, 1387.

[88] Polyakov, A. M. Quantum geometry of bosonic strings *Phys. Lett. B* **1981**, 103, 207-210.

[89] Fernet, F. Sur les courbes à double courbure *J. de Math.* **1852**, 17, 437.

[90] Kevrekidis, P. G. The Discrete Nonlinear Schrèodinger Equation: Mathematical Analysis, Numerical Computations and Physical Perspectives; Springer, 2009; Vol. 232.

[91] Hu, S.; Jiang, Y.; Niemi, A. Energy functions for stringlike continuous curves, discrete chains, and space-filling one dimensional structures *J. Phys. Rev. D* **2013**, 87, 105011.

[92] Krokhotin, A.; Liwo, A.; Maisuradze, G. G.; Niemi, A. J.; Scheraga, H. A Kinks, loops, and protein folding, with protein A as an example *J. Chem Phys.* **2014**, 140, 025101.

[93] Krokhotin, A.; Lundgren, M.; Niemi, A. J.; Peng, X. Soliton driven relaxation dynamics and protein collapse in the villin headpiece *Journal of Physics: Condensed Matter* **2013**, 25, 325103.

[94] Bashford, D.; Cohen, F.; Karplus, M.; Kuntz, I.; Weaver, D. Diffusion-collision model for the folding kinetics of myoglobin *Proteins: Structure, Function, and Bioinformatics* **1988**, 4, 211-227.

[95] Karplus, M.; Weaver, D. L. Protein folding dynamics: The diffusion-collision model and experimental data *Protein Science* **1994**, 3, 650-668.

[96] Krokhotin, A.; Lundgren, M.; Niemi, A. Solitons and collapse in the λ- repressor protein J. *Phys. Rev. E* **2012**, 86, 021923.

[97] López-Hernéndez, E.; Serrano, L. Structure of the transition state for folding of the 129 aa protein CheY resembles that of a smaller protein, CI-2 *Folding and Design* **1996**, 1, 43-55.

[98] Krokhotin, A.; Niemi, A. J.; Peng, X. Soliton concepts and protein structure *Phys. Rev. E* **2012**, 85, 031906.

[99] Sibson, R. SLINK: An optimally efficient algorithm for the single-link cluster method *The Computer Journal* **1973**, 16, 30-34.

[100] Nölting, B.; Golbik, R.; Fersht, A. R. Submillisecond events in protein folding *Proc. Natl. Acad. Sci. USA* **1995**, 92, 10668-10672.

[101] Nölting, B.; Golbik, R.; Neira, J. L.; Soler-Gonzalez, A. S.; Schreiber, G.; Fersht, A. R. The folding pathway of a protein at high resolution from microseconds to seconds *Proc. Natl. Acad. Sci. USA* **1997**, 94, 826-830.

[102] Peierls, R. The size of a dislocation *Proc. Phys. Soc.* **1940**, 52, 34-37.

[103] Nabarro, F. Dislocations in a simple cubic lattice *Proc. Phys. Soc.* **1947**, 59, 256.

[104] Thomas, N.; Thornhill, R. The physics of biological molecular motors *Journal of Physics D: Applied Physics* **1998**, 31, 253.

[105] Lopez, C. F.; Darst, R. K.; Rossky, P. J. Mechanistic elements of protein cold denaturation *J. Phys. Chem.B* **2008**, 112, 5961-5967.

[106] Radford, S. E.; Dobson, C. M.; Evans, P. A. The folding of hen lysozyme involves partially structured intermediates and multiple pathways *Nature* **1992**, 358, 302-307.

[107] Feng, H.; Zhou, Z.; Bai, Y. A protein folding pathway with multiple folding intermediates at atomic resolution *Proc. Natl. Acad. Sci. USA* **2005**, 102, 5026-5031.

[108] Noe, F.; Schutte, C.; Vanden-Eijnden, E.; Reich, L.; Weikl, T. R. Constructing the equilibrium ensemble of folding pathways from short off-equilibrium simulations *Proc. Natl. Acad. Sci. USA* **2009**, 106, 19011-19016.

CHAPTER 8

Computational Chemistry Strategies Tackling Function and Inhibition of Pharmaceutically Relevant Targets

Michele Cascella[1,2,*], Matteo Dal Peraro[3,4,*] and Marco De Vivo[5,*]

[1]Department of Chemistry, and [2]Centre for Theoretical and Computational Chemistry (CTCC), University of Oslo, Oslo, Norway; [3]Institute of Bioengineering, School of Life Sciences, École Polytechnique Fédérale de Lausanne (EPFL), Lausanne, Switzerland; [4]Swiss Institute of Bioinformatics (SIB), Lausanne, Switzerland and [5]Drug Discovery and Development, Istituto Italiano di Tecnologia, Genoa, Italy

Abstract: Computational methods relying on first principles are fundamental for dissecting basic physicochemical properties of biological systems and unveiling mechanistic details that are often silent to experiments. The tireless improvement of theoretical schemes for molecular modelling and simulations, coupled to the increasing computational power of novel architectures and integrated with available experimental inputs, allows today exploring the functioning of biological systems with unprecedented accuracy. Indeed, molecular simulations at both the quantum mechanics and molecular mechanics levels are nowadays able to dissect with high confidence the structural and dynamical features of large systems in native-like conditions, up to the point that their mode of action can be modulated in a controlled fashion. These computational chemistry strategies are particularly appealing when applied to pharmaceutically relevant targets. In this chapter, we will present recent successes of computational investigations applied to a broad variety of biochemical systems that are promising or validated targets for drug discovery. In particular, we will show how molecular modelling at the quantum mechanics level is key for revealing the mechanistic details of catalysis in bacterial and viral metallo-enzymes. We will continue by discussing how accurate molecular mechanics-based free energy calculations can provide a new quantitative description of the function of systems of relevance for multidrug resistance in bacteria. In the final part of the chapter, we will show examples where computational and medicinal chemistry is fully integrated with structural and biochemical data to study function and inhibition of target enzymes implicated in cancer and other inflammatory-related diseases. The final goal of these studies is to develop new molecular entities potentially endowed with a desired pharmacological activity. This chapter will therefore define the contribution of emerging approaches and recent advances in the field of computational chemistry for translating the atomic-level understanding of complex biological phenomena into useful information to progress in molecular medicine.

***Corresponding authors Michele Cascella, Matteo Dal Peraro and Marco De Vivo:** Department of Chemistry and Centre for Theoretical and Computational Chemistry (CTCC), University of Oslo, Oslo, Norway; Institute of Bioengineering, School of Life Sciences, École Polytechnique Fédérale de Lausanne (EPFL), Lausanne, Switzerland; Swiss Institute of Bioinformatics (SIB), Lausanne, Switzerland and Drug Discovery and Development, Istituto Italiano di Tecnologia, Genoa, Italy; E-mails: michele.cascella@kjemi.uio.no; matteo.dalperaro@epfl.ch and marco.devivo@iit.it

Zaheer Ul-Haq and Jeffry D. Madura (Eds). **10.1016/B978-1-60805-865-5.50008-3**

Keywords: Hybrid QM/MM, molecular dynamics, molecular modelling, multi-scale modelling, protein ligand interactions, structure-based drug design.

1. INTRODUCTION

Computational methods are nowadays essential in all aspects related to the design and optimization of a new drug [1]. This is because computational modelling offers the unique ability to characterize, at the atomic level, the specific function of the biochemical target, as well as key drug/target interactions. In fact, the fundamental paradigm of drug efficacy is that the drug generates its beneficial effect through its tight binding to the target(s). In this way, the drug acts by modulating, through inhibition or stimulation, the target function, ultimately causing the desired pharmacological effect. Therefore, the computational investigation of function and inhibition mechanisms of the biochemical target, as well as a meticulous characterization of the main interactions between the target and its ligands such as endogenous substrates or new small molecules, can be of great help in guiding the rational design and optimization of new drugs with improved efficacy and/or diminished side effects [2, 3].

Common computational approaches for drug design employ molecular mechanics (MM), which treats atoms and their interaction according to Newtonian physics. These methods include docking and virtual screening approaches, which are regularly applied to gain precious information on ligand binding affinity, thus helping in the selection of active compounds to initiate drug discovery efforts. However, the crucial contribution of computation along the challenging process of drug discovery has been recently reinforced by the concrete possibility to apply high-level computation in a time-affordable manner, given the rapid development of faster computer architectures and better algorithms. In this regard, molecular dynamics (MD), which usually relies on MM to define the evolution over time of a molecular system, is today consistently used to investigate (bio)chemical events that once were computationally prohibitive, such as the binding of a small molecule to its targets, which requires extended simulation times that nowadays easily reach the microseconds timescale [4-6]. Finally, due both to the extraordinary increase of computational power and the development and implementation of more efficient algorithms for wave function calculations, quantum mechanics (QM) is also today routinely applied for the characterization of the structure, dynamics, reactivity and energetics of biomolecules [3, 7-14].

In addition, the QM and MM methods can be fruitfully combined to maximize their effectiveness, generating the so-called QM/MM approach, which is particularly suitable to investigate chemical reactions happening in large model systems such as enzymes [9, 15-20]. QM/MM computations allow treating only the reactive portion of the system, *i.e.,* the catalytic site, at the QM level, while all the remaining of the system is described at the MM level (Fig. **1**) [21]. Indeed, QM/MM is routinely used nowadays for the characterization of the (free) energy landscape of enzymatic reaction mechanisms, the description of charge transfer events often detected in large biological systems, for molecular docking and drug design [22, 23] or, as a last example of a long list of possible applications, the investigation of the function of metal ions in proteins (metalloproteins) [3, 24-28]. The Nobel prize for Chemistry given in 1998 to W. Kohn for the development of Density Functional Theory and to J. A. Pople for the implementation of Quantum Chemistry methods, and the most recent Nobel prize 2013 for the seminal contributions of Karplus, Levitt and Warshel for the development of multi-scale models for complex chemical systems remark the highest consideration that these techniques have reached in the Chemistry community. Importantly, last year's Nobel Prize in chemistry has specifically honoured the impact and relevance of the QM/MM approach, which was firstly introduced by the pioneering works of Martin Karplus (Harvard), Michael Levitt (Stanford), and Arieh Warshel (USC) [21, 29, 30]. Methods such as MD, QM and QM/MM calculations can now be considered additional effective tools in the vast computational armamentarium for drug design [1].

Given these premises, in this chapter we will review some of our own recent research studies in the field of computational simulations of biological systems of pharmaceutical relevance, covering a broad variety of therapeutic areas, from viral and bacterial infection to cancer and other inflammatory-related diseases. In particular, we will start with recent successes of first-principles-based QM/MM simulations of metalloenzymes, focusing on the structural and functional role of the metal ions for catalysis in bacterial beta-lactamases [25, 31-34] and in the viral metalloenzyme ribonuclease H [35, 36]. We will continue with examples where MD was integrated with structural and biochemical data to address the mechanisms of resistance against antibiotic agents. In particular, we will present studies on protein/drug interactions in the multidrug efflux pump MexB from *Pseudomonas aeruginosa* [37], as well as MD studies on interactions of novel antimicrobial peptide dendrimer bH1 with model eukaryotic and bacterial membranes [38]. Then, we will conclude this chapter with examples where computational modelling has been used to clarify the mechanisms of covalent inhibition of the fatty amide acid hydrolase (FAAH), which represents a promising target to treat a large number of pathologies including pain and inflammation [39, 40].

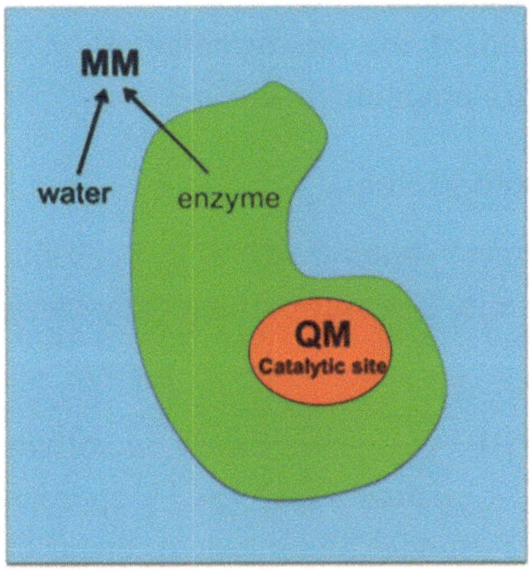

Figure 1: QM/MM approach; The model system is divided in two parts. Only the region of the catalytic site where the reaction takes place is treated at the QM level, while the remaining of the system is treated at the MM level.

This chapter will therefore demonstrate how molecular modelling and computational methods such as MD and QM/MM simulations can nowadays provide informative mechanistic insights that can effectively help the progress of de-novo enzyme design, molecular medicine and, ultimately, drug discovery.

2. METHODS IN COMPUTATIONAL MODELLING

Computational chemistry is advancing at high pace producing an array of techniques that are becoming always more reliable and accurate and able to tackle systems ranging from small isolated molecules to large biological macromolecules, like proteins and DNA. Different modelling strategies can be employed to adequately balance the competing need for accuracy and speed of the calculations. Two of the most common and well-established computational modelling approaches are based on quantum mechanics (QM) and molecular mechanics (MM), which can combine in QM/MM schemes able to provide accuracy in treating the chemistry of the system and extending the investigation to large complexes. The study of drug-target interactions is further complicated by the fact that those key interactions occur in a complex environment, where temperature and solvent effects play a crucial role. Therefore, the use of enhanced sampling methods and simplified Coarse-Grained multi-scale models allowing for more efficient exploration of the conformational space for sizable systems is becoming more and more a common practice.

2.1. Wave-Function Based Methods

Methods based on direct solution of the Schrödinger wave-function are considered the most accurate [41, 42] because they can provide the best physical description of the system.

The Hartree-Fock [43] (HF) method is based on solving the electronic problem using an iterative Self Consistent Field (SCF) approach. The electronic wave function is written as a single Slater determinant of single-electron orbital functions. Therefore, HF methods lead to a mean-field solution where electron-electron correlation is not considered. Despite its appeal, the HF-SCF method is not accurate enough to be used for accurate quantitative predictions. Over the last decades, several approaches, usually indicated as post-Hartree-Fock methods, have been developed to include electron correlation in the multi-electron wave function. These methods include introduction of perturbation terms in the Fock operator (Møller-Plesset perturbation theory [42-44]), or, more accurately, expansion of the multi-electron wave-function over a linear combination of Slater determinants, such as in Multi-Configurational Self Consistent Field (MC-SCF) [45, 46], Configuration Interaction (CI) [43], Coupled Cluster theory (CC) [42]and Complete active space SCF (CAS-SCF) [45, 46]. Both MC-SCF and CAS-SCF are considered the reference methods for the study of processes involving multiple electron states. Accurate quantum-mechanical methods are crucial in biological studies where multi-reference states or non local-electronic properties must be taken into account. For example, this is the case for photo-chemical and photo-physical properties for which these techniques are essential for their correct description [47]. Studies on vision processes initiated by rhodopsin light absorption (*i.e.,* [48-51], and on investigations into DNA induced-light damage processes (*i.e.,* [52]) provide excellent examples.

2.2. Density Functional Theory

Density functional theory (DFT) [53] is an alternative formulation of quantum mechanics where the solution of the fundamental problem addressed the particle density, so, a direct physical observable, and not the many-body wave function. To date, the commonly operative implementation of density functional theory follows the Kohn-Sham [54] approach, where the many-body problem given by a density of interacting electrons in a static nuclear electrostatic potential is mapped into that of a density of ideal non-interacting electrons. Within this formulation the major task within KS-DFT is the modelling of exchange and correlation interactions, for which an explicit analytical formula is not known. The quality of the exchange-correlation functional affects heavily the quality of the prediction by DFT calculations. Since Becke's proposition [55] of a gradient-corrected exchange functional (GGA), where both the density function and its gradient are taken into account, the reliability of the GGA DFT has been sensibly improved by developing correlation functionals with parameters acquired by fitting experimental data or created to reproduce basic physical properties. A significant improvement in DFT performances was achieved after Becke's consideration that local exchange correlation functionals could be hybridized with fractional components of HF-like exchange terms [55]. More recently, inclusion of explicit kinetic energy operators in the exchange and correlation functional yield as proposed by Tao, Perdew, Staroverov and Scuseria (TPSS) [56], led to a new generation of functionals called meta-GGA. Studies on biomolecular systems profit from DFT calculations more than from, in principle, more accurate post-HF approaches thanks to their considerably lower computational cost. Nonetheless, the choice of the correct functional for a specific system or properties may pose major issues to an inexperienced user. It is therefore crucial to rely on the large number of publications constantly assessing the performances of the various functionals (*i.e.,* [57, 58]), as a valuable tool for the correct use of DFT calculations.

One of the most important issues in DFT-based calculations is associated to its intrinsic difficulty in treating dispersion forces [59-61]. These are particularly important in biological systems, where steric packing of hydrophobic residues contributes significantly to the global stability of folded structures. In the past decade, several protocols aimed at including dispersion interactions in DFT calculations have been successfully proposed. These go from inclusion of parameterized two-body R^{-6} long-range terms derived empirically [62-65] or by atomic polarizability computed from the *in situ* atomic electron density [66], or from the instantaneous dipole moment of the exchange hole [67-70], to dispersion-corrected atom centred potentials fitted on high-level calculations [71-77], to highly parameterized meta-hybrid-GGA xc-functionals calibrated to reproduce properties of dispersion-dominated molecular sets [78-81]. Several test cases show that inclusion of dispersion interactions can change the qualitative picture of molecular structures and complexes [82, 83]; therefore, the use of any of the methods available today in the most commonly accessible codes is highly recommended.

2.3. Molecular Mechanics Approaches

Despite the most recent advances in quantum mechanical calculations for large systems, the dimensionality of biological macromolecules is computationally still too demanding. Therefore, in the past decades, simplified Molecular Mechanics (MM) approaches based on parameterized Hamiltonians have been developed. Within these approaches, the total energy of a molecular system is defined as the sum of different contributions mimicking the molecular binding action of the electronic cloud. The MM Hamiltonian [41, 45, 46] is usually composed by bonded terms describing stretching, bending, and torsional vibrational modes, and non-bonded interactions describing exchange repulsion, dispersion and electrostatic forces (Equation 1). Both stretching and bending contributions are usually expressed by simple harmonic potentials. The torsional contribution is described by a periodic function to account for multiple conformational minima. Dispersion and electrostatic interactions are taken into account through, respectively, a two-body 6-12 Lennard-Jones potential [41] and a Coulomb potentials. In the MM force field, the experimental frequencies of specific sets of molecules are fitted to reproduce *ab initio* energies. The equilibrium bond lengths and angles (r_{eq}, ϑ_{eq}) and the spring constants (k_r, k_ϑ) are therefore calibrated to reproduce those data. That is, A_{ij} and B_{ij} are derived from *Monte Carlo* simulations, [41, 45, 46] and the atomic point charges (q_i, q_j) are calculated during fitting procedures, like in the RESP procedure [84], from *ab initio* calculations.

$$E_{tot} = \sum_{bonds} k_r(r - r_{eq})^2 + \sum_{bonds} k_\vartheta(\vartheta - \vartheta_{eq})^2 + \sum_{bonds} \frac{V_n}{2}[1 + \cos(n\varphi - \gamma)] + \sum_{i<j}\left[\frac{A_{ij}}{R_{ij}^{12}} - \frac{B_{ij}}{R_{ij}^{6}}\right] + \sum_{i<j} \frac{q_i q_j}{4\pi\varepsilon_0 R_{ij}} \quad (1)$$

Established MM Force Fields [85] are defined on the basis of the specific analytical form of the potential energy function, and on the basis of the specific values of several parameters defining it. To date, most common force fields like OPLS [86], AMBER [87], GROMOS [88] and CHARMM [89] are broadly used to study biomolecular systems. Apart from the potentials developed for biological macromolecules, computational drug design and drug discovery studies profit from both the Generalized AMBER Force Field(GAFF) [90] and CHARMM General Force Field (CgFF) [91] developed for small organic compounds. The relatively cheap computational cost of MM potentials allows, for example for fast screening of large libraries of chemical compounds or the investigation of the enzymatic activation [92, 93]. Moreover, MM potentials, when coupled to enhanced-simulation techniques can be efficiently employed for lead optimization [94].

2.4. Hybrid Quantum Mechanics / Molecular Mechanics Methods

MM approaches are useful for the investigation of structural features of very large systems and for non-covalent binding of ligands to receptors. On the contrary, they cannot address reactive processes as those taking place at enzymatic active sites, unless a specific parameterization of the reaction under study is performed. Detailed mechanistic studies of any enzymatic processes thus strictly require the use of quantum-mechanical methods.

Nowadays, the most efficient approach to overcoming such limitations is the combination of Quantum Mechanics (usually DFT-based methods) with force-field-based molecular mechanics (MM) in the so-called Quantum Mechanics/Molecular Mechanics (QM/MM) scheme. The system is partitioned into two or more regions treated with different levels of theory (Fig. **1**). Originally proposed in 1976 by Warshel and Levitt [21], many different QM/MM implementations have flourished in the last two decades, being successfully employed in both biological and material science fields [2, 16-18, 24, 25, 27]. In the QM/MM formalism, the chemically relevant part of the system is described at the quantum level of theory, while the remaining portion is treated at the less expensive MM level. Several QM/MM implementations are nowadays available in QM codes. Common hybrid QM/MM schemes are the ONIOM [95] included in the GAUSSIAN suite of programs [96], or the hybrid Hamiltonian approaches included in CPMD [97, 98] and CP2K [99]. MM codes also feature interfaces to (or embed) QM algorithms (often semi-empirical), as is the case with AMBER [100, 101] or CHARMM [102]. When performing a hybrid QM/MM calculation, the most difficult step is constituted by the choice of the QM region, which is *per se* ill defined. In fact, definition of the region where the explicit electronic structure must be taken into account cannot be determined a priori, rather it must be carefully controlled case per case [103-107]. QM/MM approaches have been successfully applied to several drug design studies [3, 108-110] and enzyme reaction mechanisms [3, 18, 111-114]. More recently, hybrid approaches have been integrated into computational protocols devised for docking [115-117] and computing the binding affinity of drugs [116, 118, 119], thus providing a useful tool for *in silico* screening of lead candidates [1, 15, 120].

2.5. Conformational Sampling

On a general basis, it is important to notice that MD trajectories are intrinsically unstable, and therefore, the single events explored by MD are not, *per se*, reproducible. In fact, the relevance of MD simulations relies on the possibility of linking microscopic time-averages along trajectories to thermodynamically equilibrated ensemble averages (ergodic hypothesis). Nonetheless, such eventuality is limited by the efficiency by which MD simulations can explore the conformational space.

In fact, this hypothesis fails in case of slow diffusive processes, for which feasible sampling times are intrinsically shorter than the characteristic times, or for activated processes, for which the probability to observe a transition event between the conformational space regions describing reactant and products is very low. In these cases, free molecular dynamics may be not fully reliable and anyway not lead to an exhaustive description of the phenomena of interest, unless proper enhanced sampling techniques are combined to standard MD protocols.

Biochemical phenomena follow variational paths over the corresponding Free Energy Surface (FES); therefore, its determination is a crucial step for computational studies of their mechanisms. In most cases biochemical events require substantial activation energy even when the process is catalysed by a co-factor, an external stimulus or by the action of an enzyme. This implies that such phenomena cannot be observed by merely performing a (even very long) plain MD run. Over the years, several *enhanced sampling* strategies have been developed to compute free energy differences between reactant and product states as well as free energy changes occurring along any (bio) chemical process. Some of these procedures profit from a priori definition of one or more reaction coordinates projecting the 6N phase space to a less complex dimensionality that is presumed to be representative of the process in question. Among these methods, we mention steered and targeted molecular dynamics [121, 122], the blue moon ensemble [123], umbrella sampling [124-126] and replica exchange molecular dynamics [127] as successful examples. More recent techniques like conformational flooding [128] and metadynamics (MMD) [129-132] aim at overcoming preliminary knowledge of the reaction coordinate components, allowing an unbiased investigation of the FES. The use of enhanced sampling methods coupled to QM/MM simulations was successful in the investigation of the reaction mechanism of potential drug target enzymes [25, 27, 31, 34, 35, 39, 92, 112, 133-140].

The free energy of binding can be estimated by different methods. Here we briefly sketch the MM/PBSA protocol [141-144], used in one of the examples presented below. The free energy of binding of a ligand to a protein $\Delta G_{binding}$ is split as:

$$G_{binding} = \Delta G_{vacuo} + \Delta G_s^{Complex} - \Delta G_s^{Ligand} - \Delta G_s^{Protein}$$

where ΔG_{vacuo} is the free energy of binding *in vacuo*, and $\Delta G_s^{Complex}$, ΔG_s^{Ligand}, $\Delta G_s^{Protein}$ are the solvation energies of the protein/ligand complex, the free ligand and the free protein, respectively. The solvation energies are computed by solving the linearized Poisson-Boltzmann equation using different dielectric constants to reproduce the solvent and *in vacuo* conditions [145]. ΔG_{vacuo} is computed by estimating the enthalpic and entropic contributions separately. The enthalpic contribution is given by the interaction energy between the two fragments, while the entropy change can be computed estimating the variation of the vibrational partition function either by normal mode analysis or by quasi-harmonic approximation. Details and limits of such approach are deeply discussed in the literature [146-150].

2.6. Coarse Grained and Multi-Scale Simulations

Computational structural studies dealing with biological processes must address macromolecular systems that can have significantly different sizes and can be functional in a very broad spectrum of time scales [151, 152]. The recent advances in atomistic simulations sallow reaching limits in the millisecond time scale and or the million of atoms, if sufficient dedicated computational resources are available [4, 153]. Nonetheless, on one hand, these limits can still be far from characteristic times/sizes describing large macromolecular complexes involved in cellular processes; even more, accessible boundaries for routine calculations on molecular systems are still orders of magnitude inferior to these limits.

Time/size scale issues can be tackled by implementation of coarse-grained (CG) models, which are becoming more and more popular in the literature [154-162]. Given their increasing impact in the field, we give here a brief introduction, even though they were not directly used in the examples presented below. Different CG schemes make use of either topological or force-field like Hamiltonians based on bead models, and which can have specific or general applicability. Successful studies where CG models are applied to study different biological problems can be routinely read in the dedicated literature [163-168]. While the first CG force fields aimed at describing simple hydrophobic-hydrophilicdiphase systems (*i.e.*, [160]), nowadays it is possible to find several models able to describe also protein and protein/membrane [169, 170].

Concerning proteins, CG models are still lacking reliable transferability, which, in turn, limits their use, for example, whenever large conformational changes occur in the system. Atomistic resolution may still be crucial for quantitative investigation on phenomena like molecular recognition (for example, for a quantitative estimate of the MIC of drugs on a target protein). In recent times multi-scale modelling (MSM) techniques aiming at coupling all-atom descriptors to simplified models have become more and more popular [158, 159, 163, 171-179]. Implementation of hybrid schemes (AA/CG), where only a portion of the system is treated at the atomistic-detailed level appeared in the literature for both proteins and protein-DNA complexes [161, 162, 180, 181]. Multi-scale approaches comprise parallel schemes [176, 182, 183], adaptive resolution protocols [172, 173, 178], and different approaches for multidimensional Hamiltonians, where different portions of the system are treated at different levels of resolution at the same time [161, 177, 181].

Electrostatics is a crucial ingredient to reproduce intermolecular interactions at the CG level. Two of us have recently proposed a topological scheme to reproduce quantitatively the all-atom electrostatic potential from minimal structural CG information [184]. The proposed model reconstructs the orientation of the backbone dipoles using their statistical orientation in protein structures available in the Protein Data Bank. The protocol requires the position of the α-carbons of a protein, and the angles formed by three connected α–carbons as the only structural information. The computational costs to reconstruct the dipole orientations are negligible (scaling $\propto N$). The protocol can be easily and efficiently implemented in a MD algorithm [185]. Moreover, long-range interactions produced within such a scheme are intrinsically anisotropic, making it particularly appealing to improve CG simulations on conformational changes and secondary structure assembly [185]. Extension of the model to incorporate electrostatic interactions from side-chains can also be efficiently implemented, significantly improving the quality of protein-protein interaction studies [186].

3. ENZYMATIC CATALYSIS

Here, we report some representative QM/MM studies of pharmaceutically relevant enzymes, mainly from our own research. We aim to show how QM-based methods can help understanding the structural and energetics features of the enzymatic reactions [24, 25]. In particular, we focus on enzymes that use metal ions to efficiently perform catalysis (metalloenzymes), for which QM based methods are preferable due to the difficulties in reproducing metal coordination by classical parameterised potentials.

The role of metal ions in biochemistry is fundamental in maintaining structural stability and sustain conformational changes, and proper functionality during enzymatic catalysis. Metalloenzymes are in fact widespread proteins, ubiquitous in all life kingdoms and involved in various biosynthetic processes [187]. In here, we will first review the QM/MM investigation of the ribonuclease H (RNase H) catalytic function [134]. This enzyme catalyses nucleotidyl transfer reactions in the presence of two Mg^{2+} ions contained in the catalytic site, while representing a target for antiviral drugs [188-190]. Then, we will report on the phosphatase activity in soluble epoxide hydrolase (sEH), which is a promising target for hypertension and acute respiratory syndrome treatment [35, 135]. This enzyme carries out the phosphatase activity in its N-terminal lobe, using a single Mg^{2+} ion in the catalytic site. We will also briefly report on a recent computational investigation of topoisomerase II (topoII) [35, 135], which requires two Mg ions in the catalytic site to control the DNA topology in cells. This is a further example centred on an Mg-dependent enzyme that is a validated target for clinical antibiotics (*e.g.,* quinolones) and anticancer agents (*e.g.,* anthracyclines) [191]. As a final example, we will report on computational simulations on the metallo (Zn) β-lactamases, which are important targets for the discovery of new resistant antibiotics [31-34]. In this case, as well, we will see the difference of one *vs*. two ions for catalysis.

3.1. Ribonuclease H

Ribonuclease H (RNase H) is member of the nucleotidyl-transferase (NT) superfamily. RNase H cleaves the phosphodiester bond in the backbone of the RNA strand in RNA·DNA hybrids [192, 193]. For proper function, RNase H accommodates two metal ions in the catalytic site, which counterbalance the large negative charge on the backbone of substrate RNA and DNA strands [188]. These catalytic ions have a key role in maintaining the structural integrity of the protein/substrate complex, in promoting nucleophilic attack on the scissile phosphate, and finally in stabilizing the transition state (TS) and leaving group exit during the catalysis [194].

A divalent bimetal architecture of the catalytic site of RNase H has been described by informative high-resolution X-Ray structures, where two Mg^{2+} ions are jointly coordinated to a non-bridging oxygen of the scissile phosphate of the substrate RNA strand (Fig. **2**) [192, 193]. Notably, the enzymatic activity of RNase H changes with the nature of the metal ion and/or its concentration. RNase H has indeed optimal activity at Mg^{2+}concentration of 10-20 mM, while its function is inhibited at 50 mM [36]. This phenomenon is the so-called 'attenuation' effect. Further, while Mg^{2+} and Mn^{2+} can promote the enzymatic function when in the right range of concentration, Ca^{2+} blocks the RNase H endoribo nuclease activity [195, 196].

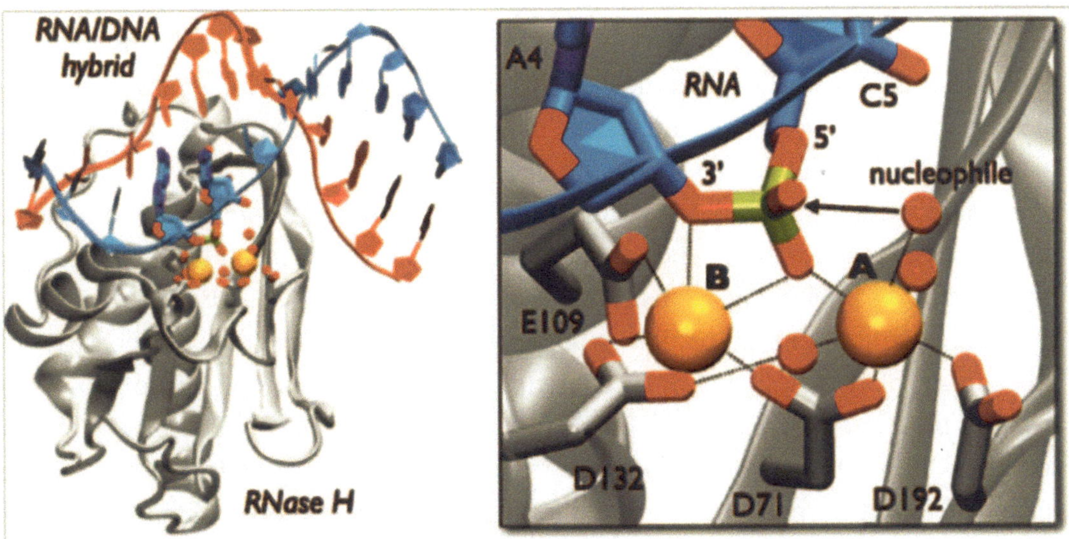

Figure 2: "RNase H structural and catalytic features. (Left) Cartoon of the complex RNase H/RNA•DNA hybrid. RNase H is in gray, DNA is in red, and RNA is in blue; orange spheres indicate the Mg^{2+} ions. (Right) Close- up of the catalytic site, including the RNA strand, and key residues and water molecules coordinated to the two Mg^{2+} ions" (Adapted from ref. [134]).

Recently, QM/MM Car-Parrinello molecular dynamics (CPMD) has clarified mechanistic details of the bimetal-aided nucleotidyl transfer reaction in RNase

H [134]. In particular, this study unravelled the nature of the enzymatic mechanism (concerted one-step or stepwise, with formation of a stable phosphorane intermediate), the energetics and formation mechanism of the nucleophilic hydroxide ion, and finally the role of the two metal cofactors in aiding the catalysis. Toward this end, two different reagent states have been considered in studying the enzymatic reaction (Fig. **3**). Namely, in one case the nucleophilic species was a water molecule (P_{WAT}), while in the other, a hydroxide ion (P_{OH-}) was the reactive nucleophile. The CP QM/MM simulations showed that P_{OH-} had the lowest free energy barrier (~10.5 kcal mol^{-1}). However, P_{WAT} was a competitive mechanism (free energy barrier of ~16 kcal mol^{-1}) if dehydration energy was also considered (~3 kcal mol^{-1}). Importantly, these free energy values are qualitatively in agreement with the kinetic data for substrate analogs for HIV-1 RNase H activity, which corroborates these mechanistic details [197].

To summarize this computational investigation on RNase H, both P_{OH-} and P_{WAT} show an in-line SN_2-like nucleophilic attack on the scissile phosphorus (Fig. **3**). This generates an associative mechanism with phosphorane-like transition states [198]. Importantly, P_{WAT} includes a meta-stable pentavalent phosphorane intermediate, which was observed so far only in the debated β-phosphoglucomutase crystal. Interestingly, the presence of such an intermediate has also been suggested by the recent study of Elsässer *et al.,* [199], which used high level QM/MM calculations to investigate the RNase A catalysis. Also, Rosta *et al* have found similar results in a more recent investigation of the reaction mechanism of RNase H, using a different flavor of the QM/MM approach, which implies the DFT/B3LYP level of theory for the QM part [200, 201]. Taken together, these QM/MM studies overall confirm the finding of ref. [134], both in terms of possible mechanisms and the associated free energy. Finally, another essential aspect of the reaction mechanism is that the two Mg^{2+} ions act in a cooperative fashion. They operate simultaneously to catalyse both nucleophile formation and leaving group stabilization. Thus, both P_{OH-} and P_{WAT} show a phosphorane-like transition state where the associative character of the transition state (TS) is supported by the two ions that get closer to each other in the TS geometry [134].

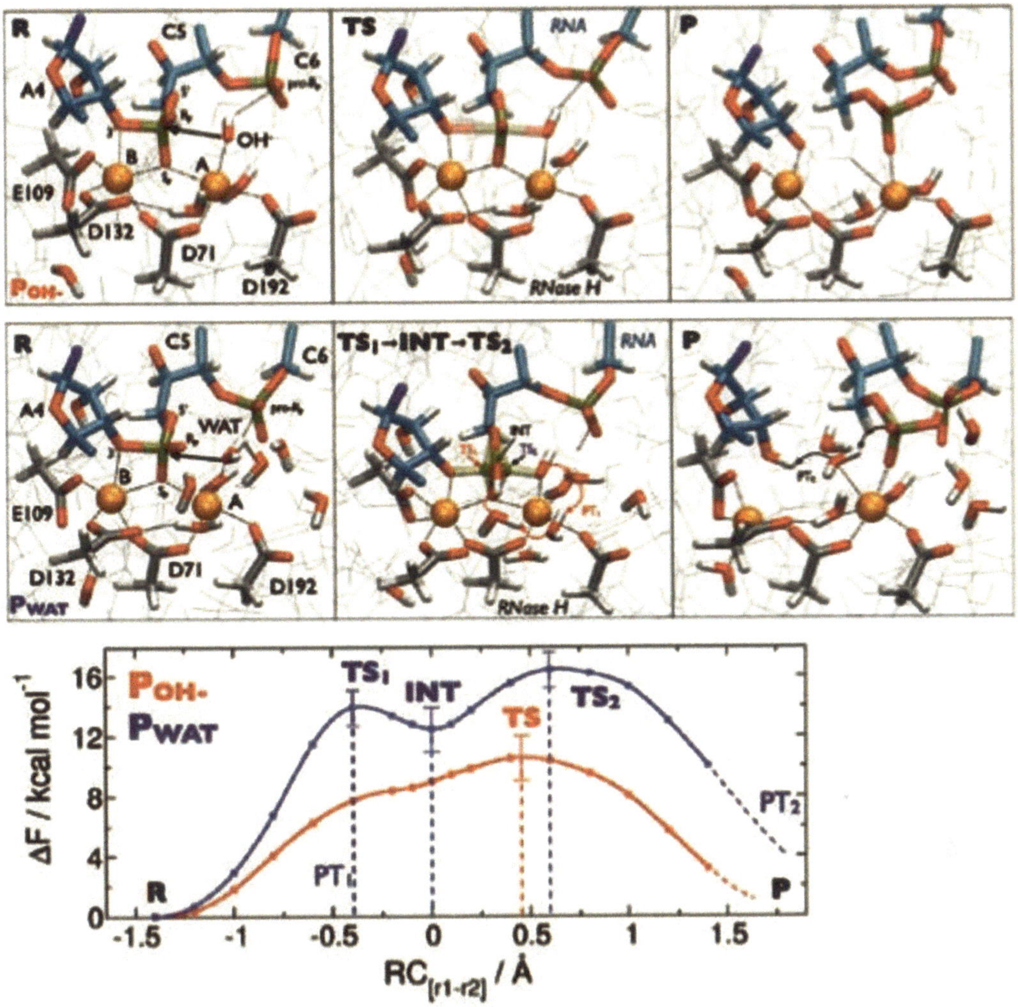

Figure 3: "UPPER SCHEME: Structural evolution of the reaction. Selected snapshots taken for the QM/MM dynamics of the two investigated pathways for RNase H catalysis (only QM atoms are shown explicitly, the rest of the system is shown in thinner lines). (Top) OH- pathway: the nucleophilic group is one hydroxide ion, R. The phosphorane-like TS is shown in the middle. Then, inversion of the phosphate stereo configuration and formation of the 5′-phosphate and 3′-hydroxy function of the RNA strand are shown in P. (Bottom) WAT pathway: the nucleophilic group is a water molecule, R. The nucleophilic attack leads to TS1, where a proton shuttle (PT1) involves 3 water molecules that bridge the scissile phosphate and WAT (red labels). The protonation of the scissile phosphate stabilizes the phosphorane group, causing the formation of the meta-stable intermediate INT (black label). Then, TS2 (blue label) leads to the final product P, in which the cleavage of the RNA strand is definitely completed, and the protonation of the 3′-hydroxy function of the RNA strand takes place (PT2). LOWER GRAPH: Free energy profiles of the two investigated pathways for RNase H catalysis (bottom)" (Adapted from ref. [134]).

3.2. Phosphatase Activity in Soluble Epoxide Hydrolase

Initially, the observed catalytic activity of soluble epoxide hydrolase (sEH), namely the hydrolysis of epoxy fatty acids, occurs in the large C-terminal domain, while the novel metal (Mg^{2+})-dependent phosphatase activity of sEH has been discovered in the smaller N-terminal domain. This novel metal (Mg^{2+})-dependent phosphatase activity of the dual-domain protein she opened a new branch of fatty acid metabolism and providing a new site for drug discovery [202-204].

Based on crystallographic data [205-207], a two-step reaction has been proposed for the two phosphoryl transfers in the sEH phosphatase activity. The first step is the nucleophilic attack of Asp9 on the phosphate group of the phosphoester substrate, with protonation of the leaving group through either Asp11 or an intervening water molecule. Secondly, a water molecule closes the catalytic cycle hydrolysis *via* a nucleophilic attack at the scissile phosphorus atom of the phosphoenzyme intermediate (Fig. **4**). These reactions have been clarified by two CP QM/MM computational studies [35, 135] that provided a first-principles-based interpretation of the experimental findings, providing great detail for the two necessary steps (Step 1 and Step 2, Fig. **4**).

Perhaps, the most interesting detail revealed by these studies is how the Mg^{2+} ion helps the reaction efficiency. In particular, these studies explain the crucial role of metal-substrate connecting water-bridges (WBs) for efficient transfer of the protons necessary for nucleophile formation (water deprotonation) and leaving group stabilization during the two phosphoryl transfers that constitute the catalytic cycle. Indeed, both steps show an in-line nucleophilic substitution with a rather dissociative character, especially marked in Step 2. A planar metaphosphate-like transition state that nicely resembles crystal structures of TS analogues is detected, while no evidence of a phospharane species in the TS regions is observed. The computed free-energy barriers were in fair agreement with experimental data, indicating Step1 (~19 kcal mol^{-1}) as the rate-determining step of the catalytic cycle (Fig. **5**). The most important contribution to enhancing catalytic efficiency is made by the nucleophile and leaving group stabilization *via* WB-mediated proton shuttles, mostly induced by the electrostatic effects of the metal ion.

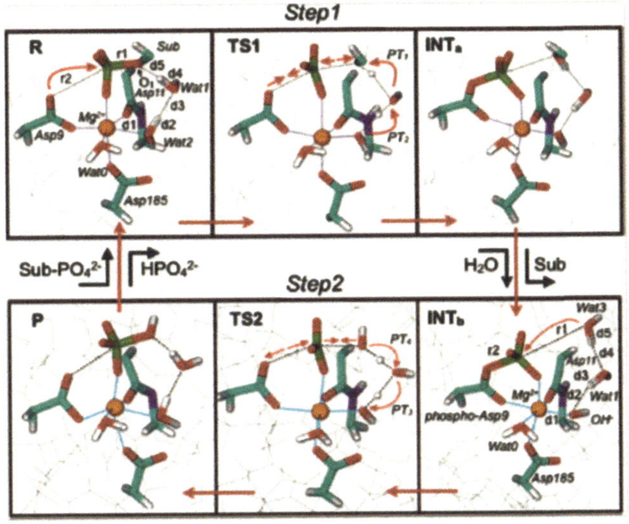

Figure 4: "Selected snapshots taken from our computer simulations of the two investigated phosphoryl transfers comprising the catalytic cycle of the phosphate activity in soluble epoxide hydrolase. (Top) Nucleophilic attack of Asp9 at the Mg^{2+}-coordinated phosphoryl group, with substrate cleavage and phosphoenzyme intermediate formation INTa. In the middle, the transition state structure TS1 shows the concomitant proton shuttle (labelled PT1 and PT2) from a Mg^{2+}-coordinated water molecule to the leaving group oxygen *via* a bridging solvent water. (Bottom) Second phosphoryl transfer from the phospho-Asp9 to one attacking solvent water, leading to the product state, with now a second proton shuttle (labelled PT3 and PT4 in TS2) traveling in the reverse direction to create the nucleophile OH^-" (Adapted from ref. [135]).

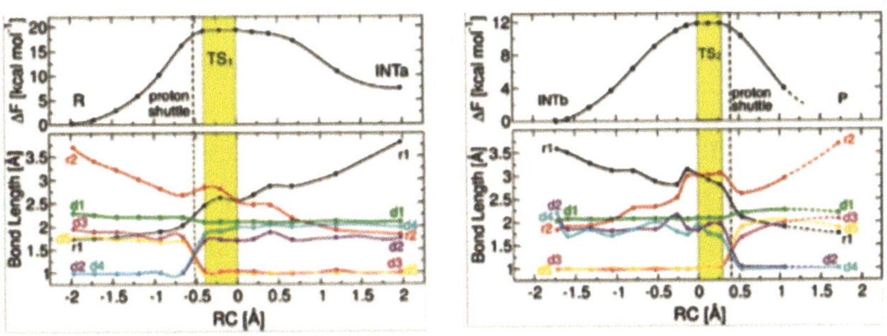

Figure 5: "Left: Free energy profile (top) and selected average bond distances (bottom) along the first catalytic step of phosphoenzyme formation (INTa). Bond distance labels as in Fig. **2**; notably, *r1* and *r2* are the breaking and forming P-O bond lengths, respectively. The proton shuttle occurs at RC ≈ -0.5A (vertical dashed line), just before the system reaches the TS plateau (orange region). Note the shortening of the Mg^{2+}-ligand distance, d1, upon proton donation and transfer along the H-bond wire (d2/d3 and d4/d5 crossing). Right: Free energy profile (top) and selected average bond distances (bottom) along the second catalytic step. Bond distance labels as in Fig. **2** (INTb panel). Here, the proton shuttle (dashed vertical line) occurs in the reverse direction (note the d2/d3 and d4/d5 crossing) after the TS plateau (orange region)" (Adapted from ref. [135]).

3.3. One *vs.* Two Metal Ions for Enzymatic Phosphoryl Transfers

The comparison of the QM/MM results on sEH (one Mg^{2+} cation) and those on RNase H (two Mg^{2+}cations) for the metal dependence in phosphoryl transfer reactions is quite instructive. In fact, based on these QM/MM studies of metalloenzymes [28, 35, 134, 135], different mechanisms (associative *vs.* dissociative) for phosphoryl-transfers seem to be induced according to the metal(s) geometry and stoichiometry during catalysis. During sEH catalysis, the metaphosphate group that is transferred is stabilized by its apical coordination to the only Mg^{2+} ion present in the catalytic site. Instead, in the RNase H enzymatic reaction, the two Mg^{2+} stabilize the attacking and leaving groups, while the metaphosphate group is in between the two ions, showing a phosphorane-like TS. Therefore, this comparison supports the hypothesis, reported for the first time in ref. [134], that two ions can more easily facilitate the formation of a meta-stable intermediate, as in the case of RNase H.

A second aspect of paramount importance in both the phosphatase activity of she [35, 135] and the endonuclease activity of RNase H is the key role of water molecules in solvating the metal centre. These waters facilitate the migration of protons involved in the phosphoryl transfer reaction, in both sEH and RNase H. This mechanism is vital to appreciating the catalytic strategy used by the enzyme to create better attacking and leaving groups. It shows the critical role played by water molecules in enzymatic mechanisms (*e.g.*, phosphoryl transfers).

A recent investigation of the metalloenzyme topoisomerase II (topoII) [208] further highlights the functional role of two metal ions for phosphodiester bond cleavage. In fact, recent X-ray structures of topoII have shown that two Mg^{2+} ions are likely placed in the catalytic site [209]. Here, the DNA strand is cleaved and re-joined to allow DNA topology control. Hybrid Born-Oppenheimer QM/MM MD simulations have been used to reconstruct a catalytically competent state, where the two ions spontaneously relax into a two-metal-ion architecture, as that in RNase H [134]. This position of the two Mg ions seems therefore similar to several other two-metal-ion phosphodiesterases, suggesting that topoII likely cleaves the substrate DNA with a mechanism that might be analogous to RNase H.

3.4. Metallo β-Lactamases

Metallo β-lactamases (MβLs) hydrolyze all kinds of β-lactam antibiotics, including the latest generation of carbapenems. These enzymes are increasingly spreading among pathogenic bacteria, showing also increasing resistance to most of the current clinical inhibitors on the market. It is therefore urgent to develop new effective MβLs inhibitors [210, 211].

The MβLs contain in theory active site either one or two zinc ions, which are essential for hydrolysis. In the case of two zinc ions into the active pocket, the first metal site (Zn1) is coordinated tetrahedrally by three histidines (His116, His118 and His196), and the nucleophilic hydroxide (Fig. **6**) [212]. The second metal (Zn2) coordinates the nucleophile, a water molecule and a ligand triad of protein residues, namely Asp120, His263 and Cys221. Recently, computational methods have unraveled the enzymatic mechanism, providing also structural and energetics details of the enzymatic reaction.

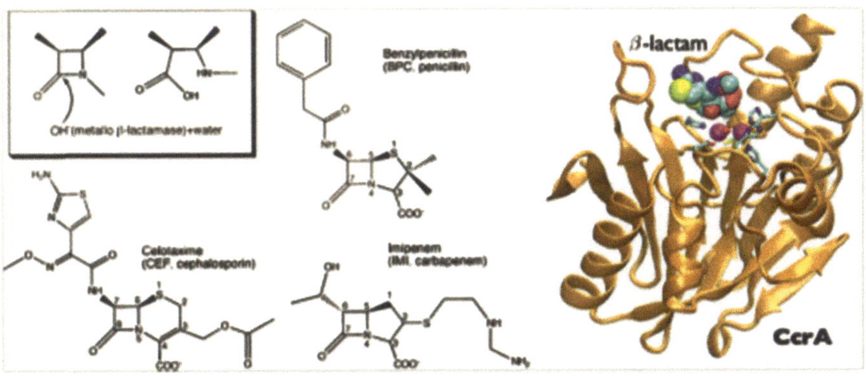

Figure 6: "(Left) β-*Lactam* hydrolysis catalysed by MβL is sketched in the inset. Also shown are substrates used in this work. (Right) Binding mode of CcrA from *B. fragilis.* Secondary structures are represented in orange cartoons; purple spheres indicate the Zn^{2+} cofactors present at the active site, whereas coordinated metal ligands are depicted in stick representation and the β-*lactam* in space-filled balls" (Adapted from ref. [34]).

Classical MD simulations and full QM calculations were used to investigate both the mono-Zn MβL from *Bacillus cereus* (BcII) and di-Zn from *Bacteroides fragilis* (CcrA) in complex with different types of β-lactams (*e.g.,* benzylpenicillin, imipenem, and cefotaxime) [31-33, 212, 213]. These studies have also revealed a few key interactions for binding recognition. Results showed that one water molecule can bridge the β-lactam carboxylate group and the metal centre. Interestingly, when two Zn metals are bound at the active site, one water is bound to Zn2, completing its coordination shell. Thus, a water-mediated salt-bridge is maintained between the β-lactam carboxylate moiety and Lys224 - a conserved residue in most MβL enzymes. It is remarkable how these few common and minimal features can be sufficient to allow the accommodation of different β-lactams. In these studies [31, 34], Car-Parrinello (DFT/BLYP) QM/MM calculations were used to investigate the hydrolysis of a commonly used cephalosporin (cefotaxime), which is actively hydrolysed either by mono-Zn and di-Zn MβL enzymes. While these simulations showed that in both cases the metal-dependent nucleophilic attack of a metal-bound hydroxide is promoted, the chemistry and kinetics of the two reactions are strongly affected by the Zn architecture and stoichiometry. In fact, when Zn2 was present, the simulations were in favor of a concerted single-step mechanism. Finally, based on the high amino acid sequence conservation among subclass B1 MβLs, this reaction mechanism was proposed for the entire B1 subclass, where β-lactams might follow similar catalytic pathways. Remarkably, this might explain why monobactams such as aztreonam, which lack the common β-lactam bicyclic core and carboxylate, are not efficiently hydrolysed by MβLs (Fig. 7).

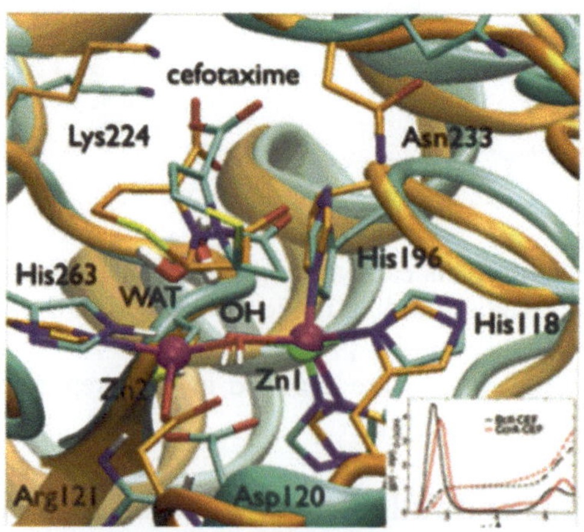

Figure 7: "Superimposition of selected MD snapshots of monozinc BcII (cyan skeleton and cartoons) and dizinc CcrA (orange) in complex with cefotaxime. The inset shows the O-O water radial distribution function (solid lines) and the coordination number (dashed lines) for the nucleophile in both conformations" (Adapted from ref. [34]).

Regardless the specific enzymatic mechanism, these examples are used herein to highlight how computational methods can shed new lights on the functional role of metals in enzymes. How the two metal ions behave along the reaction coordinate of the enzymatic reaction is indeed a very fascinating aspect of quantum enzymology. This exciting field of research is still in its infancy, and much remains to be clarified on the role of metals for catalysis. One question, among many of interest, is how different stoichiometry and physicochemical features of metal ions can lead to either inhibition or acceleration of catalysis. In this regard, in the next decades we foresee a prominent role of computational methods integrated to experimental data to address key aspects of enzymatic catalysis.

4. ANTIBIOTIC RESISTANCE

As briefly introduced in the previous paragraph on metallo-β-lactamases, the systematic and widespread use of antibiotic drugs has a relatively short history in medicine, as it dates to the first campaigns of intense use of penicillin and cephalosporin in the first decades of the past century. Since then, it soon became clear that pathogens were able to rapidly evolve mechanisms of resistance against antibiotics. In fact, the first strain of *Staphylococcus aureus* resistant to penicillin appeared only few years after broad use of penicillin began [214].

The intense use and, at times, misuse of antibiotics has lead to the evolution of bacterial strains that are now resistant to a broad spectrum of drugs. Severe health threatening strains are today known for both Gram+ and Gram- pathogens. A prototypic case is provided by methicillin resistant *S. aureus* (MRSA), which is resistant not only to methicillin, but also to several other classes of drugs, like to aminoglycosides, macrolides, tetracycline, chloramphenicol, and lincosamides [214].

Cases of induced cross-resistance by interfering bacterial strains were also reported [214], as is the case of vancomicine resistant *Enterococcus* (VRE) and MRSA, which lead to strains of *S. aureus* resistant to vancomicine. Pan-drug resistant bacterial lines have emerged in *Pseudomonas aeruginosa* and *Acinetobacter baumanii* [215]. Regrettably, since the eighties, the evolution of multidrug resistance (MDR) in bacteria has been accompanied by a continuous decrease in the approval rate of new antibiotics, in particular against Gram- bacteria [216, 217].

4.1. Antibiotic Recognition in MexB - a RND Multidrug Efflux Pump from *Pseudomonas Aeruginosa*

MDR is a global phenotype that can be attributed to multiple molecular origins. Among several other factors, expression of multidrug efflux pumps plays a key role in MDR. Evolutionary pressure has lead to the selection of bacteria mutating the native physiological exporters into efflux machineries able to extrude very different substrates from the cell.

Multidrug transporters are active against a wide range of chemically unrelated molecules, although there is a slightly preference for relatively lipophilic, planar molecules of molecular weight less than ≈800 Da [218]. Substrates are also usually, but not exclusively, weakly cationic [218].

Multidrug efflux pumps have been identified in all five active molecular transporter super families, and therefore, the MDR phenotype must have evolved independently several times [219]. For a detailed analysis of these families, please refer to the several reviews available in the literature (*i.e.,* [220-222]).

The complex dynamics associated with the efflux mechanisms poses a major complication in understanding drug-recognition and transport. Computational models based on MD simulations can therefore provide useful complementary information integrating experimental knowledge. For a general assessment on MD studies applied to efflux pump, please refer to ref. [223].

Most Gram- bacteria are intrinsically resistant to a large variety of lipophilic antibiotics [224]. This property is attributed to expression of efflux pumps of the Resistance-nodulation division (RND) superfamily, as first demonstrated by experiment on inactivation of the prototypic AcrAB/TolC RND efflux pump in *E. coli* [220].

RND efflux systems are large macromolecular complexes that extend through both the inner and outer membranes of Gram- bacteria. They are able to capture and extrude substrates from the periplasmic region between the two membranes, using electrochemical gradients at the inner membrane as the driving force. RND macromolecular complexes are formed by an inner membrane protein complex(IM) that binds to an outer membrane channel, This last belonging to the outer membrane factor family (OMF) [225]. A periplasmic membrane fusion protein (MFP) is required to stabilize the complex and make it functional [226]. So far, the single components for only two RND multidrug efflux complexes have been structurally determined. They are: *i)* the AcrB/AcrA/TolC complex form *E. coli* [227-229], and *ii)* the MexB/MexA/OprM complex from *P. aeruginosa* [230-235].

The IM and MFP proteins of these two systems are close homologous, with percentages of identity between their sequences of 69% (IM) and 57% (MFP). On the contrary TolC and OprM share only 19% sequence identity, even though the overall fold is conserved.

Despite the larger structural and biochemical information available for the *E. coli* system, the pump from *P. aeruginosa* has more interest for medical/pharmaceutical research because of the significantly more severe pathogenic impact of the bacterium [236]. From the pharmacological point of view, the IM complex is the most important component of RND pumps (Fig. **8**). In fact, drug recognition, binding, and active extrusion occur in this unit. The crystal structure of MexB (3.0 Å resolution) [235] reports an obligate homo-trimer assembly. Each unit has a transmembrane domain composed by 12 helices, and a large periplasmic domain. The periplasmic domain can be further divided into four subdomains building the *pore domain*, and two more subdomains, which constitute the docking region for the OMF. The three monomers are in close contact with each other both in the transmembrane and in the periplasmic region. Moreover, Each periplasmic subunit contains a long β-hairpin that protrudes into one other neighbouring subunit. The transmembrane region contains well-conserved tritrable amino acids, which are necessary for the proper function of the protein, most probably mediating proton translocation [237].

MexB shares the same overall fold of its close homologue AcrB, therefore, it is assumed that the general mechanism of substrate extrusion postulated for AcrB is conserved also in MexB [238-240].

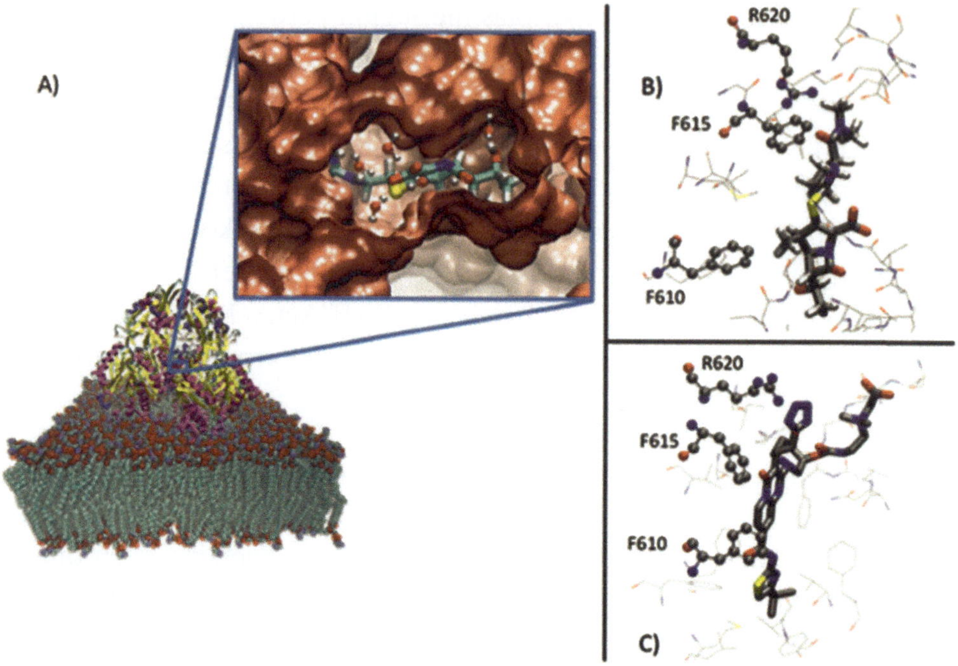

Figure 8: "Structure of RND transporter MexB (PDB: 2V50 [235]). Panel A: computational model embedded in model membrane bilayer. The inset reports the computational model structure of meropenem in the PP as from ref. [37], highlighting the residual solvating water present in the pocket. Panels B, C report a comparison between the binding mode for meropenem found by computational studies (panel B) [37] and the one from experimental co-crystallisation of a pyridopyrimidine derivative inhibitor (PDB:3W9J, [233]) (panel C)".

The first structures of AcrB reported different conformations for the three monomers [238]. Following the nomenclature by Seeger *et al.,* [239] we call these three states *Open (O), Tight (T), and Loose (L).* According to the consensus hypothesis, the three conformations represent consecutive states in a peristaltic motion, also called *functional rotation* in ref. [238]. The postulated motion starts by early recognition of a substrate at a low affinity site in the L monomer. Then, a first conformational transition switching from the L to the T conformation leads to tight binding of the substrate in a second high-affinity binding pocket. A second global conformational change converts the T into the O conformation, and the consequent release of the substrate towards the OMF channel. Structural relaxation after substrate unbinding leads back to the initial L conformation. Conformational changes should be favoured by the proton flow through the transmembrane domain.

In 2013, Nakashima *et al.,* solved the structure of MexB in complex with a pyridopyrimidine derivative [233], reporting the first structural insights for MexB/inhibitor interactions (Fig. **8**). Authors also crystallised the free MexB, finding a structure largely similar to that of Sennhauser *et al.,* [235]. The binding geometry of the same pyridopyrimidine derivative to AcrB was also resolved, and showed relevant variations in the conformation of the ligand. These findings could highlight the presence of subtle differences in the mechanisms of drug binding and translocation between the two pumps.

MexB. MexB and AcrB share two crucial regions first identified in AcrB as affinity sites of substrates. There regions are described in the literature as *i) distal binding pocket* (DP), a phenylalanine-rich pocket [238, 241], and *ii) proximal binding pocket* (PP) [241, 242], which is located toward the protein opening toward the periplasm.

In a recent computational work, Collu *et al.,* provided for the first time structural information on binding of antibiotics to MexB in these two regions [37] (Fig. **8**). Identification of binding modes in both DP and PP for meropenem and imipenem in MexB occurred by repeated flexible docking calculations using the ATTRACT software [243]. In order to reduce the computational costs, and to avoid excessive false-positive outcomes, the docking protocol was applied to the truncated periplasmic domain of MexB only. The best poses obtained by docking were used as starting configurations for all-atom molecular dynamics (MD) simulations for the four complexes containing the two antibiotics in the two pockets. The truncated complexes were solvated with roughly 45,000 water molecules for a total of 160,000 atoms per system. 50 ns long MD simulations for each antibiotic/MexB complex were performed using the ff99SB AMBER force field [90, 244].

Combining MD simulations with MM/PBSA calculations [141, 143, 144], it results that meropenem preferentially binds to DP (binding free energy = -8.1 kcal mol^{-1}) than to PP. On the contrary, imipenem has only poor affinity for both the two pockets (0.6 kcal mol^{-1} and 0.4 kcal mol^{-1}, respectively). This finding agrees with experimental data reporting a 4 to 8-fold increase in the minimum inhibitory concentration (MIC) of meropenem upon overexpression of MexB in *P. aeruginosa* whereas the MIC for imipenem is unaffected [245-249].

The qualitatively different behaviour of the two antibiotics can be rationalised in terms of dehydration properties upon binding to DP. In fact, both meropenem and imipenem are progressively dehydrated passing from the bulk to the PP to the DP. Analysis of the interactions with the solvent revealed that only imipenem formed artificially long-lifetime interactions with water molecules in DP. In particular, a significant fraction of hydrating waters (~38%) had average residence times of more than 1 ns, while, in the bulk, all hydrating waters exchange with characteristic times lower than 50 ps.

Moreover, the position of imipenem in DP recalls that of doxorubicin in mutated AcrB F610A [250]. In this particular mutant, doxorubicin, a good substrate for wild type AcrB, is instead very poorly transported [251], as also found by independent MD studies by Ruggerone and co-workers [250]. In contrast to imipenem, meropenem localizes in a region of DP in proximity of the external channel, thus assuming a favourable conformation for extrusion.

Data discussed here were acquired in the T monomer of MexB, the one supposedly tightly binding the substrate. The docking protocol used for this study did not capture relevant binding geometries of either imipenem or meropenem in the monomer in the L conformation. In fact, the L monomer of MexB appears in the crystal structure to be characterized by a closed PP, thus, not allowing binding of substrates in the absensce of a local conformational modification [235]. The analogous PP in the L monomer of AcrB is instead capable of binding chemically different substrates [241, 242]. So large differences between the AcrB and MexB asymmetric structures occurring in the L monomer only were unexpected before crystallization of MexB. The most recent structure of MexB confirms this peculiar difference between the two proteins. This may be an indication of the presence of subtle differences in the early recognition process. Further studies on the structure of MexB may shed light on this very intriguing issue.

4.2. Novel Antimicrobial Peptide Dendrimer Selectively Targeting Bacterial Membranes

Bacterial antibiotic resistance, especially in Gram- pathogens like *P. aeruginosa*, is nowadays life-threatening for several individuals, especially for people affected by degenerative chronic diseases like cystic fibrosis, or for immuno-compromised patients [252-254]. New strategies aimed at tacking pan-resistant strains of bacteria identified the bacterial cell wall as the best region to attack with antibiotic agents. Superior organisms naturally produce several antibiotic molecules, typically constituted by peptides or post-translationally modified peptide sequences. Rational design of cyclic peptides or peptide dendrimers may therefore constitute a significant route toward definition of new antimicrobial drugs. Recently, the dendrimer **bH1** (Fig. **9**) [255, 256], a new type of membrane disrupting antimicrobial peptide (AMP) [257], was proposed. This dendrimer is characterized by alternation of linear natural amino acids and branching 2,3-diaminopropionic acid (*Dap*). **bH1** shows potent antimicrobial activity against both Gram- *E. coli* (MIC = 1 μM) and *P. aeruginosa* (MIC = 5 μM). On the contrary, it has practically no selectivity from erythrocytes, used as test eukaryotic cells, with a minimal hemolysis concentration larger than 500 μM.

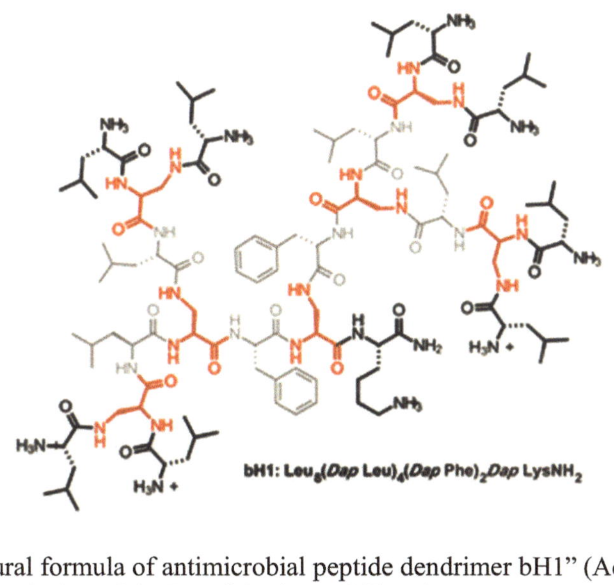

Figure 9: "Structural formula of antimicrobial peptide dendrimer bH1" (Adapted from ref. [38]). - Reproduced by permission of the Royal Society of Chemistry (http://pubs.rsc.org/en/content/articlelanding/2013/cc/c3cc44912b#!divAbstract).

Selective interaction of **bH1** with the lipopolysaccharide-coated (LPS) outer membrane of PA *versus* the eukaryotic cell membrane was studied by all atom MD simulations of the dendrimer [258-261] interacting with membrane models constituted of *i)* a 1-palmitoyl-2-oleoyl-sn-glycero-3-phosphocholine (POPC) bilayer (model of the eukaryotic membrane), or *ii)* of an asymmetric LPS on top of a 1,2-dipalmytoyl-3-phosphatydyl-ethanolamine (DPPE) layer as a model of the PA outer membrane. Force field parameters and thermally equilibrated coordinates for the two membrane systems were taken from previously published works [250, 262].

bH1 invariably approached the membrane surfaces in a relatively short time (~10 ns), showing an intrinsic tendency to adhere to the membrane surface. In this phase, most dendrimer-membrane contacts involve the eight positively charged ammonium groups at the termini of the dendrimer branches. In **bH1**/POPC, the only interaction observed was between the denrimer amino-terminal group and the lipid phosphates, no evolution of the system was observed within the subsequent 0.5 μs of MD.

The structure of LPS in the outer membrane of PA is characterized by a multi-layered assembly, where regions formed by hydrophilic saccharides alternate with regions characterized by high concentration of negatively-charged phosphate groups, stabilised by the presence of partially hydrated alkaline or alkaline-earth counter-ions [263]. Different from the **bH1**/POPC model, the **bH1**/LPS system showed a fast evolution dynamics. After an initial metastable phase at the water/surface interface, where the positively charged Leu_2*Dap* terminal dendron and the two solvent-exposed 0-α-D-glucose sugars of LPS interact with each other, (Fig. **10**, panel E), MD simulations report penetration into LPS of a first Leu_2*Dap* end by transient formation of a hydrogen bond with 6-α-D-glucose. After initial insertion, the first branch moves through deeper layers of sugars until it stabilizes at a distance of roughly 18 Å from the water-LPS interface. In this region LPS the three anionic phosphate groups of phosphorylated 2-(2-hydroxyethyl)-6-deoxy-D-manno-heptose anchor the cationic amino termini of the inserting dendron branch (Fig. **10**, panel H).

The rest of the 0.5 μs-long MD simulations reported no further penetration of this first dendrimer branch. Meanwhile, the other dendrons of **bH1** progressively inserted into LPS, always following a similar interaction pattern as for the first Leu_2*Dap* dendron. Penetration of each of the four terminal Leu_2*Dap* dendrons from the outer to the inner region of LPS occurs in a characteristic time of 100 ns by a mainly diffusive mechanism constituted by formation and rupture of labile contacts with the saccharide present in the different LPS layers. Transition from the water to the LPS membrane, on the contrary, is a fast step that requires some initial activation.

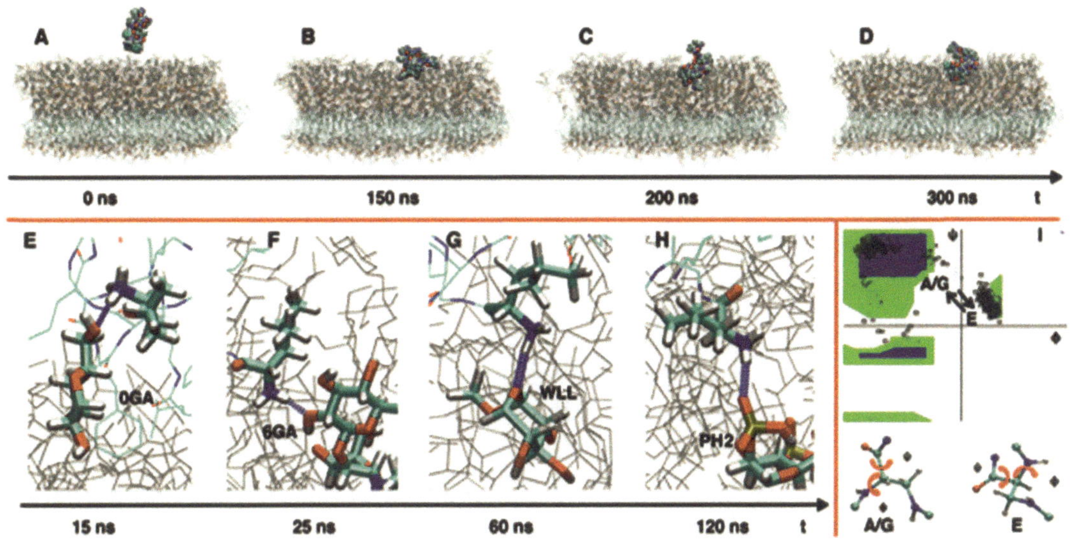

Figure 10: "Interaction of the bH1dendrimer with the LPS-DPPE membrane. Upper panels: (A-D) - insertion of bH1 into the membrane as a function of time. bH1 is shown in van der Waals spheres. Lower panels (E-H) - interactions of Leu2Dap with various residues in the LPS as a function of time. Residues involved in H-bonding are drawn in licorice. Colored lines represent the rest of bH1, gray lines represent the rest of the LPS. Blue dashed lines show hydrogen bonds. Panel I: Ramachandran plot for Dap. Black squares represent conformations explored during MD of the bH1/LPS system. While Leu2Dap is in water (like geometry A in the left panel), it explores extended conformations. Early contact with the LPS (E) induces coiling. Complete insertion into the LPS (F to G) is accompanied by recovery of the extended conformation, kept along the rest of the simulation (H, B, C and D). (Residue abbreviations: 0-α-D-glucose = 0GA, 6-α-D-glucose =WLL, 2-(2-hydroxyethyl)-6-deoxy-D-manno-heptose =PH2)" (Adapted from ref. [38]). - Reproduced by permission of the Royal Society of Chemistry (http://pubs.rsc.org/en/content/articlelanding/2013/cc/c3cc44912b#!divAbstract).

We observed that a conformational change at the branching *Dap* unit of the terminal Leu$_2$*Dap* moiety is concomitant to initial penetration into LPS. In fact, both in the aqueous phase and in the membrane, *Dap* resides in the allowed β-sheet region of the Ramachandran plot (Fig. **10**). Crossing the water/LPS interface is associated to transient migration of *Dap* to the right-handed α-helix region of the Ramachandran plot.

The required conformational change is correlated to the asymmetry of the Leu_2 *Dap* dendron. In fact, the β-branch is longer by one $-CH_2-$ group and thus, it has higher flexibility than the α-branch. During MD, this longer branch is always the first to penetrate the LPS membrane, followed by *Dap* and the shorter α-branch. Therefore, MD simulations suggest that early penetration of **bH1** into LPS is favoured to some extent by larger conformational flexibility of the penetrating units. In turn, the branching residue must be able to allow multiple conformations and local coiling in order to optimize the competing interactions between the aqueous and membrane phases, so to facilitate the early steps of the penetration.

The simulated model for the LPS membrane has two highly negatively charged layers in its structure. Within the simulated time, **bH1** migrated toward the first of these charged regions, situated at ~18 Å from the LPS surface. Penetration to the second layer, which sits approximately 8 Å deeper might occur at longer timescales. In a parallel experiment, **bH1** was exposed to 5(6)-carboxyflourescein (CF) loaded large unilamellar vesicles composed of phosphatidylglycerol as head groups. **bH1** induced CF release at low concentration (1-30 μg/mL), implying an action of disruption of the membrane. On the contrary, unilamellar vesicles with phosphatidylcholine head groups released CF only after treatment with **bH1** at concentrations at least as high as 200 μg/mL, showing that **bH1** only weakly interacts with Zwitterionic head-groups [38], thus stressing the relevance of electrostatic complementarity for the action of **bH1**.

The presented work is a case example of how MD simulation can be helpful in dissecting the molecular origin of complex phenomena like selective membrane recognition and disruption. MD runs showed that not only electrostatic, but also local flexibility is required to optimize antimicrobial potency of this kind of systems. In fact, subtle changes in amino acid sequences may be deleterious for the activity, even though the total charge of the dendrimer is unaffected. The presented simulations aimed at identifying the origin of molecular selectivity for membranes, and did not address the fate of the membrane upon dendrimer binding. In fact, the simulation time was relatively short, and the end of them LPS bilayer was still well organized and practically unaffected by the presence of the dendrimer. Studies on the antimicrobial function of **bH1** require exploration at longer timescales, also including possible aggregation of multiple **bH1** units into LPS.

5. MD AND QM-MM METHODS FOR DRUG DESIGN: THE CASE OF THE ENZYME FAAH

As discussed in the previous paragraphs, the activity of one or more enzymes can be related to the development of a disease and therefore be targeted by drug discovery programs. In fact, most small-molecule drugs produce their beneficial pharmacological effects through the modulation of a targeted enzyme function. In this context, computational methods are often used to explain in detail how selected inhibitors are able to modulate the enzyme function of interest, which can be potentially crucial in designing more effective drugs.

Here, we report on recent computational investigations of a promising enzyme target for drug discovery, namely the fatty acid amide hydrolase (FAAH). FAAH is a key enzyme involved in the endocannabinoid metabolism, which is fundamental for human health and crucial in the regulation of pathophysiological processes such as pain and inflammation [264]. FAAH is an intracellular serine hydrolase that acts with a specific mechanism of hydrolytic degradation of endocannabinoids. Therefore, inhibition of the enzyme FAAH increases the level of endogenous cannabinoids [265], which is considered a promising strategy to treat an ever-increasing number of pathologies, spanning from pain to inflammatory-related diseases such as cancer [266].

Over the last decades, a wealth of structural data on FAAH allowed a detailed understanding of the structural features of the enzyme catalytic site [267, 268]. Briefly, FAAH binding site includes a catalytic triad (Ser241-Ser217-Lys142) that performs the hydrolysis of the endocannabinoid substrate, while an oxyanion hole (Gly239-Gly240-Ser241) stabilizes the substrate for catalysis. Structural, kinetic and computational studies on FAAH catalysis have suggested a catalytic mechanism that involves a complex multi-event reaction sequence that leads the endocannabinoid substrate to hydrolysis and release, closing the overall catalytic cycle [269]. Then, additional structural data have elucidated the mechanism of inhibition of FAAH by potent enzyme inhibitors. In this respect, of particular relevance was the co-crystallization in FAAH of some potent covalent inhibitors, which have been shown to block the FAAH activity through the formation of a covalent bond with the nucleophilic Ser241 [270].

Most of these covalent inhibitors of FAAH are potent electrophilic compounds characterized by the presence of an activated carbonyl group. These include trifluoromethylketones, α-keto amides, α-keto esters and α-ketoheterocycles, such as OL-135 [271, 272]. Nevertheless, most of these compounds have low target selectivity and efficacy *in vivo*. Later, a class of FAAH covalent inhibitors with a promising drug-like profile was designed based on an N-cyclohexylcarbamic acid O-aryl ester template, including URB597, a highly potent FAAH inhibitor both *in vitro* (IC50 = 4.6 nM) and *in vivo* (ED50 = 0.15 mg/kg, in rat). Interestingly, QM/MM calculations were used to describe, at the atomic level, the reaction between FAAH and some of these carbamic acid aryl ester inhibitors [27]. The carbamoylation of the active nucleophile Ser241 by compounds of this class, including the reference compound URB597, suggested a selected reactive orientation of the inhibitor, which was later confirmed by the crystallographic resolution of the FAAH-URB597 carbamoylated structure [270]. Importantly, this represents a significant example of how QM/MM-based modelling can contribute to the rational explanation of mechanism of action of potent enzyme inhibitors [273].

The ability of FAAH to cleave amides and esters at similar rates suggested, however, that not only carbamates but also ureas could act as good carbamoylating agents. Indeed, Pfizer and Cravatt's lab recently discovered a novel class of potent FAAH inhibitors that are based on cyclic piperidine and piperazine aryl ureas, which are cleaved by Ser241 forming a covalent enzyme-inhibitor adduct. The presence of the piperidine or piperazine moiety of these compounds was indeed hypothesised to favour the covalent interaction of the inhibitor with Ser241. The distortion of the urea functionality at the FAAH active site seems prompted by the flexibility of piperidine- and piperazine-based compounds, with consequent formation of a covalent bond between the inhibitor and Ser241.

To investigate this functional hypothesis, Palermo *et al.,* [39] performed an extensive computational analysis centred on piperidine-based PF750 and piperazine-based JNJ1661010 inhibitors, which are two lead compounds used to generate clinical candidates (Fig. **11**). These two potent compounds were compared to an inactive acyclic 1-cyclohexyl-3-naphthalen-2-ylurea through the use of both MD simulations and QM/MM computations. This comparative MD and QM/MM study [39] indeed highlighted a different conformational flexibility of these three representative compounds in water and in complex with FAAH supporting the hypothesis that FAAH is able to induce a distortion only of the amide bond of the active piperidine and piperazine compounds. Indeed, MD simulations indicated that, within FAAH's binding site, the piperidine and piperazine inhibitors adopt a specific conformation that is characterized by a twist of the amide bond and incomplete pyramidalization at the amide bond nitrogen (Fig. **11**). Therefore, distorted amides undergo nucleophilic attack more easily, compared to their planar analogues. This was further established *via* QM/MM calculations, which shown a higher reactivity of the distorted amides toward nucleophilic attack, relative to those of their planar analogues, indicated by a lower ΔE_{LUMO_HOMO} for distorted conformations with respect to the planar analogues (Fig. **12**). This, coupled to a lack of distortion of the amide bond of the inactive compound, might explain the inability of planar compounds to inhibit FAAH. The essential role of flexibility of the protein/ligand complex was also demonstrated by a more recent MD-based study, which has shown the key role of flexibility of the highly flexible substrate anandamide when in complex with FAAH (Fig. **13**) [40].

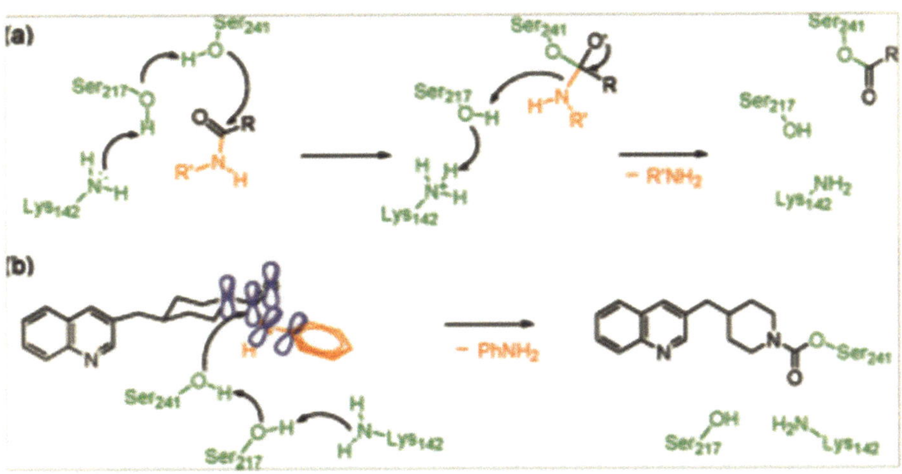

Fig. 11: contd….

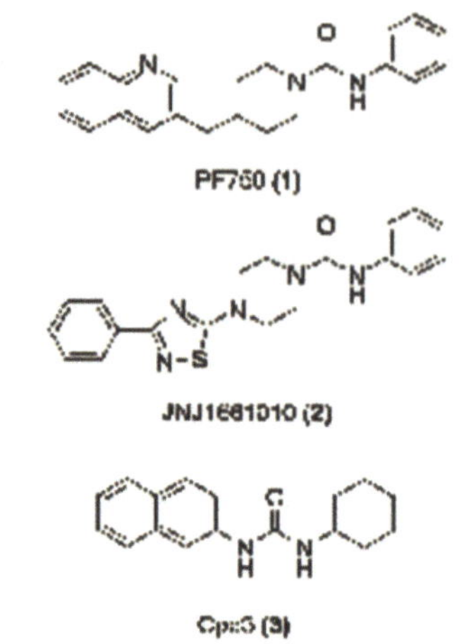

Figure 11: "(**a**) Mechanism of Substrate Hydrolysis by FAAH (shown for a generic amide substrate) and (**b**) Proposed Mechanism of FAAH Inhibition by the Piperidine and PiperazineUreas. On the right, Piperidine urea 1 (IC50 = 16.2 nM), piperazine urea 2 (IC50 = 33 nM), and acyclic urea 3 (IC50 < 30000 nM)" (Adapted from ref. [39]).

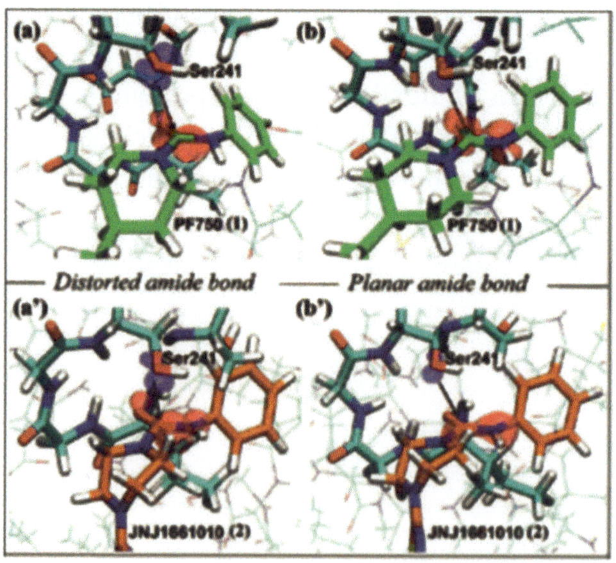

Figure 12: "Shape of the frontier orbitals for the FAAH in complex with compound PF-750 (top) and FAAH in complex with compound JNJ1661010 (bottom). Two representative snapshots characterized by distorted (a and a') and planar (b and b') amide bonds are shown for 1 and 2, respectively. FAAH residues Ile238, Gly239, Gly240, Ser241, and Ser217 are coloured cyan; 1 is coloured green, and 2 is coloured orange. The HOMO of the Ser241 nucleophile is coloured blue, while the LUMO of the electrophile (*i.e.,* the 1 and 2 carbonyl) is coloured red" (Adapted from ref. [39]).

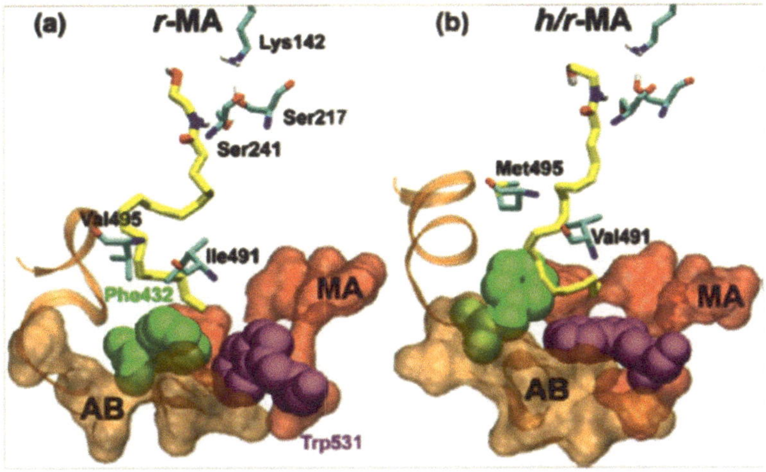

Figure 13: "Representative catalytically significant conformations of the r-MA (**a**) and h/r-MA (**b**). Residues belonging to the MA and the AB channels are shown in red and orange, respectively. Key residues of the AB channel, as well as the catalytic triad residues, are shown in cyan sticks. Phe432 (green) and Trp531 (violet) at the MA/AB interface are also shown. Anandamide is represented with yellow sticks. In both systems, anandamide assumes a curved conformation, with its acyl chain located at the MA/AB interface" (Adapted from ref. [40]).

Overall, the study of Palermo *et al.,* [39] supports the fact that the enzyme-induced twist of the amide bond likely facilitates the amide bond hydrolysis and formation of the covalent inhibitor-enzyme adduct. On the contrary, the rigidity of the planar urea moiety in the acyclic derivative seems to prevent its good fit into the catalytic site, which might partially explain its lack of inhibitory activity. This rational explanation of the inhibitory mechanism of cyclic piperidine and piperazine aryl ureas agrees with results directly obtained with experiments, and can be therefore used as a simple indication of the propensity of new urea-based compounds to act as a covalent inhibitor of FAAH. This is a representative and significant example of how MD and QM/MM methods, once prohibitive for practical drug design, have nowadays the potential to become a routinely used tool for drug design.

6. PERSPECTIVES

In this chapter, we presented few meaningful examples of computational modelling studies on pharmaceutically relevant targets. Our scope was to highlight how molecular modelling is helpful for basic understanding of the atomic-level interactions between a ligand (which could be either the endogenous substrate or a small molecule inhibitor), and its target protein. Examples spanned from quantum enzymology, where quantum mechanics was used to decipher specific metal-aided enzymatic mechanisms and related free energy balances in bacterial and viral metalloenzymes, to extensive molecular dynamics simulations of protein/drug interactions in multidrug efflux pumps, bacterial membranes and in the endocannabinoid-degrading enzyme FAAH. Given the vast applicability of these techniques, our list of examples is far from being exhaustive. Several other examples can be found in the literature (see for example [2, 3, 7, 10, 17, 18, 24-26, 108, 112]) and the interested reader is encouraged to delve into other reviews and books that might focus on the application of molecular modelling to pharmacologically relevant targets that concern the specific therapeutic area of interest.

A detailed comprehension of how the target protein works, and the individuation of key ligand-target interactions are key to more practical applications in structure-based drug design, ultimately helping the challenging process of drug discovery. Given the continuous progression of computer power and the improvement of algorithms for computations, molecular modelling will certainly spread its applications to more and more challenging questions and model systems. For example, the direct integration of molecular modelling with the increasingly broad range of experimental data (what is commonly called integrative modelling) has recently shown the potential to enhance our mechanistic understanding of biological [274]. Within this context, molecular simulations have reached nowadays a level of predictivity such that they can be used straightforwardly in parallel and/or in integration with the experiment, providing valuable synergic information within what can be called an "integrative dynamic modelling" framework [275]. For example, combination of MD studies and biochemical data led recently to the identification of a previously unknown enzymatic function for cellular-retinaldehyde-binding-protein CRALBP, a crucial retinoid transporter, whose missense mutations are associated to severe autosomal degenerative diseases of the retina [276, 277]. Moreover, the combination of low-resolution spatial data and MD simulations has been the key to have additional insights into macromolecular assembly, such as in the case of the *Yersinia enterocolitica* injectisome [278, 279], the pore-forming toxin aerolysin [275, 280], and the bacterial PhoP/PhoQ two-component regulatory system [281]. Extensive multi-scale simulations of these challenging model systems were able to identify new mechanisms of function, signing a first step in the direction of devising possible inhibition strategies.

The challenge is now to bridge the understanding of these fundamental mechanisms regulating relevant pathophysiological processes with the discovery of new drugs. It our view, computational insights on ligand-receptor interactions, and protein function will more and more impact the rational design of better inhibitors, as a promising starting point for drug discovery efforts.

ACKNOWLEDGEMENTS

MC acknowledges the support of the Norwegian Research Council through the CoE Centre for Theoretical and Computational Chemistry (CTCC) Grant Nos. 179568/V30 and 171185/V30. MDP acknowledges the support of the Swiss National Science Foundation (Grant Nos. 200020_138013, 200021_122120). MDV thanks the Italian Association for Cancer Research (AIRC) for the financial support through the grant MFAG n. 14140.

CONFLICT OF INTEREST

The authors confirm that this chapter contents have no conflict of interests.

REFERENCES

[1] Jorgensen, W. L. The many roles of computation in drug discovery. *Science* **2004**, *303*, 1813-1818.
[2] Cavalli, A.; Carloni, P.; Recanatini, M. Target-related applications of first principles quantum chemical methods in drug design. *Chem Rev* **2006**, *106*, 3497-3519.
[3] Lodola, A.; De Vivo, M. The increasing role of QM/MM in drug discovery. *Adv Protein Chem Struct Biol* **2012**, *87*, 337-362.
[4] Borhani, D. W.; Shaw, D. E. The future of molecular dynamics simulations in drug discovery. *J Comput Aided Mol Des* **2012**, *26*, 15-26.
[5] Colizzi, F.; Perozzo, R.; Scapozza, L.; Recanatini, M.; Cavalli, A. Single-molecule pulling simulations can discern active from inactive enzyme inhibitors. *J Am Chem Soc* **2010**, *132*, 7361-7371.
[6] Durrant, J. D.; McCammon, J. A. Molecular dynamics simulations and drug discovery. *BMC Biol* **2011**, *9*, 71.

[7] De Vivo, M. Bridging quantum mechanics and structure-based drug design. *Front Biosci* **2011**, *16*, 1619-1633.

[8] Peters, M. B.; Raha, K.; Merz, K. M. Quantum mechanics in structure-based drug design. *Curr Opin Drug Disc* **2006**, *9*, 370-379.

[9] Raha, K.; Peters, M. B.; Wang, B.; Yu, N.; WollaCott, A. M.; Westerhoff, L. M.; Merz, K. M. The role of quantum mechanics in structure-based drug design. *Drug Discov Today* **2007**, *12*, 725-731.

[10] Zhou, T.; Huang, D.; Caflisch, A. Quantum mechanical methods for drug design. *Curr Top Med Chem* **2010**, *10*, 33-45.

[11] Cavalli, A.; De Vivo, M.; Recanatini, M. Density functional study of the enzymatic reaction catalyzed by a cyclin-dependent kinase. *Chem Commun* **2003**, 1308-1309.

[12] Bernardi, F.; Bottoni, A.; De Vivo, M.; Garavelli, M.; Keseru, G.; Naray-Szabo, G. A hypothetical mechanism for HIV-1 integrase catalytic action: DFT modelling of a bio-mimetic environment. *Chem Phys Lett* **2002**, *362*, 1-7.

[13] Bottoni, A.; Miscione, G. P.; De Vivo, M. A theoretical DFT investigation of the lysozyme mechanism: computational evidence for a covalent intermediate pathway. *Proteins* **2005**, *59*, 118-130.

[14] Vummaleti, S. V. C.; Branduardi, D.; Masetti, M.; De Vivo, M.; Motterlini, R.; Cavalli, A. Theoretical insights into the mechanism of carbon monoxide (CO) release from CO-releasing molecules. *Chem-Eur J* **2012**, *18*, 9267-9275.

[15] Acevedo, O.; Jorgensen, W. L. Advances in quantum and molecular mechanical (QM/MM) simulations for organic and enzymatic reactions. *Acc Chem Res* **2010**, *43*, 142-151.

[16] Friesner, R. A.; Guallar, V. *Ab initio* quantum chemical and mixed quantum mechanics/molecular mechanics (QM/MM) methods for studying enzymatic catalysis. *Annu Rev Phys Chem* **2005**, *56*, 389-427.

[17] Senn, H. M.; Thiel, W. QM/MM methods for biomolecular systems. *Angew Chem-Int Edit* **2009**, *48*, 1198-1229.

[18] Warshel, A. Computer simulations of enzyme catalysis: methods, progress, and insights. *Annu Rev Biophys Biom* **2003**, *32*, 425-443.

[19] Field, M. J.; Bash, P. A.; Karplus, M. A combined quantum-mechanical and molecular mechanical potential for molecular-dynamics simulations. *J Comput Chem* **1990**, *11*, 700-733.

[20] van der Kamp, M. W.; Mulholland, A. J. Combined quantum mechanics/molecular mechanics (QM/MM) methods in computational Enzymology. *Biochemistry* **2013**, *52*, 2708-2728.

[21] Warshel, A.; Levitt, M. Theoretical studies of enzymatic reactions - dielectric, electrostatic and steric stabilization of carbonium-ion in reaction of Lysozyme. *J Mol Biol* **1976**, *103*, 227-249.

[22] Xu, M.; Lill, M. A. Induced fit docking, and the use of QM/MM methods in docking. *Drug Discov Today* **2013**, *10*, e411-418.

[23] Mucs, D.; Bryce, R. A. The application of quantum mechanics in structure-based drug design. *Expert Opin Drug Dis* **2013**, *8*, 263-276.

[24] Carloni, P.; Rothlisberger, U.; Parrinello, M. The role and perspective of a initio molecular dynamics in the study of biological systems. *Acc Chem Res* **2002**, *35*, 455-464.

[25] Dal Peraro, M.; Ruggerone, P.; Raugei, S.; Gervasio, F. L.; Carloni, P. Investigating biological systems using first principles Car-Parrinello molecular dynamics simulations. *Curr Opin Struct Biol* **2007**, *17*, 149-156.

[26] Mulholland, A. J. Modelling enzyme reaction mechanisms, specificity and catalysis. *Drug Discov Today* **2005**, *10*, 1393-1402.

[27] Lodola, A.; Mor, M.; Rivara, S.; Christov, C.; Tarzia, G.; Piomelli, D.; Mulholland, A. J. Identification of productive inhibitor binding orientation in fatty acid amide hydrolase (FAAH) by QM/MM mechanistic modelling. *Chem Commun* **2008**, 214-216.

[28] De Vivo, M.; Cavalli, A.; Carloni, P.; Recanatini, M. Computational study of the phosphoryl transfer catalyzed by a cyclin-dependent kinase. *Chemistry* **2007**, *13*, 8437-8444.

[29] Levitt, M. The birth of computational structural biology. *Nat Struct Biol* **2001**, *8*, 392-393.

[30] Jorgensen, W. L. Foundations of biomolecular modeling. *Cell* **2013**, *155*, 1199-1202.

[31] Dal Peraro, M.; Llarrull, L. I.; Rothlisberger, U.; Vila, A. J.; Carloni, P. Water-assisted reaction mechanism of monozinc beta-lactamases. *J Am Chem Soc* **2004**, *126*, 12661-12668.

[32] Dal Peraro, M.; Vila, A. J.; Carloni, P. Structural determinants and hydrogen-bond network of the mononuclear zinc(II)-beta-lactamase active site. *J Biol Inorg Chem* **2002**, *7*, 704-712.

[33] Dal Peraro, M.; Vila, A. J.; Carloni, P. Substrate binding to mononuclear metallo-beta-lactamase from Bacillus cereus. *Proteins* **2004**, *54*, 412-423.

[34] Dal Peraro, M.; Vila, A. J.; Carloni, P.; Klein, M. L. Role of Zinc content on the catalytic efficiency of B1 metallo beta-lactamases. *J Am Chem Soc* **2007**, *129*, 2808-2816.

[35] De Vivo, M.; Ensing, B.; Klein, M. L. Computational study of phosphatase activity in soluble epoxide hydrolase: High efficiency through a water bridge mediated proton shuttle. *J Am Chem Soc* **2005**, *127*, 11226-11227.

[36] Ho, M. H.; De Vivo, M.; Dal Peraro, M.; Klein, M. L. Understanding the effect of magnesium ion concentration on the catalytic activity of ribonuclease H through computation: does a third metal binding site modulate endonuclease catalysis? *J Am Chem Soc* **2010**, *132*, 13702-13712.

[37] Collu, F.; Vargiu, A. V.; Dreier, J.; Cascella, M.; Ruggerone, P. Recognition of imipenem and meropenem by the RND-transporter MexB studied by computer simulations. *J Am Chem Soc* **2012**, *134*, 19146-19158.

[38] Ravi, H. K.; Stach, M.; Soares, T. A.; Darbre, T.; Reymond, J. L.; Cascella, M. Electrostatics and flexibility drive membrane recognition and early penetration by the antimicrobial peptide dendrimer bH1. *Chem Commun* **2013**, *49*, 8821-8823.

[39] Palermo, G.; Branduardi, D.; Masetti, M.; Lodola, A.; Mor, M.; Piomelli, D.; Cavalli, A.; De Vivo, M. Covalent inhibitors of fatty acid amide hydrolase: a rationale for the activity of piperidine and piperazine aryl ureas. *J Med Chem* **2011**, *54*, 6612-6623.

[40] Palermo, G.; Campomanes, P.; Neri, M.; Piomelli, D.; Cavalli, A.; Rothlisberger, U.; De Vivo, M. Wagging the tail: essential role of substrate flexibility in FAAH catalysis. *J Chem Theory Comput* **2013**, *9*, 1202-1213.

[41] Leach, A. R. Molecular modelling: Principles and applications. *Pearson Education EMA, UK* **2001**.

[42] Helgaker, T. U.; Olsen, J.; Jorgensen, P. Molecular electronic-structure theory. *Wiley-Blackwell,* **2013**.

[43] Szabo, A.; Ostlund, N. S. Modern quantum chemistry. *Dover Publications, INC, New York* **1996**.

[44] Møller, C.; Plesset, M. S. Note on an approximation treatment for many-electron systems. *Phys. Rev.* **1934**, *46*, 618-622.

[45] Jensen, F. Introduction to computational chemistry. *John Wiley & Sons, UK* **1999**.

[46] Cramer, C. J. Essentials of computational chemistry. Theories and models, *John Wiley & Sons* **2004**.

[47] Garbuio, V.; Cascella, M.; Pulci, O. Excited state properties of liquid water. *J Phys-Condens Mat* **2009**, *21,* 033101.

[48] Rohrig, U. F.; Frank, I.; Hutter, J.; Laio, A.; VandeVondele, J.; Rothlisberger, U. QM/MM Car-Parrinello molecular dynamics study of the solvent effects on the ground state and on the first excited singlet state of acetone in water. *ChemPhysChem* **2003**, *4*, 1177-1182.

[49] Rohrig, U. F.; Guidoni, L.; Laio, A.; Frank, I.; Rothlisberger, U. A molecular spring for vision. *J Am Chem Soc* **2004**, *126*, 15328-15329.

[50] Strambi, A.; Coto, P. B.; Frutos, L. M.; Ferre, N.; Olivucci, M. Relationship between the excited state relaxation paths of Rhodopsin and Isorhodopsin. *J Am Chem Soc* **2008**, *130*, 3382-3388.

[51] Cascella, M.; Bärfuss, S.; Stocker, A. Cis-retinoids and the chemistry of vision. *Arch Biochem Biophys* **2013**, *539*, 187-195.

[52] Sobolewski, A. L.; Domcke, W.; Hattig, C. Tautomeric selectivity of the excited-state lifetime of guanine/cytosine base pairs: The role of electron-driven proton-transfer processes. *Proc Natl Acad Sci USA* **2005**, *102*, 17903-17906.

[53] Hohenberg, P.; Kohn, W. Inhomogeneous electron gas. *Phys Rev B* **1964**, *136*, 864-871.

[54] Kohn, W.; Sham, L. J. Self-consistent equations including exchange and correlation effects. *Physical Review A* **1965**, *140*, 1133-1138.

[55] Becke, A. D. Density-functional exchange-energy approximation with correct asymptotic-behavior. *Physical Review A* **1988**, *38*, 3098-3100.

[56] Tao, J.; Perdew, J. P.; Staroverov, V. N.; Scuseria, G. E. Climbing the density functional ladder: nonempirical meta-generalized gradient approximation designed for molecules and solids. *Phys Rev Lett* **2003**, *91*, 146401.

[57] Isegawa, M.; Peverati, R.; Truhlar, D. G. Performance of recent and high-performance approximate density functionals for time-dependent density functional theory calculations of valence and Rydberg electronic transition energies. *J Chem Phys* **2012**, *137, 129901.*

[58] Su, N. Q.; Adamo, C.; Xu, X. A comparison of geometric parameters from PBE-based doubly hybrid density functionals PBE0-DH, PBE0-2, and xDH-PBE0. *J Chem Phys* **2013**, *139, 174106.*

[59] Hohenstein, E. G.; Chill, S. T.; Sherrill, C. D. Assessment of the performance of the M05-2X and M06-2X Exchange-Correlation Functionals for Noncovalent interactions in biomolecules. *J Chem Theory Comput* **2008**, *4*, 1996-2000.

[60] Lu, Y. X.; Zou, J. W.; Fan, J. C.; Zhao, W. N.; Jiang, Y. J.; Yui, Q. S. *Ab initio* calculations on halogen-bonded complexes and comparison with density functional methods. *J Comput Chem* **2009**, *30*, 725-732.

[61] Zhao, Y.; Truhlar, D. G. Benchmark databases for nonbonded interactions and their use to test density functional theory. *J Chem Theory Comput* **2005**, *1*, 415-432.

[62] Bernasconi, M.; Chiarotti, G. L.; Focher, P.; Parrinello, M.; Tosatti, E. Solid-state polymerization of acetylene under pressure: *Ab initio* simulation. *Phys Rev Lett* **1997**, *78*, 2008-2011.

[63] Grimme, S. Accurate description of van der Waals complexes by density functional theory including empirical corrections. *J Comput Chem* **2004**, *25*, 1463-1473.

[64] Meijer, E. J.; Sprik, M. A density-functional study of the intermolecular interactions of benzene. *J Chem Phys* **1996**, *105*, 8684-8689.

[65] Williams, R. W.; Malhotra, D. van der Waals corrections to density functional theory calculations: Methane, ethane, ethylene, benzene, formaldehyde, ammonia, water, PBE, and CPMD. *Chem Phys* **2006**, *327*, 54-62.

[66] Tkatchenko, A.; Scheffler, M. Accurate molecular van Der Waals interactions from ground-state electron density and free-atom reference data. *Phys Rev Lett* **2009**, *102*, 073005.

[67] Becke, A. D.; Johnson, E. R. A density-functional model of the dispersion interaction. *J Chem Phys* **2005**, *123*, 154101.

[68] Becke, A. D.; Johnson, E. R. Exchange-hole dipole moment and the dispersion interaction. *J Chem Phys* **2005**, *122*, 154104.

[69] Johnson, E. R.; Becke, A. D. A post-Hartree-Fock model of intermolecular interactions: Inclusion of higher-order corrections. *J Chem Phys* **2006**, *124*, 174104.

[70] Steinmann, S. N.; Corminboeuf, C. A system-dependent density-based dispersion correction. *J Chem Theory Comput* **2010**, *6*, 1990-2001.

[71] Aeberhard, P. C.; Arey, J. S.; Lin, I. C.; Rothlisberger, U. Accurate DFT descriptions for weak interactions of molecules containing sulfur. *J Chem Theory Comput* **2009**, *5*, 23-28.

[72] Cascella, M.; Lin, I. C.; Tavernelli, I.; Rothlisberger, U. Dispersion corrected atom-centered potentials for phosphorus. *J Chem Theory Comput* **2009**, *5*, 2930-2934.

[73] Lin, I. C.; Coutinho-Neto, M. D.; Felsenheimer, C.; von Lilienfeld, O. A.; Tavernelli, I.; Rothlisberger, U. Library of dispersion-corrected atom-centered potentials for generalized gradient approximation functionals: Elements H, C, N, O, He, Ne, Ar, and Kr. *Phys Rev B* **2007**, *75*, 205131.

[74] Lin, I. C.; Rothlisberger, U. Describing weak interactions of biomolecules with dispersion-corrected density functional theory. *Phys Chem Chem Phys* **2008**, *10*, 2730-2734.

[75] Tavernelli, I.; Lin, I. C.; Rothlisberger, U. Multicenter-type corrections to standard DFT exchange and correlation functionals. *Phys Rev B* **2009**, *79*, 045106.

[76] von Lilienfeld, O. A.; Tavernelli, I.; Rothlisberger, U.; Sebastiani, D. Optimization of effective atom centered potentials for London dispersion forces in density functional theory. *Phys Rev Lett* **2004**, *93*, 153004.

[77] von Lilienfeld, O. A.; Tavernelli, I.; Rothlisberger, U.; Sebastiani, D. Performance of optimized atom-centered potentials for weakly bonded systems using density functional theory. *Phys Rev B* **2005**, *71*, 195119.

[78] Peverati, R.; Truhlar, D. G. Improving the accuracy of hybrid meta-GGA density functionals by range separation. *J Phys Chem Lett* **2011**, *2*, 2810-2817.

[79] Zhao, Y.; Schultz, N. E.; Truhlar, D. G. Design of density functionals by combining the method of constraint satisfaction with parametrization for thermochemistry, thermochemical kinetics, and noncovalent interactions. *J Chem Theory Comput* **2006**, *2*, 364-382.

[80] Zhao, Y.; Truhlar, D. G. Hybrid meta density functional theory methods for thermochemistry, thermochemical kinetics, and noncovalent interactions: The MPW1B95 and MPWB1K models and comparative assessments for hydrogenbonding and van der Waals interactions. *J Phys Chem A* **2004**, *108*, 6908-6918.

[81] Zhao, Y.; Truhlar, D. G. Exploring the limit of accuracy of the global hybrid meta density functional for main-group thermochemistry, kinetics, and noncovalent interactions. *J Chem Theory Comput* **2008**, *4*, 1849-1868.

[82] Simona, F.; Hai, N. T. M.; Broekmann, P.; Cascella, M. From structure to function: characterization of Cu(I) adducts in leveler additives by DFT calculations. *J Phys Chem Lett* **2011**, *2*, 3081-3084.

[83] Steinmann, S. N.; Corminboeuf, C. Comprehensive benchmarking of a density-dependent dispersion correction. *J Chem Theory Comput* **2011**, *7*, 3567-3577.

[84] Bayly, C. I.; Cieplak, P.; Cornell, W.; Kollman, P. A. A well-behaved electrostatic potential based method using charge restraints for deriving atomic charges: the RESP model. *J Phys Chem* **1993**, *97*, 10269-10280.

[85] Jorgensen, W. L.; Tirado-Rives, J. Potential energy functions for atomic-level simulations of water and organic and biomolecular systems. *Proc Natl Acad Sci USA* **2005**, *102*, 6665-6670.

[86] Jorgensen, W. L.; Maxwell, D. S.; Tirado-Rives, J. Development and testing of the OPLS all-atom force field on conformational energetics and properties of organic liquids. *J Am Chem Soc* **1996**, *118*, 11225-11236.

[87] Duan, Y.; Wu, C.; Chowdhury, S.; Lee, M. C.; Xiong, G.; Zhang, W.; Yang, R.; Cieplak, P.; Luo, R.; Lee, T.; Caldwell, J.; Wang, J.; Kollman, P. A point-charge force field for molecular mechanics simulations of proteins based on condensed-phase quantum mechanical calculations. *J Comput Chem* **2003**, *24*, 1999-2012.

[88] Christen, M.; Hunenberger, P. H.; Bakowies, D.; Baron, R.; Burgi, R.; Geerke, D. P.; Heinz, T. N.; Kastenholz, M. A.; Krautler, V.; Oostenbrink, C.; Peter, C.; Trzesniak, D.; van Gunsteren, W. F. The GROMOS software for biomolecular simulation: GROMOS05. *J Comput Chem* **2005**, *26*, 1719-1751.

[89] MacKerell, A. D.; Brooks, B.; Brooks, C. L.; Nilsson, L.; Roux, B.; Won, Y.; Karplus, M. CHARMM: The energy function and its parameterization with an overview of the program. *The Encyclopedia of Computational Chemistry. Ed P. v. R. S. e. al. John Wiley & Sons, Chichester* **1998**.

[90] Wang, J.; Wolf, R. M.; Caldwell, J. W.; Kollman, P. A.; Case, D. A. Development and testing of a general amber force field. *J Comput Chem* **2004**, *25*, 1157-1174.

[91] Vanommeslaeghe, K.; Hatcher, E.; Acharya, C.; Kundu, S.; Zhong, S.; Shim, J.; Darian, E.; Guvench, O.; Lopes, P.; Vorobyov, I.; Mackerell, A. D., Jr. CHARMM general force field: A force field for drug-like molecules compatible with the CHARMM all-atom additive biological force fields. *J Comput Chem***2010**, *31*, 671-690.

[92] De Vivo, M.; Cavalli, A.; Bottegoni, G.; Carloni, P.; Recanatini, M. Role of phosphorylated Thr160 for the activation of the CDK2/Cyclin a complex. *Proteins* **2006**, *62*, 89-98.

[93] De Vivo, M.; Bottegoni, G.; Berteotti, A.; Recanatini, M.; Gervasio, F. L.; Cavalli, A. Cyclin-dependent kinases: bridging their structure and function through computations. *Future Med Chem* **2011**, *3*, 1551-1559.

[94] Jorgensen, W. L.; Thomas, L. L. Perspective on free-energy perturbation calculations for chemical equilibria. *J Chem Theory Comput* **2008**, *4*, 869-876.

[95] Vreven, T.; Byun, K. S.; Komaromi, I.; Dapprich, S.; Montgomery, J. A.; Morokuma, K.; Frisch, M. J. Combining quantum mechanics methods with molecular mechanics methods in ONIOM. *J Chem Theory Comput* **2006**, *2*, 815-826.

[96] Frisch, M. J.; Trucks, G. W.; Schlegel, H. B.; Scuseria, G. E.; Robb, M. A.; Cheeseman, J. R.; Scalmani, G.; Barone, V.; Mennucci, B.; Petersson, G. A.; Nakatsuji, H.; Caricato, M.; Li, X.; Hratchian, H. P.; Izmaylov, A. F.; Bloino, J.; Zheng, G.; Sonnenberg, J. L.; Hada, M.; Ehara, M.; Toyota, K.; Fukuda, R.; Hasegawa, J.; Ishida, M.; Nakajima, T.; Honda, Y.; Kitao, O.; Nakai, H.; Vreven, T.; Montgomery Jr., J. A.; Peralta, J. E.; Ogliaro, F. o.; Bearpark, M. J.; Heyd, J.; Brothers, E. N.; Kudin, K. N.; Staroverov, V. N.; Kobayashi, R.; Normand, J.; Raghavachari, K.; Rendell, A. P.; Burant, J. C.; Iyengar, S. S.; Tomasi, J.; Cossi, M.; Rega, N.; Millam, N. J.; Klene, M.; Knox, J. E.; Cross, J. B.; Bakken, V.; Adamo, C.; Jaramillo, J.; Gomperts, R.; Stratmann, R. E.; Yazyev, O.; Austin, A. J.; Cammi, R.; Pomelli, C.; Ochterski, J. W.; Martin, R. L.; Morokuma, K.; Zakrzewski, V. G.; Voth, G. A.; Salvador, P.;

Dannenberg, J. J.; Dapprich, S.; Daniels, A. D.; Farkas, Ö.; Foresman, J. B.; Ortiz, J. V.; Cioslowski, J.; Fox, D. J. *Gaussian, Inc., Wallingford CT, 2009.*

[97] Laio, A.; VandeVondele, J.; Rothlisberger, U. D-RESP: Dynamically generated electrostatic potential derived charges from quantum mechanics/molecular mechanics simulations. *J Phys Chem B* **2002**, *106*, 7300-7307.

[98] Laio, A.; VandeVondele, J.; Rothlisberger, U. A Hamiltonian electrostatic coupling scheme for hybrid Car-Parrinello molecular dynamics simulations. *J Chem Phys* **2002**, *116*, 6941-6947.

[99] Laino, T.; Mohamed, F.; Laio, A.; Parrinello, M. An efficient real space multigrid OM/MM electrostatic coupling. *J Chem Theory Comput* **2005**, *1*, 1176-1184.

[100] Seabra, G. D.; Walker, R. C.; Elstner, M.; Case, D. A.; Roitberg, A. E. Implementation of the SCC-DFTB method for hybrid QM/MM simulations within the amber molecular dynamics package. *J Phys Chem A* **2007**, *111*, 5655-5664.

[101] Walker, R. C.; Crowley, M. F.; Case, D. A. The implementation of a fast and accurate QM/MM potential method in Amber. *J Comput Chem* **2008**, *29*, 1019-1031.

[102] Woodcock, H. L.; Hodoscek, M.; Gilbert, A. T. B.; Gill, P. M. W.; Schaefer, H. F.; Brooks, B. R. Interfacing Q-chem and CHARMM to perform QM/MM reaction path calculations. *J Comput Chem* **2007**, *28*, 1485-1502.

[103] Derat, E.; Bouquant, J.; Humbel, S. On the link atom distance in the ONIOM scheme. An harmonic approximation analysis. *J Mol Struc-Theochem* **2003**, *632*, 61-69.

[104] Ferre, N.; Olivucci, M. The amide bond: pitfalls and drawbacks of the link atom scheme. *J Mol Struc-Theochem* **2003**, *632*, 71-82.

[105] Gao, J. L.; Amara, P.; Alhambra, C.; Field, M. J. A generalized hybrid orbital (GHO) method for the treatment of boundary atoms in combined QM/MM calculations. *J Phys Chem A* **1998**, *102*, 4714-4721.

[106] Pu, J. Z.; Gao, J. L.; Truhlar, D. G. Generalized hybrid orbital (GHO) method for combining *ab initio* Hartree-Fock wave functions with molecular mechanics. *J Phys Chem A* **2004**, *108*, 632-650.

[107] Cascella, M.; Cuendet, M. A.; Tavernelli, I.; Rothlisberger, U. Optical spectra of Cu(II)-Azurin by hybrid TDDFT-Molecular dynamics simulations. *J Phys Chem B* **2007**, *111*, 10248-10252.

[108] Friesner, R. A. Combined quantum and molecular mechanics (QM/MM). *Drug Discov Today* **2004**, *1*, 253-260.

[109] Gleeson, M. P.; Gleeson, D. QM/MM Calculations in drug discovery: A useful method for studying binding phenomena? *J Chem Inf Model* **2009**, *49*, 670-677.

[110] Gleeson, M. P.; Gleeson, D. QM/MM as a tool in fragment based drug discovery. A cross-docking, rescoring study of kinase inhibitors. *J Chem Inf Model* **2009**, *49*, 1437-1448.

[111] Claeyssens, F.; Harvey, J. N.; Manby, F. R.; Mata, R. A.; Mulholland, A. J.; Ranaghan, K. E.; Schutz, M.; Thiel, S.; Thiel, W.; Werner, H. J. High-accuracy computation of reaction barriers in enzymes. *Angew Chem-Int Ed* **2006**, *45*, 6856-6859.

[112] Mulholland, A. J. Computational enzymology: modelling the mechanisms of biological catalysts. *Biochem Soc Trans* **2008**, *36*, 22-26.

[113] Rosta, E.; Klahn, M.; Warshel, A. Towards accurate *ab initio* QM/MM calculations of free-energy profiles of enzymatic reactions. *J Phys Chem B* **2006**, *110*, 2934-2941.

[114] Ho, M. H.; De Vivo, M.; Dal Peraro, M.; Klein, M. L. Unraveling the catalytic pathway of metalloenzyme farnesyltransferase through QM/MM computation. *J Chem Theory Comput* **2009**, *5*, 1657-1666.

[115] Cho, A. E.; Guallar, V.; Berne, B. J.; Friesner, R. Importance of accurate charges in molecular docking: Quantum mechanical/molecular mechanical (QM/MM) approach. *J Comput Chem* **2005**, *26*, 915-931.

[116] Khandelwal, A.; Lukacova, V.; Comez, D.; Kroll, D. M.; Raha, S.; Balaz, S. A combination of docking, QM/MM methods, and MD simulation for binding affinity estimation of metalloprotein ligands. *J Med Chem* **2005**, *48*, 5437-5447.

[117] Sander, T.; Lijefors, T.; Balle, T. Prediction of the receptor conformation for iGluR2 agonist binding: QM/MM docking to an extensive conformational ensemble generated using normal mode analysis. *J Mol Graph Model* **2008**, *26*, 1259-1268.

[118] Kaukonen, M.; Soderhjelm, P.; Heimdal, J.; Ryde, U. QM/MM-PBSA method to estimate free energies for reactions in proteins. *J Phys Chem B* **2008**, *112*, 12537-12548.

[119] Khandelwal, A.; Balaz, S. QM/MM linear response method distinguishes ligand affinities for closely related metalloproteins. *Proteins* **2007**, *69*, 326-339.

[120] Jorgensen, W. L. Efficient Drug lead discovery and optimization. *Acc Chem Res* **2009**, *42*, 724-733.

[121] Gullingsrud, J. R.; Braun, R.; Schulten, K. Reconstructing potentials of mean force through time series analysis of steered molecular dynamics simulations. *J Comput Phys* **1999**, *151*, 190-211.

[122] Jarzynski, C. Equilibrium free-energy differences from nonequilibrium measurements: A master-equation approach. *Phys Rev E* **1997**, *56*, 5018-5035.

[123] Carter, E. A.; Ciccotti, G.; Hynes, J. T.; Kapral, R. Constrained reaction coordinate dynamics for the simulation of rare events. *Chem Phys Lett* **1989**, *156*, 472-477.

[124] Major, D. T.; Gao, J. L. An integrated path integral and free-energy perturbation-umbrella sampling method for computing kinetic isotope effects of chemical reactions in solution and in enzymes. *J Chem Theory Comput* **2007**, *3*, 949-960.

[125] Marsili, S.; Barducci, A.; Chelli, R.; Procacci, P.; Schettino, V. Self-healing umbrella sampling: A non-equilibrium approach for quantitative free energy calculations. *J Phys Chem B* **2006**, *110*, 14011-14013.

[126] Torrie, G. M.; Valleau, J. P. Non-physical sampling distributions in Monte-Carlo free-energy estimation - umbrella sampling. *J Comput Phys* **1977**, *23*, 187-199.

[127] Sugita, Y.; Okamoto, Y. Replica-exchange molecular dynamics method for protein folding. *Chem Phys Lett* **1999**, *314*, 141-151.

[128] Grubmuller, H. Predicting slow structural transitions in macromolecular systems - conformational flooding. *Phys Rev E* **1995**, *52*, 2893-2906.

[129] Laio, A.; Parrinello, M. Escaping free-energy minima. *Proc Natl Acad Sci USA* **2002**, *99*, 12562-12566.

[130] Barducci, A.; Bussi, G.; Parrinello, M. Well-tempered metadynamics: A smoothly converging and tunable free-energy method. *Phys Rev Lett* **2008**, *100*.

[131] Ensing, B.; Laio, A.; Parrinello, M.; Klein, M. L. A recipe for the computation of the free energy barrier and the lowest free energy path of concerted reactions. *J Phys Chem B* **2005**, *109*, 6676-6687.

[132] Ensing, B.; De Vivo, M.; Liu, Z. W.; Moore, P.; Klein, M. L. Metadynamics as a tool for exploring free energy landscapes of chemical reactions. *Acc Chem Res* **2006**, *39*, 73-81.

[133] Cascella, M.; Micheletti, C.; Rothlisberger, U.; Carloni, P. Evolutionarily conserved functional mechanics across pepsin-like and retroviral aspartic proteases. *J Am Chem Soc* **2005**, *127*, 3734-3742.

[134] De Vivo, M.; Dal Peraro, M.; Klein, M. L. Phosphodiester cleavage in ribonuclease H occurs *via* an associative two-metal-aided catalytic mechanism. *J Am Chem Soc* **2008**, *130*, 10955-10962.

[135] De Vivo, M.; Ensing, B.; Dal Peraro, M.; Gomez, G. A.; Christianson, D. W.; Klein, M. L. Proton shuttles and phosphatase activity in soluble epoxide hydrolase. *J Am Chem Soc* **2007**, *129*, 387-394.

[136] Petersen, L.; Ardevol, A.; Rovira, C.; Reilly, P. J. Mechanism of cellulose hydrolysis by inverting GH8 endoglucanases: A QM/MM metadynamics study. *J Phys Chem B* **2009**, *113*, 7331-7339.

[137] Petersen, L.; Ardevol, A.; Rovira, C.; Reilly, P. J. Molecular mechanism of the glycosylation step catalyzed by Golgi alpha-mannosidase II: A QM/MM metadynamics investigation. *J Am Chem Soc* **2010**, *132*, 8291-8300.

[138] Raugei, S.; Cascella, M.; Carloni, P. A proficient enzyme: Insights on the mechanism of orotidine monophosphate decarboxylase from computer simulations. *J Am Chem Soc* **2004**, *126*, 15730-15737.

[139] Sulpizi, M.; Laio, A.; VandeVondele, J.; Cattaneo, A.; Rothlisberger, U.; Carloni, P. Reaction mechanism of caspases: Insights from QM/MM Car-Parrinello simulations. *Proteins* **2003**, *52*, 212-224.

[140] Lodola, A.; Branduardi, D.; De Vivo, M.; Capoferri, L.; Mor, M.; Piomelli, D.; Cavalli, A. A catalytic mechanism for cysteine N-Terminal nucleophile hydrolases, as revealed by free energy simulations. *PLoS One* **2012**, *7*, e32397.

[141] Hou, T. J.; Wang, J. M.; Li, Y. Y.; Wang, W. Assessing the performance of the MM/PBSA and MM/GBSA methods. 1. The accuracy of binding free energy calculations based on molecular dynamics simulations. *J Chem Inf Model* **2011**, *51*, 69-82.

[142] Kollman, P. A.; Massova, I.; Reyes, C.; Kuhn, B.; Huo, S. H.; Chong, L.; Lee, M.; Lee, T.; Duan, Y.; Wang, W.; Donini, O.; Cieplak, P.; Srinivasan, J.; Case, D. A.; Cheatham, T. E. Calculating structures and free energies of complex molecules: Combining molecular mechanics and continuum models. *Acc Chem Res* **2000**, *33*, 889-897.

[143] Kuhn, B.; Gerber, P.; Schulz-Gasch, T.; Stahl, M. Validation and use of the MM-PBSA approach for drug discovery. *J Med Chem* **2005**, *48*, 4040-4048.

[144] Kuhn, B.; Kollman, P. A. Binding of a diverse set of ligands to avidin and streptavidin: An accurate quantitative prediction of their relative affinities by a combination of molecular mechanics and continuum solvent models. *J Med Chem* **2000**, *43*, 3786-3791.

[145] Karplus, M.; Kushick, J. N. Method for estimating the configurational entropy of macromolecules. *Macromolecules* **1981**, *14*, 325-332.

[146] Andricioaei, I.; Karplus, M. On the calculation of entropy from covariance matrices of the atomic fluctuations. *J Chem Phys* **2001**, *115*, 6289-6292.

[147] Baron, R.; Hunenberger, P. H.; McCammon, J. A. Absolute single-molecule entropies from quasi-harmonic analysis of microsecond molecular dynamics: correction terms and convergence properties. *J Chem Theory Comput* **2009**, *5*, 3150-3160.

[148] Baron, R.; van Gunsteren, W. F.; Hünenberger, P. H. Estimating the configurational entropy from molecular dynamics simulations: Anharmonicity and correlation corrections to the quasi-harmonic approximation. *Trends Phys Chem* **2006**, *11*, 87-122.

[149] Carlsson, J.; Aqvist, J. Absolute and relative entropies from computer simulation with applications to ligand binding. *J Phys Chem B* **2005**, *109*, 6448-6456.

[150] Chang, C. E.; Chen, W.; Gilson, M. K. Evaluating the accuracy of the quasiharmonic approximation. *J Chem Theory Comput* **2005**, *1*, 1017-1028.

[151] Karplus, M.; McCammon, J. A. Molecular dynamics simulations of biomolecules. *Nat Struct Biol* **2002**, *9*, 646-652.

[152] Klein, M. L.; Shinoda, W. Large-scale molecular dynamics simulations of self-assembling systems. *Science* **2008**, *321*, 798-800.
[153] Zhao, G. P.; Perilla, J. R.; Yufenyuy, E. L.; Meng, X.; Chen, B.; Ning, J. Y.; Ahn, J.; Gronenborn, A. M.; Schulten, K.; Aiken, C.; Zhang, P. J. Mature HIV-1 capsid structure by cryo-electron microscopy and all-atom molecular dynamics. *Nature* **2013**, *497*, 643-646.
[154] Go, N.; Taketomi, H. Respective roles of short-range and long-range interactions in protein folding. *Proc Natl Acad Sci USA* **1978**, *75*, 559-563.
[155] Levitt, M.; Warshel, A. computer-simulation of protein folding. *Nature* **1975**, *253*, 694-698.
[156] Marrink, S. J.; Risselada, H. J.; Yefimov, S.; Tieleman, D. P.; de Vries, A. H. The MARTINI force field: Coarse grained model for biomolecular simulations. *J Phys Chem B* **2007**, *111*, 7812-7824.
[157] Monticelli, L.; Kandasamy, S. K.; Periole, X.; Larson, R. G.; Tieleman, D. P.; Marrink, S. J. The MARTINI coarse-grained force field: Extension to proteins. *J Chem Theory Comput* **2008**, *4*, 819-834.
[158] Noid, W. G.; Chu, J. W.; Ayton, G. S.; Krishna, V.; Izvekov, S.; Voth, G. A.; Das, A.; Andersen, H. C. The multiscale coarse-graining method. I. A rigorous bridge between atomistic and coarse-grained models. *J Chem Phys* **2008**, *128*.
[159] Noid, W. G.; Liu, P.; Wang, Y.; Chu, J. W.; Ayton, G. S.; Izvekov, S.; Andersen, H. C.; Voth, G. A. The multiscale coarse-graining method. II. Numerical implementation for coarse-grained molecular models. *J Chem Phys* **2008**, *128*.
[160] Shelley, J. C.; Shelley, M. Y.; Reeder, R. C.; Bandyopadhyay, S.; Moore, P. B.; Klein, M. L. Simulations of phospholipids using a coarse grain model. *J Phys Chem B* **2001**, *105*, 9785-9792.
[161] Villa, E.; Balaeff, A.; Schulten, K. Structural dynamics of the lac repressor-DNA complex revealed by a multiscale simulation. *Proc Natl Acad Sci USA* **2005**, *102*, 6783-6788.
[162] Zhou, J.; Thorpe, I. F.; Izvekov, S.; Voth, G. A. Coarse-grained peptide modeling using a systematic multiscale approach. *Biophys J* **2007**, *92*, 4289-4303.
[163] Ayton, G. S.; Noid, W. G.; Voth, G. A. Multiscale modeling of biomolecular systems: in serial and in parallel. *Curr Opin Struc Biol* **2007**, *17*, 192-198.
[164] Bond, P. J.; Wee, C. L.; Sansom, M. S. P. Coarse-grained molecular dynamics simulations of the energetics of helix insertion into a lipid bilayer. *Biochemistry* **2008**, *47*, 11321-11331.
[165] Carpenter, T.; Bond, P. J.; Khalid, S.; Sansom, M. S. P. Self-assembly of a simple membrane protein: Coarse-grained molecular dynamics simulations of the influenza M2 channel. *Biophys J* **2008**, *95*, 3790-3801.
[166] Chu, J. W.; Voth, G. A. Allostery of actin filaments: Molecular dynamics simulations and coarse-grained analysis. *Proc Natl Acad Sci USA* **2005**, *102*, 13111-13116.
[167] Thogersen, L.; Schiott, B.; Vosegaard, T.; Nielsen, N. C.; Tajkhorshid, E. Peptide aggregation and pore formation in a lipid bilayer: A combined coarse-grained and all atom molecular dynamics study. *Biophys J* **2008**, *95*, 4337-4347.
[168] Tozzini, V. Coarse-grained models for proteins. *Curr Opin Struc Biol* **2005**, *15*, 144-150.
[169] Treptow, W.; Marrink, S. J.; Tarek, M. Gating motions in voltage-gated potassium channels revealed by coarse-grained molecular dynamics simulations. *J Phys Chem B* **2008**, *112*, 3277-3282.
[170] Yefimov, S.; van der Giessen, E.; Onck, P. R.; Marrink, S. J. Mechanosensitive membrane channels in action. *Biophys J* **2008**, *94*, 2994-3002.
[171] Arkhipov, A.; Yin, Y.; Schulten, K. Four-scale description of membrane sculpting by BAR domains. *Biophys J* **2008**, *95*, 2806-2821.

[172] Delle Site, L.; Abrams, C. F.; Alavi, A.; Kremer, K. Polymers near metal surfaces: Selective adsorption and global conformations. *Phys Rev Lett* **2002**, *89*.

[173] Ensing, B.; Nielsen, S. O.; Moore, P. B.; Klein, M. L.; Parrinello, M. Energy conservation in adaptive hybrid atomistic/coarse-grain molecular dynamics. *J Chem Theory Comput* **2007**, *3*, 1100-1105.

[174] Heyden, A.; Truhlar, D. G. Conservative algorithm for an adaptive change of resolution in mixed atomistic/coarse-grained multiscale simulations. *J Chem Theory Comput* **2008**, *4*, 217-221.

[175] Izvekov, S.; Parrinello, M.; Burnham, C. J.; Voth, G. A. Effective force fields for condensed phase systems from *ab initio* molecular dynamics simulation: A new method for force-matching. *J Chem Phys* **2004**, *120*, 10896-10913.

[176] Lyman, E.; Ytreberg, F. M.; Zuckerman, D. M. Resolution exchange simulation. *Phys Rev Lett* **2006**, *96*.

[177] Neri, M.; Anselmi, C.; Cascella, M.; Maritan, A.; Carloni, P. Coarse-grained model of proteins incorporating atomistic detail of the active site. *Phys Rev Lett* **2005**, *95*.

[178] Praprotnik, M.; Delle Site, L.; Kremer, K. Multiscale simulation of soft matter: From scale bridging to adaptive resolution. *AnnuRev Phys Chem* **2008**, *59*, 545-571.

[179] Sherwood, P.; Brooks, B. R.; Sansom, M. S. P. Multiscale methods for macromolecular simulations. *Curr Opin Struc Biol* **2008**, *18*, 630-640.

[180] Shi, Q.; Izvekov, S.; Voth, G. A. Mixed atomistic and coarse-grained molecular dynamics: Simulation of a membrane-bound ion channel. *J Phys Chem B* **2006**, *110*, 15045-15048.

[181] Villa, E.; Balaeff, A.; Mahadevan, L.; Schulten, K. Multiscale method for simulating protein-DNA complexes. *Multiscale Model Sim* **2004**, *2*, 527-553.

[182] Liu, P.; Shi, Q.; Lyman, E.; Voth, G. A. Reconstructing atomistic detail for coarse-grained models with resolution exchange. *J Chem Phys* **2008**, *129,* 114103.

[183] Liu, P.; Voth, G. A. Smart resolution replica exchange: An efficient algorithm for exploring complex energy landscapes. *J Chem Phys* **2007**, *126*, 045106

[184] Cascella, M.; Neri, M. A.; Carloni, P.; Dal Peraro, M. Topologically based multipolar reconstruction of electrostatic interactions in multiscale simulations of proteins. *J Chem Theory Comput* **2008**, *4*, 1378-1385.

[185] Alemani, D.; Collu, F.; Cascella, M.; Dal Peraro, M. A nonradial coarse-grained potential for proteins Produces naturally sable secondary structure elements. *J Chem Theory Comput* **2010**, *6*, 315-324.

[186] Spiga, E.; Alemani, D.; Degiacomi, M. T.; Cascella, M.; Dal Peraro, M. Electrostatic-consistent coarse-grained potentials for molecular simulations of proteins. *J Chem Theory Comput* **2013**, *9*, 3515-3526.

[187] Dupureur, C. M. Roles of metal ions in nucleases. *Curr Opin Chem Biol* **2008**, *12*, 250-255.

[188] Yang, W.; Lee, J. Y.; Nowotny, M. Making and breaking nucleic acids: two-Mg2+-ion catalysis and substrate specificity. *Mol Cell* **2006**, *22*, 5-13.

[189] Broccoli, S.; Rallu, F.; Sanscartier, P.; Cerritelli, S. M.; Crouch, R. J.; Drolet, M. Effects of RNA polymerase modifications on transcription-induced negative supercoiling and associated R-loop formation. *Mol Microbiol* **2004**, *52*, 1769-1779.

[190] Klumpp, K.; Hang, J. Q.; Rajendran, S.; Yang, Y.; Derosier, A.; Wong Kai In, P.; Overton, H.; Parkes, K. E.; Cammack, N.; Martin, J. A. Two-metal ion mechanism of RNA cleavage by HIV RNase H and mechanism-based design of selective HIV RNase H inhibitors. *Nucleic Acids Res* **2003**, *31*, 6852-6859.

[191] Bailly, C. Contemporary challenges in the design of topoisomerase II inhibitors for cancer chemotherapy. *Chem Rev***2012**, *112*, 3611-3640.

[192] Nowotny, M.; Gaidamakov, S. A.; Crouch, R. J.; Yang, W. Crystal structures of RNase H bound to an RNA/DNA hybrid: substrate specificity and metal-dependent catalysis. *Cell* **2005**, *121*, 1005-1016.

[193] Nowotny, M.; Yang, W. Stepwise analyses of metal ions in RNase H catalysis from substrate destabilization to product release. *Embo J* **2006**, *25*, 1924-1933.

[194] Sissi, C.; Palumbo, M. Effects of magnesium and related divalent metal ions in topoisomerase structure and function. *Nucleic Acids Res* **2009**, *37*, 702-711.

[195] Rosta, E.; Yang, W.; Hummer, G. Calcium inhibition of ribonuclease H1 two-metal ion catalysis. *J Am Chem Soc***2014**, *136*, 3137-3144.

[196] Mordasini, T.; Curioni, A.; Andreoni, W. Why do divalent metal ions either promote or inhibit enzymatic reactions? The case of BamHI restriction endonuclease from combined quantum-classical simulations. *J Biol Chem* **2003**, *278*, 4381-4384.

[197] Shaw-Reid, C. A.; Feuston, B.; Munshi, V.; Getty, K.; Krueger, J.; Hazuda, D. J.; Parniak, M. A.; Miller, M. D.; Lewis, D. Dissecting the effects of DNA polymerase and ribonuclease H inhibitor combinations on HIV-1 reverse-transcriptase activities. *Biochemistry* **2005**, *44*, 1595-1606.

[198] Branduardi, D.; De Vivo, M.; Rega, N.; Barone, V.; Cavalli, A. Methylphosphate dianion hydrolysis in solution characterized by path collective variables coupled with DFT-based enhanced sampling simulations. *J Chem Theory Comput* **2011**, *7*, 539-543.

[199] Elsasser, B.; Valiev, M.; Weare, J. H. A Dianionic Phosphorane intermediate and transition states in an associative A(N)+D-N mechanism for the RibonucleaseA hydrolysis reaction. *J Am Chem Soc* **2009**, *131*, 3869-3871.

[200] Rosta, E.; Nowotny, M.; Yang, W.; Hummer, G. Catalytic mechanism of RNA backbone cleavage by ribonuclease H from quantum mechanics/molecular mechanics simulations. *J Am Chem Soc* **2011**, *133*, 8934-8941.

[201] Rosta, E.; Woodcock, H. L.; Brooks, B. R.; Hummer, G. Artificial reaction coordinate "tunneling" in free-energy calculations: the catalytic reaction of RNase H. *J Comput Chem* **2009**, *30*, 1634-1641.

[202] Moghaddam, M. F.; Grant, D. F.; Cheek, J. M.; Greene, J. F.; Williamson, K. C.; Hammock, B. D. Bioactivation of leukotoxins to their toxic diols by epoxide hydrolase. *Nature Medicine* **1997**, *3*, 562-566.

[203] Node, K.; Huo, Y. Q.; Ruan, X. L.; Yang, B. C.; Spiecker, M.; Ley, K.; Zeldin, D. C.; Liao, J. K. Anti-inflammatory properties of cytochrome P450 epoxygenase-derived eicosanoids. *Science* **1999**, *285*, 1276-1279.

[204] Schmelzer, K. R.; Kubala, L.; Newman, J. W.; Kim, I. H.; Eiserich, J. P.; Hammock, B. D. Soluble epoxide hydrolase is a therapeutic target for acute inflammation. *Proc Natl Acad Sci USA* **2005**, *102*, 9772-9777.

[205] Cronin, a.; Mowbray, S.; Durk, H.; Homburg, S.; Fleming, I.; Fisslthaler, B.; Oesch, F.; Arand, M. The N-terminal domain of mammalian soluble epoxide hydrolase is a phosphatase. *Proc Natl Acad Sci USA* **2003**, *100*, 1552-1557.

[206] Gomez, G. A.; Morisseau, C.; Hammock, B. D.; Christianson, D. W. Structure of human epoxide hydrolase reveals mechanistic inferences on bifunctional catalysis in epoxide and phosphate ester hydrolysis. *Biochemistry* **2004**, *43*, 4716-4723.

[207] Newman, J. W.; Morisseau, C.; Hammock, B. D. Epoxide hydrolases: their roles and interactions with lipid metabolism. *Prog Lipid Res* **2005**, *44*, 1-51.

[208] Palermo, G.; Stenta, M.; Cavalli, A.; Dal Peraro, M.; De Vivo, M. Molecular simulations highlight the role of metals in catalysis and inhibition of type II topoisomerase. *J Chem Theory Comput* **2013**, *9*, 857-862.

[209] Schmidt, B. H.; Burgin, A. B.; Deweese, J. E.; Osheroff, N.; Berger, J. M. A novel and unified two-metal mechanism for DNA cleavage by type II and IA topoisomerases. *Nature* **2010**, *465*, 641-644.

[210] Fisher, J. F.; Meroueh, S. O.; Mobashery, S. Bacterial resistance to beta-lactam antibiotics: Compelling opportunism, compelling opportunity. *Chem Rev* **2005**, *105*, 395-424.

[211] Walsh, T. R.; Toleman, M. A.; Poirel, L.; Nordmann, P. Metallo-beta-lactamases: the quiet before the storm? *Clin Microbiol Rev* **2005**, *18*, 306-325.

[212] Dal Peraro, M.; Spiegel, K.; Lamoureux, G.; De Vivo, M.; DeGrado, W. F.; Klein, M. L. Modeling the charge distribution at metal sites in proteins for molecular dynamics simulations. *J Struct Biol* **2007**, *157*, 444-453.

[213] Dal Peraro, M.; Vila, A. J.; Carloni, P. Protonation state of Asp120 in the binuclear active site of the metallo-beta-lactamase from *Bacteroides fragilis*. *Inorg Chem* **2003**, *42*, 4245-4247.

[214] de Lencastre, H.; Oliveira, D.; Tomasz, A. Antibiotic resistant *Staphylococcus aureus:* a paradigm of adaptive power. *Curr Opin Microbiol* **2007**, *10*, 428-435.

[215] Livermore, D. M. The need for new antibiotics. *Clin Microbiol Infect* **2004**, *10 Suppl 4*, 1-9.

[216] Bandow, J. E.; Metzler-Nolte, N. New ways of killing the beast: prospects for inorganic-organic hybrid nanomaterials as antibacterial agents. *ChemBioChem* **2009**, *10*, 2847-2850.

[217] Wenzel, M.; Kohl, B.; Munch, D.; Raatschen, N.; Albada, H. B.; Hamoen, L.; Metzler-Nolte, N.; Sahl, H. G.; Bandow, J. E. Proteomic response of *Bacillus subtilis* to lantibiotics reflects differences in interaction with the cytoplasmic membrane. *Antimicrob Agents Chemother* **2012**, *56*, 5749-5757.

[218] Schinkel, A. H.; Wagenaar, E.; Mol, C. A.; van Deemter, L. P-glycoprotein in the blood-brain barrier of mice influences the brain penetration and pharmacological activity of many drugs. *J Clin Invest* **1996**, *97*, 2517-2524.

[219] Saier, M. H., Jr.; Paulsen, I. T. Phylogeny of multidrug transporters. *Semin Cell Dev Biol* **2001**, *12*, 205-213.

[220] Nikaido, H. Multidrug efflux pumps of gram-negative bacteria. *J Bacteriol* **1996**, *178*, 5853-5859.

[221] Li, X. Z.; Nikaido, H. Efflux-mediated drug resistance in bacteria. *Drugs* **2004**, *64*, 159-204.

[222] Paulsen, I. T.; Brown, M. H.; Skurray, R. A. Proton-dependent multidrug efflux systems. *Microbiol Rev* **1996**, *60*, 575-&.

[223] Collu, F.; Cascella, M. Multidrug resistance and efflux pumps: insights from molecular dynamics simulations. *Curr Top Med Chem* **2013**, *13*, 3165-3183.

[224] Vaara, M. Antibiotic-supersusceptible mutants of *Escherichia coli* and *Salmonella typhimurium*. *Antimicrob Agents Chemother* **1993**, *37*, 2255-2260.

[225] Paulsen, I. T.; Park, J. H.; Choi, P. S.; Saier, M. H., Jr. A family of gram-negative bacterial outer membrane factors that function in the export of proteins, carbohydrates, drugs and heavy metals from gram-negative bacteria. *FEMS microbiol lett* **1997**, *156*, 1-8.

[226] Dinh, T.; Paulsen, I. T.; Saier, M. H., Jr. A family of extracytoplasmic proteins that allow transport of large molecules across the outer membranes of gram-negative bacteria. *J Bacteriol* **1994**, *176*, 3825-3831.

[227] Murakami, S.; Nakashima, R.; Yamashita, E.; Yamaguchi, A. Crystal structure of bacterial multidrug efflux transporter AcrB. *Nature* **2002**, *419*, 587-593.

[228] Mikolosko, J.; Bobyk, K.; Zgurskaya, H. I.; Ghosh, P. Conformational flexibility in the multidrug efflux system protein AcrA. *Structure* **2006**, *14*, 577-587.

[229] Koronakis, V.; Sharff, A.; Koronakis, E.; Luisi, B.; Hughes, C. Crystal structure of the bacterial membrane protein TolC central to multidrug efflux and protein export. *Nature* **2000**, *405*, 914-919.

[230] Akama, H.; Kanemaki, M.; Yoshimura, M.; Tsukihara, T.; Kashiwagi, T.; Yoneyama, H.; Narita, S.; Nakagawa, A.; Nakae, T. Crystal structure of the drug discharge outer membrane protein, OprM, of *Pseudomonas aeruginosa:* dual modes of membrane anchoring and occluded cavity end. *J Biol Chem* **2004**, *279*, 52816-52819.

[231] Akama, H.; Matsuura, T.; Kashiwagi, S.; Yoneyama, H.; Narita, S.; Tsukihara, T.; Nakagawa, A.; Nakae, T. Crystal structure of the membrane fusion protein, MexA, of the multidrug transporter in *Pseudomonas aeruginosa*. *J Biol Chem* **2004**, *279*, 25939-25942.

[232] Higgins, M. K.; Bokma, E.; Koronakis, E.; Hughes, C.; Koronakis, V. Structure of the periplasmic component of a bacterial drug efflux pump. *Proc Natl Acad Sci USA* **2004**, *101*, 9994-9999.

[233] Nakashima, R.; Sakurai, K.; Yamasaki, S.; Hayashi, K.; Nagata, C.; Hoshino, K.; Onodera, Y.; Nishino, K.; Yamaguchi, A. Structural basis for the inhibition of bacterial multidrug exporters. *Nature* **2013**, *500*, 102-106.

[234] Phan, G.; Benabdelhak, H.; Lascombe, M. B.; Benas, P.; Rety, S.; Picard, M.; Ducruix, A.; *Etc*hebest, C.; Broutin, I. Structural and dynamical insights into the opening mechanism of *P. aeruginosa* OprM channel. *Structure* **2010**, *18*, 507-517.

[235] Sennhauser, G.; Bukowska, M. A.; Briand, C.; Grutter, M. G. Crystal structure of the multidrug exporter MexB from *Pseudomonas aeruginosa*. *J Mol Biol* **2009**, *389*, 134-145.

[236] Poole, K.; Krebes, K.; McNally, C.; Neshat, S. Multiple antibiotic resistance in *Pseudomonas aeruginosa*: evidence for involvement of an efflux operon. *J Bacteriol* **1993**, *175*, 7363-7372.

[237] Guan, L.; Nakae, T. Identification of essential charged residues in transmembrane segments of the multidrug transporter MexB of *Pseudomonas aeruginosa*. *J Bacteriol* **2001**, *183*, 1734-1739.

[238] Murakami, S.; Nakashima, R.; Yamashita, E.; Matsumoto, T.; Yamaguchi, A. Crystal structures of a multidrug transporter reveal a functionally rotating mechanism. *Nature* **2006**, *443*, 173-179.

[239] Seeger, M. A.; Schiefner, A.; Eicher, T.; Verrey, F.; Diederichs, K.; Pos, K. M. Structural asymmetry of AcrB trimer suggests a peristaltic pump mechanism. *Science* **2006**, *313*, 1295-1298.

[240] Sennhauser, G.; Amstutz, P.; Briand, C.; Storchenegger, O.; Grutter, M. G. Drug export pathway of multidrug exporter AcrB revealed by DARPin inhibitors. *PLoS biol* **2007**, *5*, e7.

[241] Nakashima, R.; Sakurai, K.; Yamasaki, S.; Nishino, K.; Yamaguchi, A. Structures of the multidrug exporter AcrB reveal a proximal multisite drug-binding pocket. *Nature* **2011**, *480*, 565-569.

[242] Eicher, T.; Cha, H. J.; Seeger, M. A.; Brandstatter, L.; El-Delik, J.; Bohnert, J. A.; Kern, W. V.; Verrey, F.; Grutter, M. G.; Diederichs, K.; Pos, K. M. Transport of drugs by the multidrug transporter AcrB involves an access and a deep binding pocket that are separated by a switch-loop. *Proc Natl Acad Sci USA* **2012**, *109*, 5687-5692.

[243] May, A.; Zacharias, M. Energy minimization in low-frequency normal modes to efficiently allow for global flexibility during systematic protein-protein docking. *Proteins* **2008**, *70*, 794-809.

[244] Cheatham, T. E., 3rd; Cieplak, P.; Kollman, P. A. A modified version of the Cornell *et al.,* force field with improved sugar pucker phases and helical repeat. *J Biomol Struct Dyn* **1999**, *16*, 845-862.

[245] Eguchi, K.; Ueda, Y.; Kanazawa, K.; Sunagawa, M.; Gotoh, N. The mode of action of 2-(thiazol-2-ylthio)-1beta-methylcarbapenems against *Pseudomonas aeruginosa*: the impact of outer membrane permeability and the contribution of MexAB-OprM efflux system*J Antibiot* **2007**, *60*, 129-135.

[246] Livermore, D. M. Multiple mechanisms of antimicrobial resistance in *Pseudomonas aeruginosa*: our worst nightmare? *Clin Infect Dis* **2002**, *34*, 634-640.

[247] Mesaros, N.; Glupczynski, Y.; Avrain, L.; Caceres, N. E.; Tulkens, P. M.; Van Bambeke, F. A combined phenotypic and genotypic method for the detection of Mex efflux pumps in *Pseudomonas aeruginosa*. *J Antibmicrob Chemother* **2007**, *59*, 378-386.

[248] Ong, C. T.; Tessier, P. R.; Li, C.; Nightingale, C. H.; Nicolau, D. P. Comparative *in vivo* efficacy of meropenem, imipenem, and cefepime against *Pseudomonas aeruginosa* expressing MexA-MexB-OprM efflux pumps. *Diag Microbiol Infect Dis* **2007**, *57*, 153-161.

[249] Riera, E.; Cabot, G.; Mulet, X.; Garcia-Castillo, M.; del Campo, R.; Juan, C.; Canton, R.; Oliver, A. *Pseudomonas aeruginosa* carbapenem resistance mechanisms in Spain: impact on the activity of imipenem, meropenem and doripenem. *J Antibmicrob Chemother* **2011**, *66*, 2022-2027.

[250] Vargiu, A. V.; Collu, F.; Schulz, R.; Pos, K. M.; Zacharias, M.; Kleinekathofer, U.; Ruggerone, P. Effect of the F610A mutation on substrate extrusion in the AcrB transporter: explanation and rationale by molecular dynamics simulations. *J Am Chem Soc* **2011**, *133*, 10704-10707.

[251] Bohnert, J. A.; Schuster, S.; Seeger, M. A.; Fahnrich, E.; Pos, K. M.; Kern, W. V. Site-directed mutagenesis reveals putative substrate binding residues in the *Escherichia coli* RND efflux pump AcrB. *J Bacteriol* **2008**, *190*, 8225-8229.

[252] Aris, R. M.; Gilligan, P. H.; Neuringer, I. P.; Gott, K. K.; Rea, J.; Yankaskas, J. R. The effects of panresistant bacteria in cystic fibrosis patients on lung transplant outcome. *Am J Resp Crit Care* **1997**, *155*, 1699-1704.

[253] Harris, A.; Torres-Viera, C.; Venkataraman, L.; DeGirolami, P.; Samore, M.; Carmeli, Y. Epidemiology and clinical outcomes of patients with multiresistant *Pseudomonas aeruginosa*. *Clin Infect Dis* **1999**, *28*, 1128-1133.

[254] Hsueh, P. R.; Teng, L. J.; Yang, P. C.; Chen, Y. C.; Ho, S. W.; Luh, K. T. Persistence of a multidrug-resistant *Pseudomonas aeruginosa* clone in an intensive care burn unit. *J Clin Microbiol* **1998**, *36*, 1347-1351.

[255] Reymond, J. L.; Darbre, T. Peptide and glycopeptide dendrimer apple trees as enzyme models and for biomedical applications. *Org Biomol Chem* **2012**, *10*, 1483-1492.

[256] Stach, M.; Maillard, N.; Kadam, R. U.; Kalbermatter, D.; Meury, M.; Page, M. G. P.; Fotiadis, D.; Darbre, T.; Reymond, J. L. Membrane disrupting antimicrobial peptide dendrimers with multiple amino termini. *MedChemComm* **2012**, *3*, 86-89.

[257] Bruschi, M.; Pirri, G.; Giuliani, A.; Nicoletto, S. F.; Baster, I.; Scorciapino, M. A.; Casu, M.; Rinaldi, A. C. Synthesis, characterization, antimicrobial activity and LPS-interaction properties of SB041, a novel dendrimeric peptide with antimicrobial properties. *Peptides* **2010**, *31*, 1459-1467.

[258] Karatasos, K.; Adolf, D. B.; Davies, G. R. Statics and dynamics of model dendrimers as studied by molecular dynamics simulations. *J Chem Phys* **2001**, *115*, 5310-5318.

[259] Lyulin, S. V.; Darinskii, A. A.; Lyulin, A. V. Energetic and conformational aspects of dendrimer overcharging by linear polyelectrolytes. *Phys Rev E* **2008**, *78*.

[260] Pavan, G. M.; Barducci, A.; Albertazzi, L.; Parrinello, M. Combining metadynamics simulation and experiments to characterize dendrimers in solution. *Soft Matter* **2013**, *9*, 2593-2597.

[261] Welch, P.; Muthukumar, M. Dendrimer-polyelectrolyte complexation: A model guest-host system. *Macromolecules***2000**, *33*, 6159-6167.

[262] Kirschner, K. N.; Lins, R. D.; Maass, A.; Soares, T. A. A Glycam-based force field for simulations of lipopolysaccharide membranes: Parametrization and validation. *J Chem Theory Comput* **2012**, *8*, 4719-4731.

[263] Nascimento, A.; Pontes, F. J. S.; Lins, R. D.; Soares, T. A. Hydration, ionic valence and cross-linking propensities of cations determine the stability of lipopolysaccharide (LPS) membranes. *Chem Commun* **2014**, *50*, 231-233.

[264] Kinsey, S. G.; Long, J. Z.; O'Neal, S. T.; Abdullah, R. A.; Poklis, J. L.; Boger, D. L.; Cravatt, B. F.; Lichtman, A. H. Blockade of endocannabinoid-degrading enzymes attenuates neuropathic pain. *J Pharmacol Exp Ther* **2009**, *330*, 902-910.

[265] Labar, G.; Michaux, C. Fatty acid amide hydrolase: from characterization to therapeutics. *Chem Biodivers* **2007**, *4*, 1882-1902.

[266] Petrosino, S.; Di Marzo, V. FAAH and MAGL inhibitors: therapeutic opportunities from regulating endocannabinoid levels. *Curr Opin Investig Drugs* **2010**, *11*, 51-62.

[267] Bracey, M. H.; Hanson, M. A.; Masuda, K. R.; Stevens, R. C.; Cravatt, B. F. Structural adaptations in a membrane enzyme that terminates endocannabinoid signaling. *Science* **2002**, *298*, 1793-1796.

[268] Bertolacci, L.; Romeo, E.; Veronesi, M.; Magotti, P.; Albani, C.; Dionisi, M.; Lambruschini, C.; Scarpelli, R.; Cavalli, A.; De Vivo, M.; Piomelli, D.; Garau, G. A binding site for nonsteroidal anti-inflammatory drugs in fatty acid amide hydrolase. *J Am Chem Soc* **2013**, *135*, 22-25.

[269] McKinney, M. K.; Cravatt, B. F. Evidence for distinct roles in catalysis for residues of the serine-serine-lysine catalytic triad of fatty acid amide hydrolase. *J Biol Chem* **2003**, *278*, 37393-37399.

[270] Mileni, M.; Kamtekar, S.; Wood, D. C.; Benson, T. E.; Cravatt, B. F.; Stevens, R. C. Crystal structure of fatty acid amide hydrolase bound to the carbamate inhibitor URB597: discovery of a deacylating water molecule and insight into enzyme inactivation. *J Mol Biol* **2010**, *400*, 743-754.

[271] Seierstad, M.; Breitenbucher, J. G. Discovery and development of fatty acid amide hydrolase (FAAH) inhibitors. *J Med Chem* **2008**, *51*, 7327-7343.

[272] Favia, A. D.; Habrant, D.; Scarpelli, R.; Migliore, M.; Albani, C.; Bertozzi, S. M.; Dionisi, M.; Tarozzo, G.; Piomelli, D.; Cavalli, A.; De Vivo, M. Identification and characterization of Carprofen as a multitarget fatty acid amide hydrolase/cyclooxygenase inhibitor. *J Med Chem* **2012**, *55*, 8807-8826.

[273] Alexander, J. P.; Cravatt, B. F. Mechanism of carbamate inactivation of FAAH: Implications for the design of covalent inhibitors and *in vivo* functional probes for enzymes. *Chem Biol* **2005**, *12*, 1179-1187.

[274] Russel, D.; Lasker, K.; Webb, B.; Velazquez-Muriel, J.; Tjioe, E.; Schneidman-Duhovny, D.; Peterson, B.; Sali, A. Putting the pieces together: integrative modeling platform software for structure determination of macromolecular assemblies. *PLoS biology* **2012**, *10*, e1001244

[275] Degiacomi, M. T.; Dal Peraro, M. Macromolecular symmetric assembly prediction using swarm intelligence dynamic modeling. *Structure* **2013**, *21*, 1097-1106.

[276] Bolze, C. S.; Helbling, R. E.; Owen, R. L.; Pearson, A. R.; Pompidor, G.; Dworkowski, F.; Fuchs, M. R.; Furrer, J.; Golczak, M.; Palczewski, K.; Cascella, M.; Stocker, A. Human Cellular Retinaldehyde-Binding Protein has secondary thermal 9-cis-retinal isomerase activity. *J Am Chem Soc* **2014**, *136*, 137-146.

[277] Helbling, R. E.; Bolze, C. S.; Golczak, M.; Palczewski, K.; Stocker, A.; Cascella, M. Cellular Retinaldehyde Binding Protein-Different binding modes and micro-solvation patterns for high-affinity 9-cis- and 11-cis-retinal substrates. *J Phys Chem B***2013**, *117*, 10719-10729.

[278] Kudryashev, M.; Stenta, M.; Schmelz, S.; Amstutz, M.; Wiesand, U.; Castano-Diez, D.; Degiacomi, M. T.; Munnich, S.; Bleck, C. K. E.; Kowal, J.; Diepold, A.; Heinz, D. W.; Dal Peraro, M.; Cornelis, G. R.; Stahlberg, H. *In situ* structural analysis of the Yersinia enterocolitica injectisome. *Elife* **2013**, *2*, e00792.

[279] Wagner, S.; Sorg, I.; Degiacomi, M.; Journet, L.; Dal Peraro, M.; Cornelis, G. R. The helical content of the YscP molecular ruler determines the length of the Yersinia injectisome. *Molecular Microbiology* **2009**, *71*, 692-701.

[280] Degiacomi, M. T.; Lacovache, I.; Pernot, L.; Chami, M.; Kudryashev, M.; Stahlberg, H.; van der Goot, F. G.; Dal Peraro, M. Molecular assembly of the aerolysin pore reveals a swirling membrane-insertion mechanism. *Nat Chem Biol* **2013**, *9*, 623-629.

[281] Lemmin, T.; Soto, C. S.; Clinthorne, G.; DeGrado, W. F.; Dal Peraro, M. Assembly of the transmembrane domain of E. coli PhoQ histidine kinase: Implications for signal transduction from molecular simulations. *PLoS Comput Biol* **2013**, *9*, e1002878.

Subject Index

Zaheer Ul-Haq and Jeffry D. Madura (Eds). **10.1016/B978-1-60805-865-5.50009-5**

H

I

K

L

Q

R

S

T

U

V

W

Kolibri 360 GmbH
Ferdinand-Jühlke-Straße 7
99095 Erfurt, Deutschland
produktsicherheit@kolibri360.de